GROUND IMPROVEMENT TECHNIQUES

GROUND IMPROVEMENT TECHNIQUES

By

Dr. P. PURUSHOTHAMA RAJ

Former Dean
College Development Council
University of Madras

LAXMI PUBLICATIONS (P) LTD

(An ISO 9001:2015 Company)

BENGALURU • CHENNAI • GUWAHATI • HYDERABAD • JALANDHAR
KOCHI • KOLKATA • LUCKNOW • MUMBAI • RANCHI
NEW DELHI

GROUND IMPROVEMENT TECHNIQUES

Printed and bound in India
Typeset at : Goswami Associates, Delhi
First Edition : 1999, *Reprint* : 2005, 2006, 2007, 2008, 2009, 2011, 2012, 2013, 2014,
Second Edition : 2016, *Reprint* : 2018, *Edition* : 2020
ISBN : 978-81-318-0594-7

PUBLISHED IN INDIA BY

(An ISO 9001:2015 Company)
113, GOLDEN HOUSE, GURUDWARA ROAD,
DARYAGANJ, NEW DELHI - 110002, INDIA
Telephone : 91-11-4353 2500, 4353 2501
Fax : 91-11-2325 2572, 4353 2528
www.laxmipublications.com info@laxmipublications.com

Branches

✆	Bengaluru	080-26 75 69 30	
✆	Chennai	044-24 34 47 26,	24 35 95 07
✆	Guwahati	0361-254 36 69,	251 38 81
✆	Hyderabad	040-27 55 53 83,	27 55 53 93
✆	Jalandhar	0181-222 12 72	
✆	Kochi	0484-237 70 04,	405 13 03
✆	Kolkata	033-22 27 43 84	
✆	Lucknow	0522-220 99 16	
✆	Mumbai	022-24 93 12 61	
✆	Ranchi	0651-220 44 64	

C—01265/019/012
Printed at: Ajit Printing Press, Delhi.

TO THE MEMORY OF MY FATHER

PREFACE TO THE SECOND EDITION

It is indeed a pleasure to present the second edition of this unique book on this subject. The present edition has been thoroughly revised and enlarged to include the latest information in the advanced field of Civil Engineering.

It is sincerely hoped that the book in its present form will be of immense use for students and Civil Engineers.

Suggestions for the improvement of this book will be gratefully acknowledged.

—Author

PREFACE TO THE FIRST EDITION

The rapid urban and industrial development pose an increasing demand for land reclamation, utilization of unstable and environmentally affected ground and safe disposal of wastes. In order to meet these demands, ground improvement techniques have been evolved which now emerged as a major part of the civil engineering practice. Basically, these techniques involve modifications of mechanical, hydraulic, physical, cementing and chemical properties of the ground.

Any technique can be accepted and generally applied provided comprehensive specifications, workable technical standards and field examples are available. In the field of ground improvement techniques adequate, standards and specifications are available and this field has been recognized as an important and rapidly expanding one. The rate of growth and the overwhelming information already available resulted in a demand for compilation of information in the subject. This book fulfils such a need.

This book is not intended for a beginner in geotechnical engineering, instead for a matured student in the subject and for a professional who is already trained in the fundamentals of geotechnical engineering. A general treatment of the subject is made with due emphasize on practical aspects with limited theoretical treatment wherever necessary. This book is not very exhaustive and aimed at the practicing engineers and post graduate students who want alternative approaches for a problem and also want to be sure that they are aware of the recent development.

The subject has been treated in nine chapters with an introductory chapter on Formation and Development of Ground followed by eight chapters dealing with specific techniques. The main emphasis in each chapter is with the practical aspects of the technique in that the method of analysis, economics, construction materials, details and durability are considered in greater depth than in a conventional geotechnical engineering textbook.

The SI system of units is used throughout the book. For further reading a list of references is provided in the bibliography.

Materials have been drawn freely, in one form or the other, from various books, and technical publications. Indian Standards which also formed the source of material for this book and the author expresses his sincere thanks. Due reference of the author/s or the source is made in the text, figures and tables. Inspite of such sincere efforts, some contributions might not have been acknowledged, such omissions are not intentional and the author regrets for the same.

The library facilities of Pondicherry Engineering College have been extensively used for which the author expresses his sincere thanks to the Principal and to the Librarian, Dr. K. Nithyanandam, and his devoted staff. The author is thankful to Dr. D. Govindarajulu and Dr. S. Kothandaraman, Dept. of Civil Engineering, PEC, for their help.

The author extends his thanks to Mr. M. Nazir Hamad Kane, for his careful work in typing the word processed manuscript. Thanks are also due to Mr. S. Kanagasabapathy and Mr. M. Gajendran for drafting of the sketches and charts.

A book of this nature could not have been possible but for the patience, understanding and continuous encouragement shown by my wife Indira.

The author welcomes suggestions from students, teachers and engineers for further improvement of the book.

—Author

Contents

LIST OF SYMBOLS

Symbol	Represents
A_c	Cross-sectional area
a	Constant/spacing of wells
a_v	Coefficient of compressibility
C	Coefficient
C	Constant
C_c	Compression index
c	Side/undrained cohesion
c_h	Coefficient of consolidation for horizontal drainage
c_t	Point cohesion
c_v	Coefficient of consolidation for vertical drainage
D	Particle size/constrained modulus/effective depth
D_8	Specific constrained modulus
d	Average diameter/drainage path
ECB	Ethylene copolymer bitumen
e_0	Initial void ratio
G_8	Specific gravity
GR	Groutability ratio
H	Thickness of layer/depth of natural water table
H_e	Average head loss
H	Head/height of drop
h_w	Head at the well
i	Hydraulic gradient
k	Coefficient of permeability
k_e	Coefficient of electro-osmotic permeability
k_h	Coefficient of horizontal permeability
k_v	Coefficient of vertical permeability
L	Distance to source from well/electrode spacing
L_c	Stone column lengtn
$LDPE$	Low density polyethvlene
$LLDPE$	Linear low density polyethylene
m_v	Coefficient of volume compressibility

N	Standard penetration resistance
NOS	Natural overburden stress
OCR	Overconsolidation ratio
P	Total load on stone column
P_c	Preconsolidation pressure
P_o	Overburden pressure
PA	Polyamide
PE	Polyethylene
PP	Polypropylene
PVC	Polyvinylchoride
q_a	Allowable bearing capacity
q_c	Cone penetration resistance
R_r	Radius of influence
r_e	Radius
r_w	Radius of well
S_c	Primary compression
S_i	Initial compression
S_8	Secondary compression
S_t	Sensitivity
SF	Safety factor
T	Time factor
T_r	Time factor for radial flow
T_v	Time factor for vertical flow
t	Time
t_p	Time when primary compression complete
U	Degree of consolidation
u	Pore pressure difference
U_r	Degree of consolidation for radial flow
U_z	Degree of consolidation for one dimensional flow
V	Voltage difference
v	Discharge velocity/vacuum at pump intake
v_8	Seepage velocity
W	Weight of explosive/weight dropped
w_L	Liquid limit
w	Unit weight of water
P	Change in pressure
t	Time increment
$\Delta\sigma$	Additional stress
σ_r	Effective radial stress
σ_v	Effective consolidation stress
ϕ	Drained angle of internal friction
η	Efficiency factor

Chapter 1

FORMATION AND DEVELOPMENT OF GROUND

1.1 INTRODUCTION

Materials that constitute earth's crust are broadly classified into two categories as rock and soil. Rock is a material strongly bonded of minerals whereas soil is an assemblage of solid particles formed by disintegration of rocks. It spreads beneath rivers and seas and on land along with all organic and inorganic materials overlying the bedrock.

The type and characteristic properties of the soils depend on its formation and deposition by transportation agents. Additional particle binding takes place due to the presence of carbonates, oxides and organic matter. The exposure of soil with time develops a weathering profile from the ground surface down.

Changes in ground, after formation, occurs due to different natural causes and man's activities other than that produced by structures. Man-made lands called reclaimed lands are formed in low lying areas and on water by land fillings.

The ground formed based on the above activities should have adequate mechanical and hydraulic properties, otherwise the ground has to be improved.

This Chapter deals with the formation of different types of soil and their further alteration due to different reasons and identifies the conditions of ground which requires ground improvement.

1.2 FORMATION OF ROCK, SOIL AND SOIL PROFILE

Soil and rock are the principal structural materials with which a civil engineer deals. Hence a knowledge of the origin, manner of occurrence, and characteristics of these materials is essential for the successful practice of civil engineering design and construction. Rock is a consolidated material composed of natural aggregates that are connected by strong bonding forces. Soil, on the other hand, is an unconsolidated material composed of natural aggregate of mineral grains which have resulted from the disintegration of rock. Further in engineering profession soil is also considered to include the reclaimed soils that is the residue of vegetable and animal life including industrial residual wastes. Thus soil is a particulate material wherein the grains are bonded together not as strongly as in rocks.

1.2.1 Principal Rock Types

The rock materials encountered at the surface of the earth or beneath the surface of soil are classified under three major groups, *viz.*, igneous, sedimentary and metamorphic rocks, based on their mode of origin. Igneous rocks are formed by the cooling of the molten magma or by the recrystallization of the older rocks under heat and pressure great enough to render them fluid. Sedimentary rocks are the products of minerals formed by physical disintegration of any kind of rock and chemical decomposition and deposits of plant and animal remains. Metamorphic rocks are those produced by internal processes such as heat, pressure and plastic flow, acting on rocks of any kind, with the limitation that the rocks concerned remained essentially solid during their transformations. Each and every rock is classified under any one of these groups based on the texture, structure and mineralogical composition of the rock.

Igneous rocks are classified primarily on the basis of texture and colour. They are generally very hard and have textures that vary from coarsely crystalline to glossy depending upon the rate of cooling. The igneous rocks are generally massive without any structural features but joints and cracks are usually found in all igneous rocks. Most of the igneous materials that cooled very rapidly may contain gas bubbles or else have a fragmental structures. The principal mineral constituents are light coloured quartz and feldspar and dark coloured hornblende, biotite, augite and olivins.

Sedimentary rocks may be further divided into three groups, *viz.*, clastic, organic, and chemical in accordance with the origin of the sediment. Rock or mineral fragments derived from pre-existing mineral belong to clastic group, those deposited from organisms living in water are grouped as organic and those precipitated by chemical activity or left to evaporation is referred to as chemical. The texture of chemical sediments is usually microscopic whereas clastic and organic may vary from coarse grained to microscopic. The predominent minerals of sedimentary rocks are quartz, calcite and clay minerals. Almost all the sedimentary rocks are classified as sandstone, limestone or shale. The other minor constituents of these rocks are iron oxides, feldspar and carbonaceous material. The most important structural characteristics of sedimentary rocks from the geological point of view is their bedding or stratification. Shales are formed predominantly from deposits of fine grained particles (silts* and clays**). It is estimated that shale covers over 50% of the rock that is exposed at the earth's surface or closest to the surface under soil cover. Sound shale can provide a good foundation material. It is not suitable as a construction material because of its tendency to break down under handling and weathering.

Metamorphic rocks are distinguished from others primarily on the basis of structure and mineralogical composition. The agents of metamorphisms such as heat, pressure, and hydrothermal solutions lead to the crystallization and orientation of minerals not found in parent rock.

Particle sizes—*Silt: 0.002 mm to 0.075 mm, **Clay: < 0.002 mm.

Metamorphic rocks are characterised by the foliated structure. However, metamorphic rocks derived from sediments possess a massive rather than a foliated structure. These rocks also show structural defects as cracks and joints. By metamorphism limestone, sandstone and shale are changed to marble, quartzite and slate. Metamorphic rocks formed from sound igneous or sedimentary rocks can be good materials for construction. Typical metamorphic minerals are chlorite, sericite, talc, hornblende and biotite. The most important engineering characteristics of metamorphic rocks are the softness of schists and the high susceptibility of all foliated rocks to weathering.

1.2.2 Origin of Soils

Up to a depth of about 20 km the earth's crust comprises of both rock and weathered rock (as soil). Soils (weathered rock) are originated from the rocks and minerals of the earth's crust. The principal minerals subject to weathering to produce soil at or near the earth's surface and available in the order of abundance are: quartz, feldspar, pyroxene, amphibole, *etc*. Solid rocks are decomposed to fragments creating soils by the continuous weathering processes in combination with crystal deformations. The type of soil developed depends on the rock types, its mineral constituents and the climatic region of the area.

Rocks containing quartz or orthoclase minerals with high silica content (*e.g.*, granite and rhyolite) mostly decompose into sands or gravelly soils with a little clay. Fine textured silty and clayey soils are formed due to decomposition of rocks (*e.g.*, gabbros and basalts), containing minerals of iron, magnesium, calcium or sodium, with little silica. Clays are not fragments of primary minerals from parent rock but secondary minerals formed by the decomposition of primary minerals. Thus the behaviour of gravel and sand are different than that of clay as the former are composed to primary minerals.

Gneiss and schist decompose into silt-sand mixtures with mica, slates and phyllites to clays, marble to limestone and quartzite to sands and gravels. As discussed earlier through metamorphism ingenous or sedimentary rocks could form metamorphic rocks and accumulated deposits of soil particles with cementing materials subjected to high pressure could form sedimentary rocks. Thus the cyclic process of transforming rock to soil and vice-versa is a continuous process occurring over millions of years through complex chemical and physical processes.

1.2.3 Types of Weathering

Breaking down of intact masses of rock into smaller pieces by physical, chemical or solution processes is called **rock weathering**. All these three processes may take place simultaneously but at different rates, depending on the climate, topography and composition of the original rock. Rock weathering is one of the important geological processes. Rock products formed by weathering are deposited as unconsolidated sediments as soils.

The process by which rock disintegrates into smaller fragments due to combination of grinding, shattering, breaking and stress changes without involving any change in the properties is called physical or mechanical weathering or disintegration. For example, stress readjustments during regional uplift accompanied by water run off, freezing of water in cracks and pores, the abrasion of gravel and boulders by mountain streams and rivers, the pounding of water waves on beaches or cliffs, and the sand blast of sand-laden desert winds cause physical weathering.

Mechanical weathering may also be caused by organic activity, such as cracking forces exerted by plants growing in the crevassis of rocks and the moving fragments towards the surface by animals or insects. Mechanical weathering predominates in dry regions and areas with rugged topography.

Chemical weathering or decomposition is the alteration of the rock minerals, due to chemical reaction, to form new minerals which have chemical and physical properties different from their parent materials. The chemical reaction of minerals occurs with water, dissolved carbondioxide and oxygen from air, organic acids from plant decay and dissolved salts present in the water. Hydrated iron oxide, carbonates and sulphates are formed when rainwater comes in contact with rock surfaces. Similarly rainwater with pH < 7 may react chemically with some rock. Further, during a geological time period even a weak acid may cause decomposition. For example, leaching may remove the cementing properties of sedimentary rocks. Chemical weathering predominates in warm, humid regions and in areas with flat topography.

Solution is the dissolving of soluble minerals from the rock, leaving the insoluble minerals behind as a residue. Solution is predominent in humid regions underlain by soluble rocks.

Thus in the weathering process, rock minerals are broken physically, changed chemically and dissolved in water. The end product is soil which consists predominantly of quartz and clay minerals with varying amounts of mica, ferromagnesium minerals, iron oxides and carbonates. This alteration process further continues with changes in environment and that are induced by man-made activities such as construction of drainage and structures, excavation, flooding and filling.

1.2.4 Major Soil Types

Based on the method of formation soils can be grouped into three broad categories—residual, sedimentary or fill. Residual soils have formed from the weathering of rocks or accumulation of organic material and practically remain at the location of origin with little or no movement of individual soil particles. Sedimentary soils are formed by individual particles which were created at one location and later transported and deposited at another location. Fill is a man-made deposit.

Residual Soils: All the three processes of weathering can cause the formation of residual soils. If the rate of rock weathering is more than the rate of erosion or transportation of weathered material, the resulting accumulated soil is the residual soil. The contributing factors for the rate of weathering and nature of weathered products are climate, time, rock source, vegetation, drainage and bacterial activity. As the weathering action decreases with depth, the deeper residual soils maintain the concentration of mineral and orientation of grains of the parent rock. Thus the thickness of residual soil accumulated at a particular location depends on the rate of rock weathering and the availability of erosive forces to carry the soil away after it is formed. Residual soils generally have wide range of particle sizes, shapes and composition based on the degree and type of weathering and the minerals of parent rock.

Residual soils may be derived from all the three basic rocks. Residual soils produced from igneous rocks, such as granite, are tan and yellow sandy silts and silty sands with mica and clay of kaolinite family. These soils from granites ordinarily form good construction material. Basalt, another igneous rock, which is rich in ferromagnesium materials produce residual soils with highly plastic montmorillonite clays with deep red to dark brown colour and are not suitable for construction or as foundation. The black cotton soils found in India belong to this group.

Residual soils derived from clastic sedimentary rocks exhibit an altered mineral composition and are finer grained than the original sediments. Further the depth of weathering of these sedimentary rocks is generally less than for igneous rocks in the same environment. The residual soils from carbonate sedimentary rocks consist of all the insoluble impurities of the rock. These soils have clay varying from kaolinite to montmorillonite and silica varying from boulder to silt-size and iron oxide.

Residual soils formed from metamorphic rocks, such as gneiss and schist, range from sandy silts to silty sands with varying amount of mica. These deposits are extremely variable in composition and extent and the minerals are found in the same order of bands as in the original rock. The residual soils derived from quartzites and marbles resemble those from granites and limestones respectively in their engineering behaviour. Because of the variation in composition, these residual soils need careful study before use.

Residual soils are found abundant in humid and warm regions which are favourable to chemical weathering of rock. Residual soils exist in many parts of the world, *viz.*, South Asia, Africa, Southeastern North America, Central America, the islands of Caribbean and South America. Sowers (1963) has reported the following typical depths of residual soils:

Southeastern United States	6 to 23 m
Angola	7.5 m
South India	7.5 to 15 m
South Africa	9 to 18 m
West Africa	10 to 20 m
Brazil	10 to 25 m

Studies on residual soils are limited compared to sedimentary soils.

Sedimentary Soils: Sedimentary soils are produced depending on the formation, transportation and deposition of sediments. Sediments are formed basically from the physical and chemical weathering of rocks on the surface. Generally, physical weathering of rocks produces silt-, sand- and gravel-sized particles whereas chemical weathering produces clay-sized particles. Transportation of sediments can be performed by the five agents, *viz.*, water, air, ice, gravity and organisms. During the process of transportation the sizes of the particles are altered and accordingly sorted depending on the agency. Lambe and Whitman (1979) has given the effects of these five transporting agents on sediments (Table 1.1). Thus, particles, so formed and transported, are deposited in water or land to form sedimentary soils. The deposition in water may be caused due to reduction in velocity, decrease in solubility and increase in electrolytic concentration.

Sedimentary soils are further classified based on the mode of transportation causing the deposit.

Soil particles which are transported by water, called water-transported soils, may be in the form of suspended particles, or, by rolling and sliding along the bottom of the stream or river. Particles transported by water range in size from boulders* to clay. Soils that are transported by river water and deposited are called **alluvial deposits.** Soil particles carried by a river while entering a lake, deposit all the coarse particles because of sudden decrease in velocity and such a deposit is

*Greater than 300 mm size.

referred to as **lake delta.** However, the fine particles reach the centre of the lake and get deposited under quiet waters. Every season alternate layers are formed and such lake deposits are called **lacustrine deposits.** These deposits are weak and compressible and pose problems for foundation. If coarse and fine grained particles are deposited in sea-water, then they are called **marine deposits.** Marine sediments are made up of terrestrial and marine contributions. The soil deposited in salt-water will have a highly flocculated structure and thus have a larger void ratio than the soil deposited in fresh-water. Marine clay deposits are generally weak, compressible and problematic soil for foundations. In water stagnated areas where the water table is fluccuating and vegetational growth possible, **swamp and marsh deposits** develop. Soils formed under this environment are soft, high in organic content and unpleasant in odour. Accumulation of partially or fully decomposed aquatic plants in swamps or marshs is termed **muck or peat.** Muck is a fully decomposed material, spongy, light in weight, highly compressible, and not suitable at all for construction purposes of any kind.

Table 1.1. Effect of Transportation on Sediments
(*Source : Lambe and Whitman, 1979*)

	Water	*Air*	*Ice*	*Gravity*	*Organisms*
Size	Reduction through solution, little abrasion in suspended load, some abrasion and impact in traction load	Considerable reduction	Considerable grinding and impact	Considerable impact	Minor abrasion effects from direct organic transportation
Shape and roundness	Rounding of sand and gravel	High degree of rounding	Angular, soled particles.	Angular, non-spherical	
Surface texture	Sand: Smooth, polished, shiny Silt: Little effect	Impact produces frosted surfaces	Striated surfaces	Striated surfaces	
Sorting	Considerable sorting	Very considerable sorting (progressive)	Very little sorting	No sorting	Limited sorting

Ice in the form of glaciers has been a very active agent of both weathering and transportation. Deposits formed due to this are generally termed as **glacial deposits.** Melting of a glacier causes deposition of all the materials and such a deposit is called a **till.** The land form or topographic surface after a glacier has receded is referred to as a **ground moraine or till plain.** Till deposits which have been overrun by glaciers contain coarser particles and form construction materials. **Eskers** are the remains of soils deposited by the surface and subsurface glacial rivers. **Drumlins** are elongated low hills of till that point in the direction of the ice travel. Large boulders picked up by a glacier, transported to a new location and dropped are called **erratics.** Glacial deposits provide poor to excellent foundation.

Gravity is generally capable of transporting materials only to a limited distance and hence there will be no appreciable change in the materials moved by the transportation process. Gravity deposits are called as **talus.** Talus includes the materials collected at the base of cliff and landslide deposits.

Wind, like water can erode transport fine grained particles by rolling or carrying them and deposit. Soils carried by wind and subsequently deposited are referred to as **aeolian deposits.** Accumulations of such wind-deposited sands are termed as **dunes.** Dunes generally occur in desert areas, and on the downwind side of bodies of water having sandy beaches. Dune sands may be used for construction purpose to a limited extent. In arid regions, fine-grained soils can be transported by wind to a great distance. Wind-blown silts and clays deposited with some cementing materials in a loose stable condition is designated as **loess.** These deposits have low density, high compressibility and susceptible for collapse on saturation.

Fill: The process of soil formation discussed in the preceding sections are due to nature. A man-made soil deposit is a **fill** and the process adopted to form the fill is called **filling.** The materials needed for a fill are obtained from a source called **borrow** or processed by blasting and are transported through land or water using suitable vehicles or machineries and deposited in the required location by dumping. The fill can be densified by a suitable compaction device (as in a core of an earthdam or a structural fill or a subgrade) or left uncompacted (as in rock toe or hydraulic fill).

1.2.5 Soil Profile

Uppermost soils develop a characteristic weathering profile from the ground surface down due to continuous exposure to environmental activities. The term **soil profile** is used to indicate a vertical section through the subsoil showing the thickness and sequence of the individual strata. The soil profile is called **simple or regular**, if the boundaries between strata are approximately parallel. If the boundaries show irregular pattern the soil profile is called **erratic**. The physico-chemical properties of soil profile up to a depth of 3 m are influenced by the environmental factors such as changes of moisture and temperature, amount and seasonal distribution of moisture and temperature, amount and seasonal distribution of rainfall, changes due to biological agents, ground water level, *etc*. The degree of soil profile development depends on the length of time of environment activity. In freshly deposited materials the profile is shallow and poorly defined and in older deposits it may be as thick as 3 to 5 m and clearly defined.

The science of soil profile analysis is called pedology or soil science which is one of the basic science of agronomy. As civil engineers are also concerned with upper portion of soil deposits as a foundation or as a source of construction material, a knowledge of the soil profile at a particular location is very important for them. Further a suitable ground improvement technique can be decided only after knowing the details of a soil profile.

The uppermost part of the ground surface is subject to the mechanical effects of weathering and to the loss of some constituents due to leaching. This region is referred to as A-horizon. The thickness may range from a few centimetres to half a metre. The lower part is referred to as the B-horizon, where part of the substance washed out of the A-horizon are precipitated and accumulated. The B-horizon is also referred to as the zone of accumulation and thickness varies from 0.50 to 0.75 m.

The properties of soils in the A and B horizons are chiefly needed for road works. For foundation and earthworks the material underlying the B-horizon is important which is referred to as C-horizon. The character of the soil beneath the B-horizon is determined only, by the raw materials from which it is derived, by the method of deposition, and by subsequent geological events. The strata beneath the B-horizon may be homogenous, stratified or erratic. Generally, the C-horizon will contain material which was first deposited by water, wind or ice in the geological cycle. For erratic deposits it may be difficult to arrive at average values. Generally the physical properties of almost every natural soil stratum vary to a considerable extent in the vertical direction and to a smaller degree in horizontal direction.

Soil profile of various regions are briefly dealt below (Sowers, 1979).

Cool and Temperate Humid Region Soil Profile: In cool temperature regions with humid climates, there is large accumulation of dead leaves, plants and other organic debris due to heavy growth of vegetation. These accumulated materials decompose slowly and produce weak acids which accelerate the weathering and the soil-moisture movement is downward to the water table.

The uppermost soil is subjected to chemical alteration producing clays of the kaolinite family, soluble carbonates and semisoluble reduced iron materials. These are leached downwards by the soil moisture resulting in sandy soils with underlain clays. However, the top soil will be dark coloured organic matter with spongy texture. The leached material accumulates in the clay layer with decreasing effect of leaching with depth. Thus the clay layer is dark coloured and contains a greater concentration of clay minerals, iron, and carbonates than the original soil. Below this clay layer lies the slightly weathered parent material followed by unaltered parent material or rock.

Hot-Humid Region and Profile: The upper parts of the deposits in this region also experience alternate wetting and drying cycle and downward leaching. The decomposition is faster and no organic acid is produced. Because of the formation of soluble carbonates, the silicate weathering proceeds in a neutral environment and the soluble colloidal silica is leached downwards. The aluminium and iron become highly oxidized and cement the quartz into a stiff rock like soilds. The iron and aluminum get accumulated due to continued weathering and leaching and form nodules or concretions with a texture of a loose but cemented gravel with a colour ranging from tan to bright red. This process is referred to as **laterization** and the well indurated material is called laterite or ferricrete. Well-developed laterities are porous, light in weight but strong and relatively incompressible. Some of the strongly cemented laterities are used as load bearing material in place of gravels. On the other hand some of the less-developed laterites may get softened upon wetting. Laterites can be identified based on the presence of silica to alumina ratio.

Dry Region Soil Profile: In this type of region there will be no presence of organic matter. Because of surface evaporation the moisture movement is generally upwards resulting in accumulation of soluble materials such as carbonates near the surface and in the partial cementing of the soil. This sort of drainage is referred to as drainage by desiccation. This phenomenon of desiccation is very much pronounced in soils of semi-arid and arid regions. The distribution of carbonates are often non-uniform. When dry, these soils are highly incompressible and strong. However, upon wetting they weaken suddenly and collapse. They could be used as construction material depending on the parent material. In extremely arid regions, a saline or alkali topsoil is formed due to soluble salts brought by capillary action. These soils are referred to as **desiccated soils** and quite often mistaken for soft rocks.

Humid-Poorly Drained Soil Profile: The wet, poorly drained soils form similar group depending on the moisture. Organic growth is usually rapid in a very wet environment and the decay process is slow leading to accumulation of organic matter. Slower rates of decay associated with stagnation of water produce **fibrous peat** while higher rates of decay with fluctuating water levels, produce nearly fibreless **muscles.** In cool regions thick peat deposits are formed due to very slow decay.

1.3 SOIL DISTRIBUTION IN INDIA

India has a combination of tropical and temperate climatic conditions and exposed to monsoon weather, almost, throughout the country. The average annual rainfall varies from 20 to 400 cm. The soils of India are generally very old and fully matured and their formation is due to seasonal climatic distribution and amount of rainfall. The chemical reactions involved are more intense and therefore rock disintegration is rapidly followed by chemical decomposition. The soils of Indo-Gangetic plains are mainly alluvial and boulders being brought down by rivers from the Himalayas. Most of the other soils of India have been formed in the areas where they are found. The soils of India may generally be classified as follows (Pichamuthu, 1970): (*i*) Alluvial soils, (*ii*) Black soils, (*iii*) Red soils, (*iv*) Laterite soils, (*v*) Forest and hill soils, (*vi*) Arid and desert soils, (*vii*) Saline and alkaline soils, and (*viii*) Peaty and marshy soils. However, keeping in view the aspects of geotechnical engineering the major soil deposits of India (Katti *et. al.*, 1975; Ranjan and Rao, 1991) which cover large areas (Fig. 1.1) are discussed below.

1.3.1 Marine Deposits

The marine deposits spread along the Indian coast are relatively narrow belt and generally derived from terrestrial sources except for the coral sands and mud near the Andaman and Nicobar Islands and the Lakshadweep. These deposits are very soft to soft normally consolidated highly compressible clays. The sensitivity range is in the order of slight to medium sensitive and essentially inorganic in composition. The thickness of deposits vary from 5 to 20 m. Marine deposits cover along the coast of West Bengal, Andhra Pradesh, Tamil Nadu, Pondicherry, Kerala, Goa, Karnataka, Maharashtra and Gujarat. The deposits generally need a pre-treatment before application of any external load.

1.3.2 Black Cotton Soils

Black cotton soil is one of the major soil deposits of India and is spread over a wide area of 300,000 sq. km. To a large extent they are found in regions having low to medium slope and poor drainage conditions. The primary bed rock is basalt or trap and in some locations quartizites, schists and sedimentary rocks are found. It is expansive in nature due to the presence of montmorillonite and illite clay minerals. Some of these black cotton soils are also found to contain high amount of carbonates. The depth of black cotton soils can be as high as 20 m. Black cotton soils extend over the states of Maharashtra, Madhya Pradesh, Karnataka, Andhra Pradesh, Tamil Nadu and Uttar Pradesh. Based on the pedological conditions the depth of crack and crack pattern vary. The soil surface is hard in summer season but becomes slushy and lose its strength substantially during rainy season. Volume changes up to a depth of 1.5 m generally occur due to seasonal moisture changes. Highly loaded structures are most susceptible to damage as a result of the volume changes. Because of the shrinking and swelling characteristics of the soil a special treatment of the soil or design approach has to be adopted.

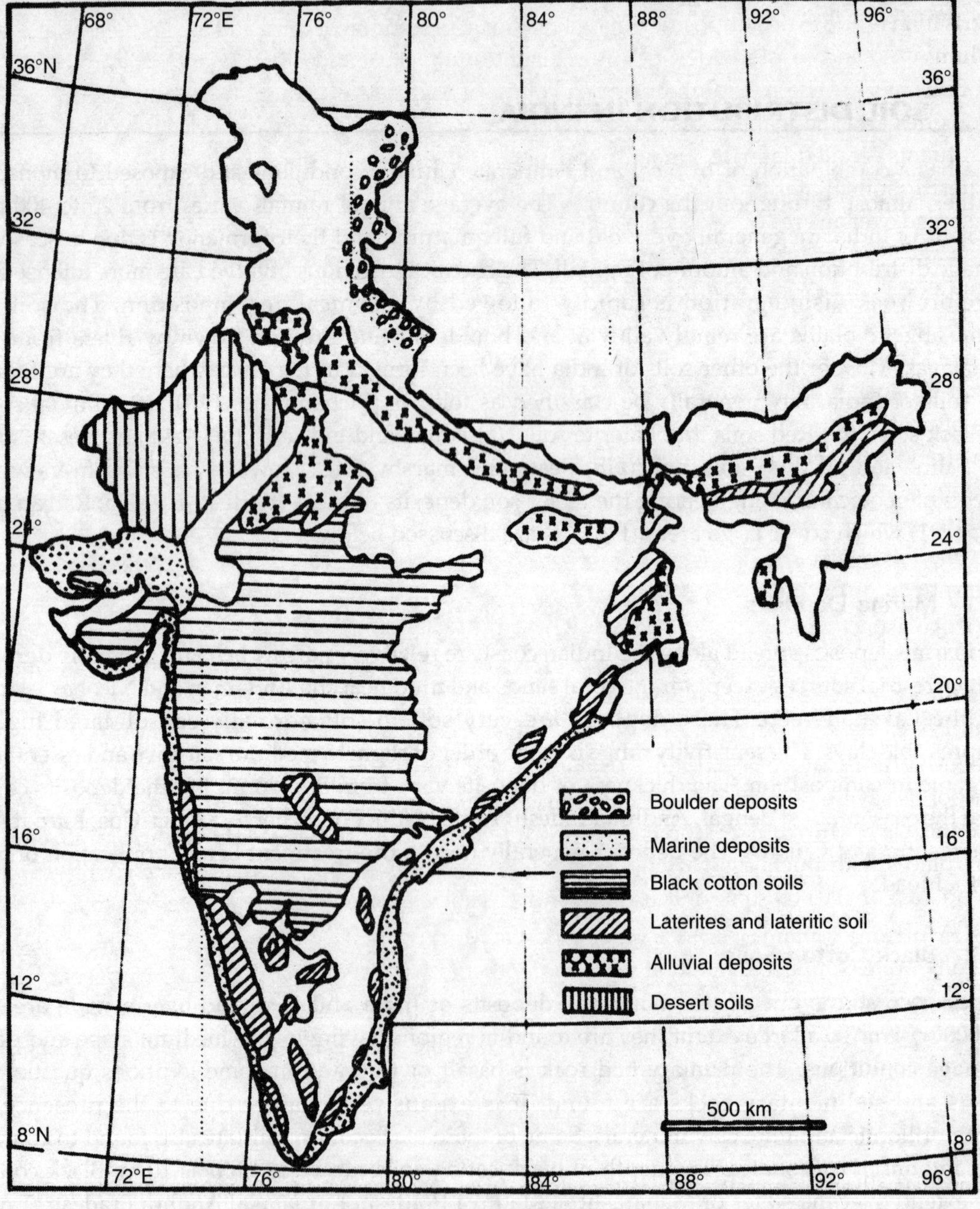

***FIGURE 1.1.* Major Soil Deposits of India.**

(Source: Katti et. al., 1975 and Ranjan and Rao, 1991.)

1.3.3 Laterites and Lateritic Soils and Murrums

Tropical regions of high moisture and high temperature establish special conditions of accelerated and intense weathering which lead to the development of an unique soil type identified as laterite. In such regions weathering activity is so intense that a tremendously thick soil (exceeding 30 m) may be developed from the parent rock through processes collectively termed as laterisation. Laterisation is due to decomposition of rock, removal in solution of silica and bases and accumulation of aluminium and sesquioxides, titanium, magnesium, clays and other amorphous products (Iyer and Pillai, 1972). The red, pink or brown colour of laterities is essentially due to the presence of iron oxides and manganese. A coarse-grained concretionary material with 90% of these lateritic constituents is called **laterite**. Fine-grained materials with lower concentrations of oxides are called as **lateritic soils**. Lateritic soils of India are residual soils and forms one of the major soil deposits and spread over an area of 100,000 sq. km. Lateritic soils extend over Kerala, Karnataka, Maharashtra, Orissa and West Bengal States. Indian laterites have been derived from different rocks under not humid conditions in places with an average annual rainfall of about 200 to 300 cm.

The characteristic feature of laterites is the high strength when it is cut and dried in Sun. Hardened soil can retain the high strength even subjected to immersion in water. The specific reason for such a behaviour is due to dehydration of iron oxides and the presence of halloysite type of clay mineral. It has been reported in literature that some of the laterites show extremely high strength comparable to that of burnt bricks. Many laterites are porous in nature and possess medium to high permeability. Calicut laterites were found to be rich in halloysite and crystalline goethite whereas Rajamundry laterites showed presence of crystalline kaolinite and metahallosite (Rao and Raymahshay, 1981). The reason for such a behaviour has been attributed to the geological environment of the areas.

An aspect which is problematic is the permeability of laterites in relation to construction of reservoirs (Katti *et. al.*, 1975). Road cuts in laterite deposits pose a serious stability problem. Further difficulties have been experienced in the assessment of lateral stresses in lateritic profiles (Iyer and Pillai, 1972).

Murrums are residual soils formed basically from weathering of basaltic rock, where monsoon is severe (Ketkar, 1970). Murrums constitute mixtures of weathered rock pieces, clayey sand and clay. Murrums may contain rock fragments in varying sizes. Murrums are found in many places with thickness varying from a few centimetres to a few metres and totally spreads over an area of 400,000 sq. km. It usually underlies black cotton soils, red or yellow soils.

Murrums contain over 50% gravel and very less percentage of clay fraction. The material is generally quite compact and soft at shallow depth and hard at deeper depth. Murrums because of the presence of gravel and sand size particles and compact exhibit very high strength. They are used extensively as road material and fill material. The bearing capacity of such soils are extremely high and do not pose any foundation problem.

1.3.4 Alluvial Deposits

The familiar alluvial deposits of India are in the Indo-Gangetic and Brahmaputra flood plains. The thickness of the deposit in these areas exceed 100 m. North of Vindya Satpura range is covered with river alluvium and deltaic type are spread near the sea. Alluvial deposits extend from Assam in the east to Punjab in west. Alluvial deposits generally present alternating layers of sandy silt and clay and in some locations organic layers are also encountered. The Bengal basin is another important alluvial deposit and is believed to have been formed in the middle of Mesozoic era. The subsoil of the upper strata of about 100 m thickness of the Bengal basin is of recent origin and is deposited by the Ganga river system (Som, 1975). These deposits are primarily taken place under typical alluvial environments. The soil around Calcutta, termed the normal Calcutta deposit, consists primarily of desiccated brownish grey/light brown silty clay/clayey silt with occasional lenses of silty fine sand followed by grey to dark grey silty clay with decomposed wood and exist up to a depth of 15 m (Som, 1975). Another deposit of Bengal basin is the river channel deposits, consisting of sandy silts to silty sands up to a depth of 30 m (Som, 1975).

1.3.5 Desert Soils

Thar desert in Rajasthan, covering about 500,000 sq. km., consists of desert soils of India. These are wind blown deposits generally present in the form of sand dunes with an average height of about 15 m. The deposits are formed under arid conditions and the dune sands are predominantly of non-plastic uniformly graded fine or silty sands. Some of the problems associated with the soil are of soil stabilisation for roads and runways and scarcity of water for any construction activity.

1.3.6 Boulder Deposits

Large boulders are carried down hills due to rivers and deposited near foot hills. Such deposits are encountered in the sub-Himalayan regions of Himachal Pradesh and Uttar Pradesh. The properties of these deposits are complex and depend on the size of boulders and the soil matrix. Generally, because of particle contact high frictional resistance should be anticipated (Ranjan and Rao, 1991).

1.4 ALTERATIONS OF GROUND AFTER FORMATION

A structure to be constructed on the ground or under the ground should be designed considering the properties of the ground at the time of formation, at the starting of the project and the possible changes which might take place to the ground during the design-life of the structure. The size and shape of a given deposit and the engineering properties of the soil in the deposit may change very significantly. Ground movements which are independent of stresses imposed by the structure can occur due to different activities. For example, ground movement may occur due to swelling and shrinking, frost action, seepage, vibration, *etc.* Many of these changes may be caused independent of man's activity and some due to construction activity itself. Hence, necessary precautions have to be taken while designing a structure keeping in view the possible anticipated changes which could occur during the design-life of the structure. A few of the factors which contribute for ground alteration are discussed below.

1.4.1 Effect of Seasonal Moisture Variation

Soils may undergo volume changes caused by seasonal moisture content variation. When a saturated soil is allowed to dry, a meniscus develops in each void at the soil surface and brings in tension in the soil water leading to a compression in the soil structure which is termed as shrinkage. Such a compression caused by shrinkage in fine-grained soils is as effective as that produced by an external load which could produce a pressure as great as 500 kN/m^2. In partially saturated soils the force causing shrinkage arises from the pressure difference across the curved air water interfaces. The effect of shrinkage depends on initial moisture content, type and amount of clay content, mode and environment of geological deposition. Shrinkage is reduced due to the presence of sand and silt-size particles. Soils with high plasticity shrink greatly causing settlement. Shrinkage takes place in horizontal and as well in vertical directions because of capillary tension which is exerted in all directions. Highly compressible clays exhibited cracks 0.5 m wide and 5 m deep (Sowers, 1979). Repeated shrinkage produces a network of shrinkage cracks in all directions.

Some soils (expansive soils like black cotton soils of India) not only shrink due to drying but also show marked swelling with increase of moisture content. Swelling is caused mainly due to repulsive forces which separate the clay particles leading to volume increase. The mechanism causing swelling may be attributed to a number of phenomena such as the elastic rebound of the soil grains, the clay mineral's affinity for water, the cation exchange capacity and electrical repulsive forces and the expansion of entrapped air. All these factors may contribute for swell in precompressed clays whereas in normally consolidated clays the affinity of water by clay particles and the electrical repulsion of clay particles may predominate. If the swelling is prevented high pressures as high as 500 kN/m^2 may be developed.

In regions which have well defined, alternately wet and dry seasons, susceptible soils swell and shink in regular cycles. Due to such seasonal volume change there will be rise and fall in the ground surface accompanied by tension cracks in the soil during dry season and closing of the cracks in wet season. The movements are larger in grass covered areas than in areas devoid of vegetation. Soils with illite or kaolinite show large initial volume decrease on drying with only a limited swelling on rewetting.

As shrinking and swelling depend on the character of the soil and the moisture changes, it is difficult to assess them qualitatively. Theoretically, a moist soil can shrink until it reaches the shrinkage limit. Further if the soil is initially drier than the shrinkage limit it should not shrink. In general, the lower the shrinkage limit, the greater the potential shrinkage. Similarly, the swell will be less when the moisture content is around plastic limit or slightly more. In the field the most reliable methods of assessing shrinkage cracks is by local enquiry and observations of shrinkage, cracking and desiccation in trial pits. Adequate field observations for several years in a region are needed to properly estimate the shrinkage and swelling. If the field condition compels to use a swelling soil, the effects of swelling may be minimised by placing the clay at the highest practicable moisture and providing surcharge over the soil. It has been found that the inherent potential to swell depends on the plasticity index (Holtz and Gibbs, 1956 ; Seed *et. al.*, 1962) as indicated in Table 1.2.

Table 1.2. Volume Change Potential
(*Source: Holtz and Gibbs, 1956*)

Volune change potential	*Plasticity Index %*		*Shrinkage limit %*
	Arid area	*Humid area*	
Low	0 to 15	0 to 30	> 12
Moderate	15 to 30	30 to 50	10 to 12
High	> 30	> 50	< 10

1.4.2 Effect of Water Seepage and Surface Erosion

Mainly in sandy soils troubles occur due to water seepage and erosion. Internal erosion can result from carrying away of fine soil particles by ground water seeping in broken sewers or culverts and in careless technique of deep excavation below the water-table. The consequent loss of ground from beneath foundations may lead to collapse of structure. Subsidence also possible due to solution of minerals from the ground as a result of water seepage.

Surface erosion may occur due to loss of materials in strong winds or erosion by flowing water. Fine particles such as fine sands and silts and dry peat are very much susceptible for erosion by the winds. Surface erosion by flowing water may be severe if structures are constructed in the bottom of the valleys particularly where monsoon rains are very heavy.

Such actions of seeping water and erosion may lead to changes in the properties of soils and in soil profile which needs careful consideration while designing a structure. Erosion can be prevented by providing adequate depth of foundation, by growing suitable vegetation, or by blanketing the erodable soil by gravel, crushed rock or clay, *etc.*

1.4.3 Effect of Vegetation

Swelling and shrinking problem is also aggravated due to the effect of the roots of vegetation. The roots of trees, plants and shrubs consume considerable amount of water from the soil leading to the soil shrinkage. Roots of isolated trees can spread to a radius greater than the height of the tree. The removal of water by the roots causes shrinkage both vertically and horizontally. The root system causes two types of problems, *viz.*, (*i*) heave of foundation on sites which have recently been cleared of trees and hedges and (*ii*) settlement in existing structures sited close to growing trees or caused by subsequent planting of trees or shrubs close to them (Tomlinson, 1986). Thus, care should be taken to assess the settlement and also the forces tending to tear the foundations.

1.4.4. Effect of Temperature Variation

Both low and high temperature cause volume change leading to heave and shrinking respectively. When daily mean temperature remains below 0°C for a long period the soil moisture near the ground surface freezes. Continued sub-zero weather leads to increase in depth of freeze which results in rise of ground surface known as frost heave. The frozen water is concentrated into ice lenses or layers that lie parallel to the ground surface. Frost heave is rarely uniform but exerts enormous upward pressure causing damage to structures constructed on them. During warm weather the frozen soil thaws causing low shear strength and ground subsidence. The zone of soil subject to freezing and thawing depends on several factors such as type of soil or rock, the cover of vegetation, exposure to the sun, surface configuration, and ground water movements. The depth below ground surface to which 0°C temperature extends is called the frost line. The depth of frost line depends on the air temperature and its period and soil ability to conduct heat.

In permanently frozen ground, called permafrost, the frost heave effects are very severe. Permafrost can also occur in distinct layers, lenses, sheets or dikes separated by thawed material. The surface of permafrost is rarely stable, but varies with cyclic changes in climate and ground water flow.

Coarse-grained soils such as gravel and sand with no fines are rarely subject to freeze causing objectional heave. Fine sands and silts have the optimum combination of fine pores and relatively high permeability that results in maximum ice formation and heave. As clays have cracks and fissures they are susceptible for frost action.

Construction of any structure on or above the frozen ground needs a careful design because of the freeze-thaw property of the frozen ground.

When soil is subjected to very high temperature severe shrinkage cracks may occur. Such conditions may arise on soil beneath foundation of boilers, kilns and furnaces.

1.4.5 Effect of Vibration

It is a common experience that a sandy soil when subjected to vibrations from such sources as moving machinery, traffic, pile driving, blasting or earthquakes usually increase the density of sand and cause subsidence of its surface. It is also known vibration is one of the economical means of compacting loose sand layers. Hence, the effect of vibration may be harmful or beneficial. It has been shown that (Terzaghi and Peck, 1967) the settlement of a sand surface subjected to a pulsating load may be many a times greater than that produced by a peak static load. Experimental studies and field experience have shown that most serious settlements in sand due to vibrations are caused by high-frequency vibrations in the range of 50 to 2500 impulses per minute (Terzaghi and Peck, 1967).

Settlement caused by vibration on clay is usually small because of the cohesive bond between clay particles. Thus vibration does not cause any serious damage in clayey soils under any circumstances. Even a soft clay settles to a moderate extent when it is repeatedly subject to vibration at natural frequency. Whatever may be the subsoil conditions it is recommended to provide adequate provisions to reduce the amplitude of vibrations (Terzaghi and Peck, 1967).

The effect of blasting is equivalent to that produced by a mild earthquake. Damages caused by blasting is not basically due to settlement but from the transient ground motions and the air blast associated with the shock. Terzaghi and Peck (1967) also mentioned that long-continued traffic could produce considerable settlement.

1.4.6 Effect of Mining Subsidence and Pumping

Ground subsidence due to mining, pumping or dredging is generally of high magnitude. An earlier method of mining, particularly coal, was by sinking 'bell-pits', practised in medieval times. The subsidence is not sudden and occurs slowly with less magnitude. As mining techniques improved gradually. "Pillar and stall workings" technique was developed followed by the present day method of coal mining by "longwall workings". In the former method, support to the roof is given by methods known variously as 'pillar and stall', 'room and pillar' or 'bord and pillar'. Galleries are made from the shaft and the roof is supported by rectangular pillars of unworked coal. In the first phase, *i.e.*, on the onward workings, only 30 to 50% of the coal is extracted. In the second phase, *i.e.*, on the 'return workings' the pillars are removed either entirely to allow full collapse of the roof or partially to give continued support on some locations to safeguard the structures on the ground surface at such locations. By the longwall workings the coal face is continuously advanced over a long front. The roof close to the face is supported by props and the cavity left by the coal extraction is partially filled by waste material. When the props are removed the roof subsides resulting in slow settlement of the ground surface. In general, the trough or basin of subsidence occurs all round the area of coal extraction. Because of the sudden collapse of the roof, movement of ground surface occurs both vertically and horizontally. The movements are not uniform and it is difficult to predict the settlement accurately. Other forms of underground mineral extraction by mining methods give rise to similar subsidence problems. The extent and depth of subsidence can be greatly reduced by carefully planned extraction accompanied by recharge.

Excessive pumping from oil wells reduces the neutral stresses in the oil-bearing rocks and increases the effective stress. The rocks consolidate as the oil is removed and the ground surface sinks correspondingly. Similarly lowering of water table during construction activity, as a temporary or permanent measure, increases the effective load on the subsoil by an amount equal to the difference between the drained weight and the submerged weight of the mass of soil located between the original and the lowered water table. This increase of the effective overburden pressure causes additional compression leading to ground subsidence.

1.4.7 Effect of Construction Operation

Ground subsidence during construction may also occur due to increasing load on surrounding soil and excavation apart from vibrations and lowering of water table.

Load applied on one area of a ground surface above a soil may cause the surface of the adjacent soil to tilt. Practical significance of this aspect depends on the soil profile and the dimensions of the loaded area. In soft clay subsoil the magnitude and distribution of the settlement can be approximately assessed on the results of appropriate soil tests. This effect on the soft clay subsoil may be greater although not necessarily detrimental. In the sandy subsoil the settlement due to load on the adjoining site can not be computed reasonably but has to be estimated based only on the records of precedents.

Settlement due to excavation depends to a large extent on the type of bracing used and the care with which bracing is installed. The magnitude of settlement on the ground surface adjacent to an excavation can not be computed but a forecast can be made based only on reliable well-documented case records maintained for that area. Even in an open cut in sand without bracing, the settlement due to excavating the cut does not extend beyond a distance equal to the depth of the cut. If the cut is properly braced the maximum settlement may not be greater than 0.5% of depth of the cut. In an open cut in soft clay, the soil located at the sides of the cut act as a surcharge contributing for a lateral yield at the bottom of the cut towards the excavation. As a consequence of these movements, the ground surface located above the yielding clay settles.

1.5 RECLAIMED SOILS

The term reclaimed soil comprises of all materials deposited on a site using various methods for different purpose. It is justifiable to name the reclaimed materials as soil, when it comes to the purposes of construction of structures on them considering them as foundation materials. Industrial and commercial development of urban areas, development of navigation channels for ports and other water front structures require a large usable land which could be possible only by reclamation. Reclamations are also needed, though may be less important, for disposal of garbage, industrial wastes, paper sludge, mine tailings, *etc*. Reclamations may be on unusable low level land areas or on large bodies of water. These reclamations especially near large bodies of water lead to unsavory odours, greater turbidity and toxity of shore waters and affect in great extent the ecology of all marine life (Iyer, 1975). The geotechnical problems to be answered in these areas are control of settlement, increase of bearing capacity and biological stability of fill. Generally reclamation followed by ground treatment is preferable and economical than designing deep foundations.

1.5.1 Types of Reclamation Materials

The materials which are used in practice for reclamation purposes fall into the following groups (Iyer, 1975): (*i*) hydraulic fills of dredged soil, (*ii*) sanitary fill, (*iii*) paper sludge, (*iv*) flyash including slag, and (*v*) rubbish and debris.

Hydraulic Fills: For large reclamations hydraulic fills are most commonly used. The soil needed for this is usually obtained by developing it from the adjacent river, lake, or ocean to place it at the desired location. Generally a well graded soil is preferred for hydraulic fill. Sand deposited by hydraulic methods will be relatively in loose condition and has to be densified. Silt and clay hydraulic fills are difficult to compact after placement. Where time permits, such fill is left to consolidate and stabilize naturally. The main problems with hydraulic fills are low density, segregation and turbidity of the area. The extent of segregation in the fill and the amount of turbidity of the adjoining water depends on the type of dredger used. Quick dumping of well-graded material has been shown to produce good results in depths up to 15 m (Iyer, 1975).

Sanitary Fills: Sanitary landfills is considered as problem of chemical stability. In such fills, high moisture content due to rainfall, snow melt or otherwise reacts with the waste materials themselves to form a polluted liquid called **leachate** which often has dreadful characteristics. The leachate from such fills may give rise to pollution of drinking waters and bad odours. Periodic collection and treatment of leachate will be helpful. Production of methane and other gases cause explosions and fire hazards. Such fills experience large settlement due to movement of fine material into large void spaces, material loss due to chemical and biological changes, creep and consolidation. Miling of waste materials and bailing them before filling will be very much helpful to reduce the volume and the time required for compaction. Further fire hazards and fly menace are reduced.

Paper Sludge: Paper mill sludge is also used as a material for landfilling. This sludge consists of kaolinite and organic cellulose fibres with ash content ranging from 32 to 59%. The density of the fill is low and the shear strength increases with degree of consolidation and attains good bearing capacity with time.

Flyash and Slag: Flyash is a more stable material with a low percentage of solubility. Steel furnace slags have been in use as reclamation material. But blast furnace slags are not recommended as they have different properties than that of steel furnace slags. Incineration residues are also being used as fill materials. All these materials are of light weight and are highly alkaline.

Rubbish and Debris: These fill materials represent a most heterogeneous material ranging from stone, concrete pieces to paper, glass and grass. These are used as bottom portion of a fill and rolled. Top portion may be a structural fill or hydraulic fill. Such fills are highly compressible and a load test has to be conducted to evaluate its properties. Table 1.3 (Iyer, 1975) gives a summary of the different types of materials used as reclaimed soils and their associated problems.

Table 1.3. Materials used as Reclaimed Soils
(*Source: Iyer, 1975*)

Sl. No.	*Type of Material*	*Method of Placing*	*Method of Control*	*Problems Associated*	*Remedial Measures*
1.	Fairly Clean sand	Hydraulic fill	Field density *N*-values	Uniform fill of moderate density	Vibratory rolling Vibroflotation
2.	Silt or Clayey sand	Hydraulic fill	Field load tests on large plates	Heterogeneous fill of large voids	Preloading
3.	Stiff Cohesive soils	Hydraulic fill	Field load tests on large plates	Skeleton of clay bales matrix of sands and clay	Preloading
4.	Soft Cohesive soil and Mine tailings	Hydraulic fill	Laboratory tests	Laminated normally or under consolidated clay	Proloading with sand drains or stirring and drying/Adopt foundations rigid/flexible articulated type
5.	Paper sludge	Rolled fill	Laboratory chemical tests	Organic, underconsolidated material, chemical action	Preloading with sand layers
6.	Fly ash	Rolled or Hydraulic fill	Laboratory chemical tests	Chemically unstable-silt size-frost corrosive to foundation	Preloading, lime treatment
7.	Slag from open furnace	Rolled fill	Laboratory tests	Expansive-causes crack	Preloading, lime treatment
8.	Loading soil under wet condition (possibly lateritic)	Compaction Deposition in water	Field tests on large plates	Heterogeneous fill, leachate, gas explosion	Preloading
9.	Sanitary fill	Rolled fill	Field test	Subsidence, large secondary compression	Preloading, packing and loading

1.5.2 Construction Methods

When large volumes of soil must be excavated and transported, hydraulic methods may be economical. In cohesionless or slightly cohesive soil fills the hydraulic excavation is most effective. Jets of water under a high pressure as high as 1500 kN/m^2 through 50 to 100 mm nozzles wash the soil from the borrow location to sluices. Then the mixture of soil and water can be pumped and transported through pipes to the required location.

The suspension can be pumped several kilometers with only dredge pumps and booster pump can be used for still long distances.

Adequate care has to be exercised in selecting a suitable site for disposal of waste. The selected area should be above the maximum level to the local water table. It is advisable to create only a shallow landfill with the wastes because in deeper fills decomposition is slower. After due selection of a proper site for the fill a proper procedure has to be decided for landfilling the wastes.

Hydraulic fill Construction: If a hydraulic pipe line discharges directly on the ground pipe at its focus, a fan-shaped mound is formed. The deposited soil is coarsest at the outer boundaries and increasingly finer at points near the pool. The natural side slopes created by a flowing coarse soil slurry are usually in the range between five and ten horizontal to one vertical. Hydraulically transported soft clay soils may develop slopes as, shallow as 50 to 1.

If it is necessary to contain the area receiving the fill, dikes are constructed or sheetpiling is installed to forming a ponding area. A starter dike of 2 m high (first stage) is placed first around the area to be filled [Fig. 1.2 (*a*)]. The hydraulic pipe is laid on the dike and the discharge sluices or valves are installed at regular intervals. While the filling is processed the solids settle with the coarsest particles close to the dike and the fine particles move to the centre and settle. Excess silt and clay in suspension are discharged with water. In order to prevent water pollution, the unwanted fines are recovered in an auxiliary diked area. In general controlling the sedimentation time, coarser fill can be obtained. When the pond is full the height of dike is raised to the required second stage level [Fig. 1.2 (*b*)], and the process repeated till the required level is reached. Heights greater than 30 m is practically feasible with efficient control.

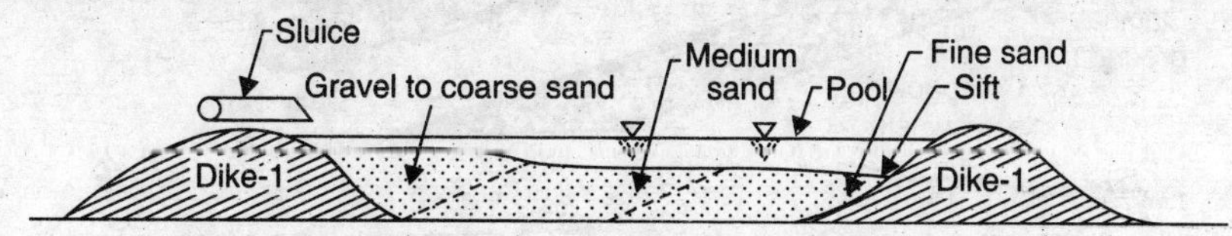

(a) Stage 1. Low dikes and initial pool

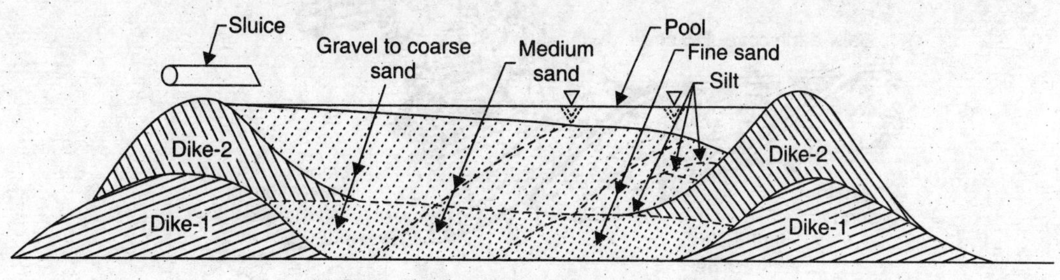

(b) Stage 2. Dikes built up form stage 1 filling

FIGURE 1.2. **Hydraulic Filling with Dikes.**

As the hydraulic fills are generally loose throughout the depth, proper densification or stabilization is needed if it is intended to support structural load. Hydraulic fill construction is generally cheap and stable land areas have been created. However, in certain cases expensive methods of densification may be required to make the fill suitable for its purpose. Hydraulic fill construction is suitable for land fill and fill along water front.

Landfilling Procedures for Waste Disposal: Wastes may fall into the categories of industrial, municipal and hazarduous. However, the step by step procedure for landfilling of these wastes is identically same-to take the waste material, place it in a suitably lined area, backfill around it as it is being placed, and eventually cover it over in an environmentally safe and secure manner. Obviously more care and greater precautions should be taken while handling highly toxic materials. The problems of wastes are enormous. Although different new methods of waste treatment are proposed, the paramount task is to provide a safe and secure landfill. The disposal of waste of all categories by the three means of landfilling procedures are discussed below.

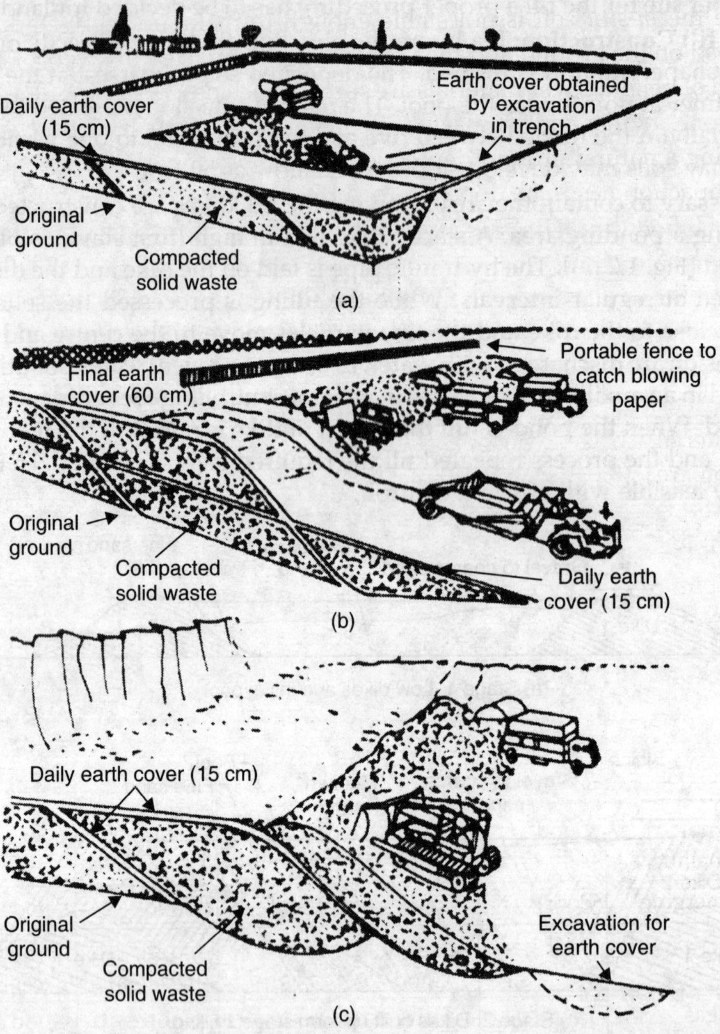

FIGURE 1.3. **Typical Sanitary Landfill Showing Three Different Methods of Operation : (*a*) Trench Method, Where Trucks Deposit the Loads into the Trench and a Bulldozer Spreads and Compacts, and Covers Them Over at the End of the Day ; (*b*) Area Method, Where a Bulldozer Spreads out the Refuse and a Scraper Hauls the Cover Material; and (*c*) Ramp Method, Where the Refuse is Compacted as Before but the Cover is Excavated from Directly in Front of the Working Face.** (*Source: Brunner and Keller, 1972*)

Trench method is suitable when adequate backfill material is present and the site has a high water table and the volume of waste to be disposed is relatively small. The chosen site should have sufficient strength and stability of side slopes to avoid a collapse. Wastes are dumped from the sides of the trench and spread into layers which is compacted simultaneously by the spreading of bulldozers. As far as possible, at the end of each day the final height of fill shculd be reached for the entire width. The necessary backfill material is obtained by extending the length of the trench or by starting a new and parallel trench. Each day's deposit of waste must be completely covered before work ends for the day (Fig. 1.3).

Area method is suitable where large volume of waste has to be handled and landfilled and at the same time the *in-situ* soil is not stable enough to maintain the well-defined slopes. The wastes are dumped on the ground and then spread and compacted in layers of 45 and 75 cm each. For every 2 to 3 m of height, cover materials are spread. Each day's covered lift is called a **cell**. Cells are stacked upon one another and compacted adequately. This method of successive cell application is used when filling a natural ground depression, ravine or canyon. If the number of staked cells exceed three, compaction has to be done in an efficient way.

The third method is called rapid method which is an adaptation of the area method. This is suitable when large volumes of waste are to the disposed and a very low water table exists at the location. Usually a single cell is made to the final height but the cover is taken from directly in front of the working face. Thus the cells in this method are thicker and shorter than the area method.

The selection of soil for covering the waste material is very important. Suitability of different soils for landfill cover is presented in Table 1.4 (Brunner and Keller, 1972). As far as possible a suitable cover material, available near the area of fill should be used. If such a soil is not available near the landfilling area, the available soil may be used irrespective of its suitability. However, the final cover over the upper most cell should be the one which could be almost near to the one recommended in Table 1.4.

Table 1.4. Suitability of General Soil Types for Use as Landfill Cover Material (*Source: Brunner and Keller, 1972*)

Function	*Clean Gravel*	*Clayey-silty Gravel*	*Clean Sand*	*Clayey-silty Sand*	*Silt*	*Clay*
Workability, placement of cover	Good	Excellent	Excellent	Good	Fair	Poor
Prevent rodents from borrowing or tunnelling	Good	Fair to Good	Good	Poor	Poor	Poor
Keep flies from emerging	Poor	Fair	Poor	Good	Good	Excellent*
Minimise moisture entering fill	Poor	Fair to Good	Poor	Good to Excellent	Good to Excellent	Excellent*
Minimise landfill gas venting through cover	Poor	Fair to Good	Poor	Good to Excellent	Good to Excellent	Excellent*
Provide pleasing appearance and control blowing paper	Excellent	Excellent	Excellent	Excellent	Excellent	Excellent
Grow vegetation	Poor	Good	Poor to Fair	Excellent	Good to Excellent	Fair to Good
Be permeable for venting decomposition gas**	Excellent	Poor	Good	Poor	Poor	Poor

*Except when cracks extend through the entire cover.

**Only if well drained.

The wastes material deposited usually have a dry density of the order of 1.57 to 6.28 kN/m^3 and moisture content of 10 to 35% with low bearing capacities, *e.g.*, 19.0 to 34 kN/m^2.

Natural consolidation of sill is very slow and can not be used without improvement. The landfilled site can be better utilized after consolidating with the aid of a surcharge fill. As most of the wastes are combustible in nature, settlements as high as 40% of the original thickness of the fill can occur. The time required for complete consolidation may take one to four years.

During the settlement process there are possibilities of gas generation within the degrading fill and creation of leachate.

1.5.3 Landfill Gas and Leachate

Most landfill sites produce methane and carbon dioxide gases. If an impermeable cover is provided over the entire landfill, methane can move horizontally beyond the boundaries of the fill then escape to the atmosphere. As the methane gas is harmful and creates health disorders it is advisable to properly vent the gas.

As carbon dioxide is 1.5 times denser than air and 2.7 times denser than methane it moves to the bottom of the landfill. Since it is readily soluble in water, it may mix with the leachate and enter into the groundwater. Dissolved carbon dioxide in water reduces the pH of water and thereby increases the hardness and mineral content of water. A sand-gravel drainage layer of a suitable geotextile can be provided above the bottom of the landfill with a proper collection system for gathering the gas (and the leachate) for proper disposal at the ground surface (Koerner, 1985).

Leachate should be properly collected above the primary liner (if provided) at the bottom of the landfill, treated adequately and disposed as per the stipulated norms. But many of the existing landfills are unlined. In an unlined fill or if the liner leaks, the leachate will escape from the cell to the surrounding soil. The basic need is not to allow the leachate escape from the landfill. This could be achieved only by providing liners or containment walls.

1.5.4 Landfill Liners and Caps

Both bottom land and side slopes of a landfill area need protection by liners. A wide range of materials are used to act as liners. In general the liners may be constructed of rigid or flexible materials. The rigid liners are shotcrete or gunite, concrete, soil-cement, bituminous-concrete, and bituminous panels. The flexible liners are compacted soils, chemically treated soils, bentonite clays and geomembranes.

Irrespective of the type of material, adequate site preparation and strengthening the soil subgrade is simply mandatory for a successful performance of a lining system. This aspect is very important as no liner, whatever may be its rigidity, can withstand large differential settlement.

Inspite of a perfectly designed and constructed lining the system can not provide a zero seepage loss system. Thus an underdrain system can be placed beneath the primary linear to collect any leakage passing through it. This collected leachate may be treated and disposed or recirculated back to the containment area. The present day practice, particularly for hazardous waste landfills, is to provide a secondary liner beneath the underdrain system. For a detailed treatment on liners, reference may be made to Koerner (1985).

When a landfill is realised to be a hazard in its current condition or whenever landfill has ceased to be usable it should be capped based on the environmental norms such that it will be safe and secure. The flow of leachates generated from landfills to the ground water system may be prevented by construction of seepage cut-off walls. Most seepage cut-off walls are constructed by slurry method and are backfilled with a mixture of soil and bentonite.

1.6 NATURAL OFFSHORE DEPOSITS

The discussions made so far are on the variety of on-land soils. As civil engineering activity is confined to the coastal region, no special interest was shown on offshore soil deposits. But in the recent years on account of the need to exploit for oil and gas reserves primarily located offshore, knowledge on offshore deposits gained engineering importance.

Coastline structures are constructed in depths of water varying from 0 to 15 m on soils that are almost similar to on-land soils. Now, offshore structures are built on water depths varying from 50 to 300 m for exploitation of offshore resources. These structures are intended for production, processing and transportation of oil. Invariably these structures are founded as sealed which are built on different types of soils and subjected to different types of loading conditions. Thus a design engineer has to know the nature and behaviour of offshore soils and the design procedure for offshore foundations. The first aspect is discussed below and for the design aspects advanced books on foundation design have to be referred to.

1.6.1 Terrigenic Soils

Soils which originate or found on land are called as **terrestrial** soils. Soils in the seas and oceans which are originated on land and then transported and deposited on the seabed are referred to as **terrigenic** soils.

In general a close relationship exists between terrestrial soils near the coast and those to the nearby offshore region. It is reported that the terrigenic soils in the Bay of Bengal are rich in kaolinite, illite and chlorite which are the minerals present in soils that originate in the Himalayas and are transported by the rivers of Ganges and Brahmaputra. Likewise, the terrigenic soils surrounding the Indian peninsular coastline are reported to be rich in montmorillonite since rivers from the Deccan carry soil into the region. The interaction of river sediments in sea and the sea environment produces some special deposits. Some such formations of deposits are given below.

Calcareous clays are terrigenic soils which are found in the seas between latitudes 30° North and 30° South. Such deposits are encountered in abundance in the Bombay high region. These clays are found to exist in the form of needle like particle sizes of silt and clay. It is attributed to the result of precipitation of calcium carbonate on account of some reaction which could have occurred in the sea water. These particles are found to be electrical inert. In some of the calcareous clays, calcium carbonate was reported to exist as a deposit at particle contacts in the form of a cementing material.

Rivers bring sediment which is deposited on the seabed. Subsequent sediments are deposited on the sediment layers which are already deposited. Each deposition induces excess pore water pressure which has less time for dissipation resulting in very less effective stress. That is the deposit in total has yet to consolidate to the hydrostatic condition and said to be under-consolidated. This is attributed to the rate of deposition of sedimentation is faster than the time required for consolidation. Such deposits have low shear strength and high compressibility which are not suitable for structures to be founded. In India deposits of highly under-consolidated clays are encountered in the sea near the Godavari and Krishna basins to a maximum depth of 100 m.

1.6.2 Pelagic Soils

Another type of soil originate entirely under the offshore environment which are called pelagic soils. Pelagic soils are reported to originate (*i*) biogenetically, from the skeletical remains of sea animals, (*ii*) chemically, through precipitation of chemicals and minerals in sea water and (*iii*) as atmospheric dust settles through water in the ocean onto the seabed. Such deposits evidently accumulate very, very slowly and may exist at very high void ratios. The material is called **ooze** which have very little shear strength and very high compressibility. These soils are yet to be considered as a geotechnical material.

1.7 GROUND IMPROVEMENT POTENTIAL

In the aforementioned sections it has been discussed in depth about the formation of soil and rock, soil distribution in India, conditions (both natural and man-made) which are formed due to reclamation. All the above ground conditions are not one and the same but different in variety of ways. The soil and rock conditions can be placed under three general categories, *viz.*, hazardous, poor and favourable (Hunt, 1986) so as to identify the gravity of the ground condition which will enable the engineer to decide a proper treatment-approach and or a design approach.

Hazardous: A regional or a local field condition is such that a regular design-approach or an economical treatment technique may not be feasible and construction in such a location may result in ultimate disaster. As far as possible such locations should be avoided.

Poor: A local condition including regional conditions which may require special design and/or special treatment for development.

Favourable: A local condition including regional conditions for which normal design and ground treatments are suitable.

1.7.1 Hazardous Ground Conditions

Construction on sites located on or in close proximity to faults, particularly in seismically active regions, may serve ground shocks. Such locations are totally unsuitable and hazardous and should be avoided.

Loose to medium dense fine sands may easily be susceptible for quicksand condition leading to liquefaction when such sites are located in seismically active regions and adjacent to deep rivers and bays. Due to liquefaction loss of ground support and lateral movement could occur. In such locations the treatment will be costly and the potential risk also is very high.

Ground collapse may take place in locations underlain by dormant or active mines or cavernous limestones. Collapse may also occur in such locations if the ground water from adjacent site is drawn enormously. Any design approach to consider this factor is not feasible and the treatment costs are high and potential risk also high.

Some of the natural slopes in glacial lacustrine clays, clay shales, colluvium, thick deposits of residual soils, dipping stratified rock masses and any deep cuts are susceptible for slope failures in the form of landslides, avalanches, flows, *etc*. Such failures will be sudden and occur more in moist climates. Inevitably such locations are hazardous for construction activity.

Flood plains and other relatively low ground may be quite often exposed to seasonal floods and may lead to ground subsidence. Such locations need elaborate flood protection measures such as construction of adequately high grades, dikes and surface drainage control system. No proper design-approach or treatment technique is feasible.

Any landfill of hazardous waste should never be selected for construction or any other activity irrespective of the age of the fill.

Thus the above ground conditions may be categorized as hazardous and such locations should be totally avoided.

1.7.2 Poor Ground Conditions

Loess, porous lightly cemented clays, low-density recent alluvium of arid-climate valleys may collapse on saturation followed by subsidence. In such site condition if it is feasible, saturation may be prevented, or precollapsed before construction by flooding or the structure may be designed with large allowable settlement and thereby make the ground usable.

Expansive clays and rocks including the black cotton soils of India undergo large volume changes with changes in moisture contents. Prevention of swelling lead to high uplift pressure and in dry weather experiences large network of cracks. Similarly behaviour could be expected on soils susceptible for freeezing which show frost heave-thaw behaviour similar to swelling-shrinkage behaviour of expansive clays. In these regions the active zone has to be identified and structures should be designed to meet the behaviour. If the depth of active zone is less in expansive clays a suitable ground treatment technique can be applied.

Soft to firm clays generally have low bearing capacity and some of the clays may be highly sensitive to disturbance and some are highly fissured. Such clays generally undergo long-term consolidation of significant magnitude. Such locations can be easily handled by design of suitable deep foundations or by suitable treatment techniques.

Organic soils are highly compressible in nature. If the depth of organic soil is less a suitable ground improvement technique is feasible or else the organic soil may be removed and replaced with a better soil. On the other hand a suitable deep foundation can be designed with the consideration of negative force on the deep foundation.

Loose sand and silts may need a proper treatment, in particular if deep foundations are planned.

Ground water location, although is not a soil or rock condition, is an important factor to be considered in dealing with any of the field problems, design methods and treatment techniques.

1.7.3 Favourable Ground Conditions

Cohesive granular soils such as sandy-clay mixtures, are relatively strong and form good supporting medium for moderately to heavily loaded foundations.

Cohesionless granular soils, such as medium dense to dense sands and sand-gravel mixtures provide excellent foundation conditions for most loading conditions.

Shallow rock without any discontinuities provides the best foundation to support any type of loading. However, its supporting capacity depends on rock quality, characteristics of discontinuities and long-term properties.

1.7.4 Alternative Approaches

As discussed above, *in-situ* ground characteristics of construction site are different from those desired and, almost always, form ideal for a designated need. In urban environment sites with favourable support conditions become scarce. In many a situations, in order to satisfy client's desire the engineer may be compelled to construct structures at locations selected for reasons other than ground support conditions. Thus, it has become increasingly important for the engineer to know the extent to which alterations have to be made for the construction of an intended structure at a stipulated site.

If unsuitable site conditions are encountered at the desired site of a proposed structure, one of the following four procedures may be adopted to ensure satisfactory performance of the structure (Mitchell, 1976):

(*i*) By pass the unsuitable soil by means of deep foundations extending to a suitable bearing material.

(*ii*) Redesign the structure and its foundations for support by the poor soil, a procedure that may not be either feasible or economical.

(*iii*) Remove that poor material and either treat it to improve and replace it or substitute for it by a suitable material.

(*iv*) Treat the soil in place or improve its properties.

1.7.5 Geotechnical Processes

Ground improvement in its broadest sense is the alteration of any property of a soil or rock to improve its engineering performance. This may be a permanent measure to improve the completed facility or a temporary process to allow the construction of a facility. Various process of ground improvement are available to increase the strength, reduce compressibility, reduce permeability or improve ground water condition. Further, if there is any foundation distress in the existing structures, in-place foundation treatment can be applied to rehabilitate the structure. The techniques involved in the attainment of the required improvement facilities is referred to as geotechnical processes.

Ground improvement techniques may be classified based on the nature of the process, involved, material used, the desired result, *etc.* Various techniques discussed in the subsequent chapters are compaction, drainage methods, precompression and vertical drains, vibration methods, grouting and injection, mechanical, cementing and chemical stabilisation, geosynthetics, and miscellaneous methods. The factors that must be considered in the selection of the best technique in any case include the following (Mitchell, 1976):

(*i*) Soil type—soil, clay, organic, *etc*.

(*ii*) Area and depth of treatment required—depend on the geometric characteristics of the soil deposit and the nature of facilities proposed for construction.

(*iii*) Type of structure and load distribution.

(*iv*) Soil properties—strength, compressibility, permeability, *etc*.

(*v*) Permissible total and differential settlements.

(*vi*) Availability of materials—stone, sand, water, admixture, stabiliser *etc*.

(*vii*) Availability of skills and equipment.

(*viii*) Environmental considerations—waste disposal, erosion, water pollution, *etc*.

(*ix*) Local experience and preferences.

(*x*) Economics.

The discussion on the following chapters on different techniques is directed primarily at their use as a permanent measure. However, some of them may be used to expedite or facilitate construction as well.

Chapter 2 COMPACTION

2.1 INTRODUCTION

Soil is exclusively used as a basic construction material from the existence of earth structures such as dams, embankments, levees, *etc*. Where the natural topography is undulated and needs to be changed to accommodate a building, highway or any other civil engineering development, soil is generally the material used for filling in low locations.

It is essential that the in-place soil should possess certain properties to withstand the forces caused by structures. Soils should have adequate strength, should be capable to resist settlement or heave, should possess proper permeability and should be durable and safe against deterioration. These desirable features can be achieved by compacting the in-place soil or by providing a compacted fill with proper selection of the fill soil type. Soil in the field is compacted by applying energy in the form of pressure (rolling), impact (ramming) or vibration. This Chapter deals with the different aspects of surface compaction with reference to ground improvement.

2.2 COMPACTION MECHANICS

Excavating a soil from a place and redepositing them for construction purposes without special care will have high average porosity, permeability and compressibility without having adequate strength and stability. It was customary even in ancient times to compact fills to be used as dams or levees Thus, it is generally a matter of economics to place the fill at the highest density that can be obtained practically in the field and that can be maintained during the entire life span, of the structure. In order to have a logical and optimal design procedure one should know the density that the material has to be compacted such that the strength, compressibility and permeability characteristics of the material can be determined. As the strength and other properties of most soils vary with the moisture at which the soil is placed, it is necessary to place the soil at a particular moisture content to obtain the design density. Therefore, for a proper design the required density and the corresponding moulding water content, particularly for cohesive soils, should be known.

2.2.1 Densification

Densification or reduction in void volume may occur in a number of ways such as re-orientation of particles, fracture of the grains or the bonds between them followed by re-orientation, and bending or distortion of the particles and their adsorbed layers. In a cohesive soil the densification is primarily attained by distortion and re-orientation which are resisted by inter-particle forces. In a cohesionless or coarse-grained soil the densification is primarily accomplished by re-orientation of grains which is resisted by the friction between the particles.

Moisture plays a major role in the process of densification. For example, capillary tension in moisture films between the grains cause an increase in contact pressure and thus a temporary increase in friction. But a continued increase in moisture content, decreases the capillary tension and reduces the resistance due to friction resulting in more densification. If the moisture content is still higher because of building up of neutral stress which prevents further reduction in void ratio, and the additional effort is wasted. Therefore, saturation is the theoretical limit for compaction at any water content.

There is no one method of compaction which is equally suitable for all types of soil. However, the energy consumed in a compaction process is due to the compactive efforts of the compaction device. The effectiveness of the energy depends on the type of particles of which the fill is composed and on the way in which the effort is applied.

2.2.2 Moisture-Density Relationship

Compaction is measured quantitatively in terms of dry density. The increase in the dry density of soil, brought by the process of compaction, depends on the moisture content of the soil and on the energy input by the process of compaction. For a given amount of compaction (*i.e.*, compactive effort) there exists for each soil a moisture content termed the "optimum moisture content" at which a maximum dry density is achieved.

The engineering behaviour of soil under compaction for a given amount of compaction primarily depends on the moisture content. At low moisture contents, the soil is stiff and difficult to compress and results in low dry densities with high air contents (Fig. 2.1). An increase in moisture content, causes the water to act as a lubricant and thus makes the soil more workable and accordingly results in higher densities with lower air contents. As the air content becomes less, the water and air in combination tend to keep the particles apart and prevent any appreciable decrease in air content. However, the total voids continue to increase with moisture content, resulting in further decrease in dry density.

In Fig. 2.1 the zero air void or the 100% saturation line is shown. It could be observed that the saturation line beyond the peak approaches the moisture-density curve but never reaches since it is never possible to expel all the entrapped air in the voids of the soil by compaction.

The maximum dry density that can be obtained either in the laboratory tests or in the field with any soil depends upon its type and varies from about 22 kN/m^3 for a well graded gravel to about 14 kN/m^3 for a heavy clay. Figure 2.2 shows the moisture-density curves for different soils. Curves for well-graded granular soils and clay-sand mixtures usually have a well-defined peak. Moisture density curves for clay soils usually have more gently rounded peaks similar to uniformly graded sands. In general, a flat moisture-density curve denotes a closely graded soil whereas a well graded soil gives a curve with pronounced peak. This is illustrated in Fig. 2.3 for two sands.

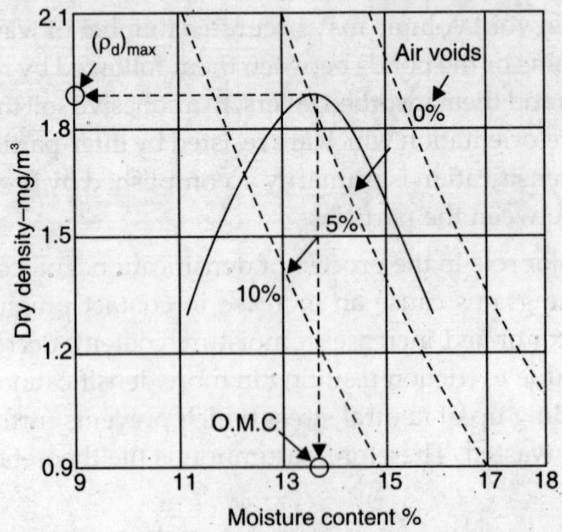

FIGURE 2.1. **Dry Density-Moisture Content Relationship.**

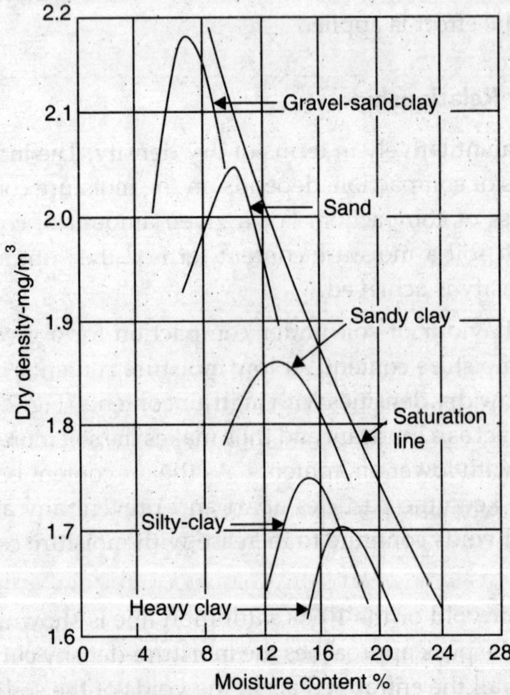

FIGURE 2.2. **Typical Compaction Curves for Different Soils.**

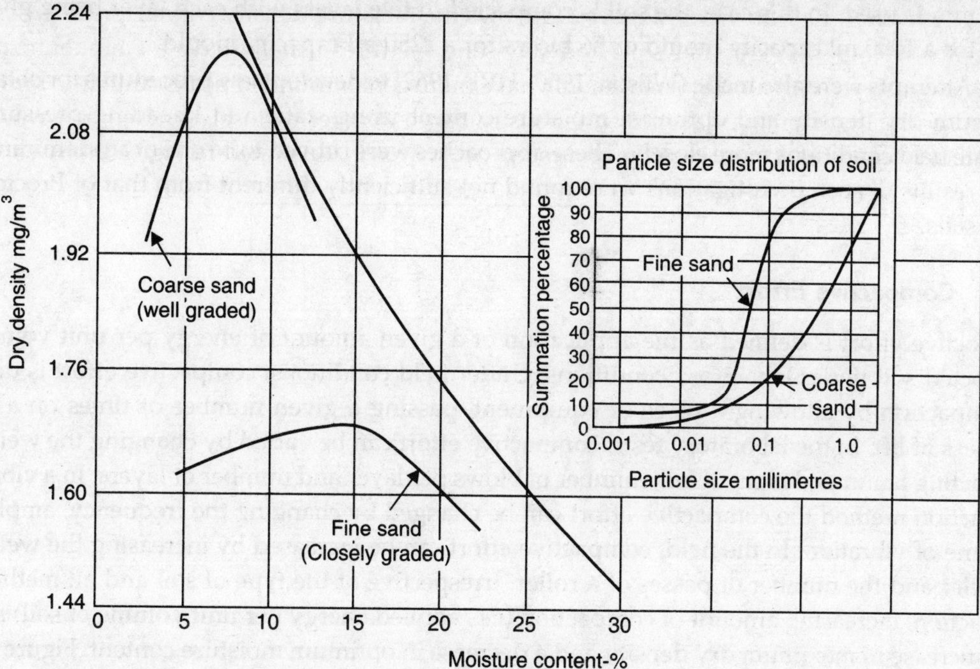

FIGURE 2.3. **Dry Density *Vs* Moisture Content Curves for Two Grades of Sands.**
(*Source: HMSO, 1974*)

2.2.3 Laboratory Compaction Tests

Different arbitrary standards have been developed to simulate the various range of amounts of effort as applied by the full sized equipment used in soil construction. However, the simplest and the most widely used are the "Proctor tests" named after R.R. Proctor who first developed the optimum moisture content-maximum dry density concept (Proctor, 1933). Two such tests are : (*i*) the Standard Proctor test (IS : 2720—Part 7, 1970) which causes adequate compaction for most applications such as backfills, highway fills and earth dams, and (*ii*) the Modified Proctor test (IS : 2720—Part 8, 1983) which is used for heavier load applications, such as airfield and highway base courses.

These tests are performed by compacting the wet soil sample into a mould in a specified number of layers. Each layer is compacted at a stipulated compactive effort. The compactive effort is measured in terms of energy per unit volume of compacted soil. The required compactive effort is attained by controlling the weight of hammer, height of drop, number of layers, and number of blows for each layer. After compacting the final layer the bulk density of the soil and moisture content are determined. Tests are repeated on fresh samples with increased moisture contents. From the values of bulk density and moisture content obtained, the dry density is calculated. A plot of dry density versus moisture content is made.

The Standard Proctor compaction test adopts 1000 ml capacity mould, 2.6 kg ram with a height of fall 310 mm, three layers and 25 blows per layer. Indian Standard (IS : 2720—Part 7, 1974) also recommends a 2250 ml capacity mould with 56 blows per layer keeping the same rammer weight and number of layers. For Modified Proctor test a rammer with a mass of 4.89 kg and a fall of 450 mm is used. In this case, the soil is compacted in five layers with each layer being given 25 blows for a 1000 ml capacity mould or 56 blows for a 2250 ml capacity mould.

Attempts were also made (Wilson, 1950; HRB, 1962) to develop new procedures for obtaining maximum dry density and optimum moisture content using static and kneading pressures, to simulate field conditions more closely. These approaches were proved to be one of academic interest as the results of such investigations were found not sufficiently different from that of Proctor's in many soils.

2.2.4 Compactive Effort

Compactive effort is defined as the application of a given amount of energy per unit volume of compacted soil under laboratory conditions. Under field conditions, compactive effort is defined as compaction by allowing a piece of equipment, passing a given number of times on a given thickness of lift. In the laboratory tests, compactive effort can be varied by changing the weight of compacting hammer, height of fall, number of blows per layer and number of layers. In a vibratory compaction method the compactive effort can be changed by changing the frequency, amplitude, and time of vibration. In the field, compactive effort can be increased by increasing the weight of the roller and the number of passes of a roller. Irrespective of the type of soil and all methods of compaction, increasing amount of compaction (*i.e.*, applied energy per unit volume of soil) results in an increase in maximum dry density and a decrease in optimum moisture content. Figure 2.4 (*a*) shows soils compacted at two different compactive efforts. A comparison of these curves reveal that above the optimum moisture content, when the percent air voids is small, increase in compaction has little effect upon the dry unit weight but below the optimum moisture content, when the percent air voids is large, the effect of increase in compaction is considerable.

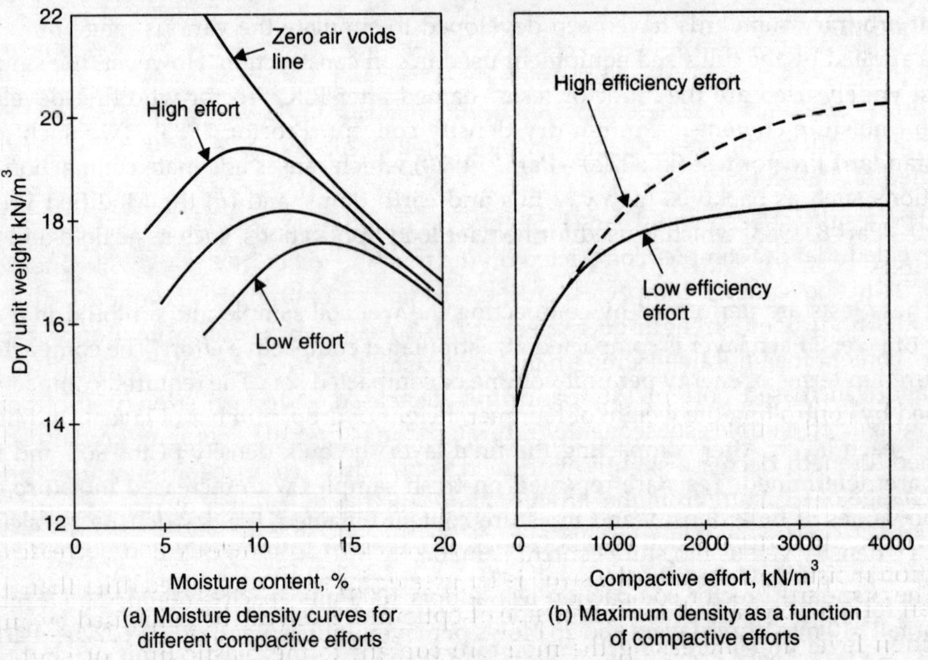

***FIGURE 2.4.* Effect of Compaction Effort.**

(*Source: Sowers, 1979*)

The relationship between compactive effort and maximum dry unit weight is shown in Fig. 2.4 (*b*). It could be seen that the relationship is not linear and only a small increase in density could be achieved by a large increase in compactive effort. Further, the method of application of compactive effort has a significant role on the density. In coarse-grained soils vibration, which reduces the friction between the grains, is particularly effective. On the other hand in cohesive soils, pressure that bends and forces the grains into new positions is better. Repeated application of small pressures can not be effective in cohesive soils compared to single large application and the duration of the effort sometimes influence the density. In coarse-grained soils with high water contents, the neutral stress that resists compaction will not build up if the effort is applied slowly.

2.2.5 Engineering Behaviour of Compacted Fine-grained Soils

The physical properties of a compacted soil depends on the soil type, moisture content, density and soil structure. In cohesive soils at low moisture contents less than optimum, an aggregated structure called flocculated structure is formed and at higher moisture contents a dispersed structure is formed with flaky particles aligned in parallel. The nature and magnitude of compaction in a fine-grained soil has a significant influence on the engineering behaviour of compacted soil. Lambe's (1958, 1962) exhaustive study on compacted soils has thrown more light on better understanding of compacted fine-grained soils.

It has been found that an increase in moulding water content causes a decrease in permeability of samples compacted dry-side of optimum moisture content whereas at wet-side of optimum moisture content there is only slight increase. Further an increase in compactive effort is found to reduce the permeability since it increases the dry density and thereby reducing the voids available for flow and increases the particle-orientation.

Comparing samples compacted on the dry-side of optimum and on the wet-side of optimum with same dry density and subsequently saturated, the sample compacted on the wet-side is found to compress more than the one compacted on the dry-side at low stress levels. But at high stress levels the sample compacted on the dry-side is observed to compress more than the sample compacted on the wet-side.

The results of Seed and Chan (1959) based on the study of stress-strain behaviour of soils have revealed that the samples compacted on the dry-side of optimum tend to be more rigid and stronger than the samples compacted on the wet-side of optimum. It is found that at constant density the undrained strength decreases with increasing moisture content and at a constant moisture content the undrained strength increases with increase in density. The increasing density brings about increased pore pressure and thus decreased undrained strength (Seed and Chan, 1959). Compacted samples subsequently saturated under constant confining pressure showed an undrained strength decrease slightly with increasing initial (compaction) moisture, presumably due to change of structure from the stronger aggregated to the weaker dispersed form.

It has been observed that swelling of clays increases with density and decreases with the compaction moisture. Generally, the swell is far greater for a soil compacted drier than the plastic limit than for one compacted on the wet-side of optimum. Swell can be minimised by limiting the compaction level and increasing the moisture content to the plastic limit or slightly above (Sowers, 1979).

Table 2.1 compares the salient properties of samples compacted on the dry-side of optimum with samples compacted on the wet-side of optimum (Lambe and Whitman, 1979). Thus an engineer designing a subgrade, a sub-base, a dam or an embankment should consider the behaviour of the soil during the entire life span of the structure.

Table 2.1. Comparison of Compaction on Dry-of-Optimum with Wet-of-Optimum
(Source: Lambe and Whitman, 1979)

Property	*Compaction*
Structure	
Particle arrangement	Dry-side more random
Water deficiency	Dry-side more deficiency, therefore more water imbibed, more swell, lower pore pressure.
Permanence	Dry-side structure more sensitive to change
Permeability	
Magnitude	Dry-side more permeable
Permanence	Dry-side permeability reduced, much more by permeation.
Compressibilty	
Magnitude	Wet-side more compressible in low stress range, dry-side in high stress range
Rate	Dry-side consolidates more rapidly
Strength	
As moulded	
Undrained	Dry-side much higher
Drained	Dry-side somewhat higher
After saturation	
Undrained	Dry-side somewhat higher if swelling prevented; Wet-side can be higher if swelling permitted.
Drained	Dry-side about the same or slightly greater
Pore water pressure at failure	Wet-side higher
Stress-strain modulus	Dry-side much greater
Sensitivity	Dry-side more apt to be sensitive.

2.3 FIELD PROCEDURE

The field procedure for constructing a controlled compacted fill is simple to practice. The fill soils are identified in a borrow pit. The soil is processed in the pit for the required moisture content before transportation. The processed soil is transported to an area being worked. It is spread in thin layers and additional water is added if necessary. Each layer is compacted to a predetermined density by a suitable compaction devices. This procedure is explained in the subsequent paragraphs.

2.3.1 Controlled Compacted Fill

When soil is used for construction purposes it is placed in layers, each layer is being compacted before being covered by a subsequent layer and compacted, till a final required elevation and shape is attained. The compacted soil should have adequate strength and stability that are as good or better than many natural soil formations. Earth structures such as dams and embankments are primarily compacted soil which could withstand all the forces to which they are subjected. Similarly earth fills can support buildings, highways, *etc.* Such fills are referred to as controlled compacted earth fill or structural earth fill. Application of controlled compacted earth fill may be categorized under five groups : (*i*) fills used to support structures and pavements, (*ii*) compaction of pavement subgrade, (*iii*) construction of earth-and-rock fill dams, (*iv*) construction of embankments, and (*v*) compaction of pavement layers, *viz.*, base and sub base courses. Only the first two aspects are relevant to this book.

Preparing sites for buildings where controlled fill is required can fall into the following three categories (Hunt. 1986):

1. Fills placed on a stripped, compacted subgrade to raise general site grade for such purposes as flood protection or to provide dock-high loading in an industrial building.
2. Fills placed for structures after excavating weak soils from lowlands (the excavation-backfill method).
3. Fills combined with benches that are constructed for hill side development.

2.3.2 Pavement Subgrade

It is important that the subgrade of a road should have adequate stability. Highly organic soils and peat as subgrade should be avoided or should be replaced with more stable materials. In very cold regions, where severe frosts occur, frost susceptible soils such as silts and chalks should be excluded or the route alignment should be changed. In case it is inevitable to change the route alignment, the material has to be replaced with non-susceptible material or a thick non-susceptible material has to be placed on the top of subgrade as fill and compacted. In either case the replaced material would function as the base. In certain areas where considerable seasonal moisture changes occur resulting in volume change in clay subgrade, especially in expansive clays like black-cotton soils, has to be minimised by a suitable ground improvement technique. In order to ensure adequate stability the sub-grade should be thoroughly compacted when the soil is at a favourable moisture content.

2.3.3 Material Selection Factors

Rock and soil are among the oldest construction materials. They are locally available, readily handled by a variety of techniques, durable and cheap. For these known reasons only the modern engineer also has been using soil and rock for construction. The field engineer must be familiar with the materials and the construction operations that are necessary to obtain the required results.

Soils have various levels of quality for use as a fill. The selection of borrow material for fill depends on the purpose of the fill, availability of the material, and climatic conditions during placement. In general organic soils and clays are not used for fill. However, clays are used in earth dams as an impervious medium called as core.

For structural foundations and floors well-graded granular soils with less than 10 to 15% fines are preferred as they have high allowable soil pressure when properly compacted and at the same time have excellent drainage. In such soils, building up of moisture from capillary forces resulting in building dampness and possible frost heaving is minimised. Generally silts and clays are avoided as adequate support capacity by compaction is not feasible. Further such soils may pose construction difficulties arising from their response to increased moisture during rainy periods. Any cohesive mixture may be used provided capillary cut-offs are installed immediately beneath floor of the structure. Strength of compacted fill from such soils are generally lower than granular soil fills. Hydraulic fills of silt-sand mixtures, uncompacted except for surface compaction at final grade, have been used to support lightly loaded structures such as houses.

Earth and rockfill dams and embankments are constructed using a wide range of materials. The major objective of a dam is to act as a water barrier whereas an embankment is required to support only its own weight and that of traffic. In general organic soils, soft clays and silts are avoided.

Fills and subgrades beneath pavements should have materials with free-draining properties so as to prevent moisture build up from capillary forces leading to frost heave, expansion of clayey soils or pumping of softened fine-grained soils under wheel loads.

2.3.4 Borrow Pits

During the process of site investigation itself, it is necessary to locate suitable borrow areas. Evaluation are generally made, about the availability on-site materials from cuts and excavations and from off-site borrow sources.

Requirements of source material for fill depends on the project type. For a building, fill may require only one type of select material. On the other hand for a dam a wide range of materials are required to use in the different parts of an earth dam to satisfy different functions.

Hydraulically excavated soil can make excellent fill when they are non-organic and coarse-grained. In case of load bearing fills stock piling and drying are usually necessary prior to placement and compaction. In order to obtain efficient operations of the excavation equipment, moisture control is often necessary in the borrow pit. Plowing and exposure to air and sunshine often help to dry the soil. Addition of moisture makes it easier to excavate hard and dry soils. If the soil is drier than optimum, addition of water is necessary for compaction. When the moisture is added at the borrow pit, it is mixed uniformly.

In order to ensure the quality of the materials, adequate; supervision of the borrow pit is necessary. This requires experienced technicians who can recognize the specified soils and take adequate steps for checking moisture in borrow pits, during placement and rolling.

2.3.5 Placement and Processing

The placement of the fill materials depends on the method of handling, and the size of the area to be filled. The materials excavated by shovels, draglines, and elevating grades are handled by trucks and wagons. The scraper spreads the materials in ribbon-like layers. Sometimes levelling of the layers are needed after the spread by the scrapes. If the material is too wet, they are cut and turned with a disk plow so that they will aerate and dry in the sun. If the material is too dry, the correct amount of water is added by sprinkling irrigation and mixed into the soil by plowing.

Different types of compaction equipment are available to use for different soils under various field conditions. Detailed treatment on compaction equipment is made in the next section.

2.3.6 Compaction Specification

Two methods of compaction specification have been in practice (Foster, 1962) to specify the compaction requirements in the various compacted layers. In one type, called the performance type of specification, the compaction requirement is stated in terms of the physical properties of the compacted layer. Typical specification of this type is the percentage of maximum dry density obtained in a standard compaction test. Sometimes both the dry density and moisture content are specified. For example, a specification of 95% of maximum dry density is frequently designated. Specification in terms of percentage of the maximum dry density in a standard test is preferable to a stated value of dry density, because it ensures that a compactive effort comparable to that of the standard laboratory test is applied in the field. This type of specification is applicable for cohesive soils. Other physical properties that are used are the percentage of air voids at a specific moisture content, void ratio and relative density. The last two are used primarily for cohesionless materials. In performance type of specification, the contractor is given wide scope in the selection of equipment and lift thickness. In the other type called the work type specification, the type of equipment, the lift thickness, the moisture content and the amount of work required to obtain the necessary density are specified. The work type of specification has been used for dams and embankment works whereas the performance type of specification has found wide spread usage in highway and airfied pavement works.

2.4 SURFACE COMPACTION

All types of soils can not be compacted satisfactorily by a single piece of equipment. Soil compaction is achieved by different means such as tamping, kneading, vibrating or impact. Appropriate equipment should be used for the type of soil and the field condition. In appropriate use of equipment for a given soil type will result not only in a poor quality of construction but also lead to construction delays. Equipments discussed in the subsequent sections are used on the surface of the ground to improve the material property to a limited depth from the ground surface and are termed as surface compaction equipments. Compacting materials to large depths by operating equipment on the ground surface is being used more recently. This new technique is called heavy tamping also named as dynamic compaction, dynamic consolidation or pounding.

Details of these equipments and their appropriate use for different soils are explained subsequently.

2.4.1 Static Compaction Equipments

These are rollers which can be used in general to compact any material. These rollers are of two types: one with two large wheels in the rear and smaller single drum in the front and the other type has large single drum in the front and the rear. Light weight rollers of 3 tons for foot-path construction whereas for road construction heavy rollers are used. Static rollers of weight 8 to 10 tons are used for works ranging from earthwork and sub-bases to bituminous road surfacing materials. Tandom roller is preferred as a finishing roller as wearing.

Static compaction rollers and compactors may be classified into the following groups (Mahesh Varma, 1979):

(*i*) Towed static smooth compactors
(*ii*) Static sheeps-foot or pad-foot compactors
(*iii*) Static three wheel self-propelled compactors
(*iv*) Static tandom compactors.

Towed Static Smooth Compactors: These are the old type of compactors. These are the first type of rolling compaction equipment. These were pulled by men or horses. During the rein of Romans these were used to smoothen roads and paths. Static rollers impose a limited pressure on the soils because of their relatively large contact areas. These rollers have a very limited effective depth of compaction, hence their use should be restricted to situations where only thin layers or a surface zone needs to be compacted.

Static Sheeps-Foot or Pad-foot Compactors: Sheeps-foot rollers, also called as tamping-foot rollers or pad-foot rollers, having projecting studs or feet on the surface of the rollers and compact by a combination of tamping and kneading action. These feet are also called as lugs which are of different shapes. The kneading action of the feet causes successive layers to be fused and thus increases the stability. These compactors consist of steel drums which can be filled with water or sand to increase the weight. As rolling proceeds, most of the roller weight is imposed through the projecting studs and results in fairly high contact pressure.

When a loose soil layer is initially rolled, the projections sink into the layer and compact the soil near the lowest portion of the layers. In subsequent passes, the zone being compacted already continuously rises until the surface is reached. This process of continuously rising effect experienced by the compactor is referred to as "walk-out" by the compactor. For most efficient compaction the roller should walk-out enough to lift the drum off the construction surface atleast 25 to 50 mm during the last few passes. Failure to walk-out signifies presence of high water content allowing shear failure of the soil. However, this does not necessarily indicate that enough compaction is not attained. A roller may walk-out when the roller weight is reduced.

These rollers are available in drum width ranging from 120 to 180 cm and drum diameter ranging from 100 to 180 cm and the loaded mass per drum ranges from about 2950 to 13600 kg. The tamping feet may be clubbed or tapered. The area of foot ranges from 36 to 50 sq.cm and of length range from 18 to 25.5 cm. These rollers are available as both self-propelled units and as rollers only.

Sheeps-foot rollers have the following advantages over other types of rollers:

(*i*) more suitable for cohesive soils
(*ii*) the feet produces kneading action
(*iii*) increase in blending of soils

(*iv*) simplification of adding water or drying as needed

(*v*) possible soil compaction over a wide range of moisture content, and

(*vi*) effective in breaking down large pieces of soft rock.

Major disadvantages are:

(*i*) relatively slow operation

(*ii*) lower compacted density

(*iii*) large entrapped air, and

(*iv*) soft zones are not revealed easily

Static Three Wheel Self-Propelled Compactors: These compactors have three rolls, a small split steering roll in the front and two large drive rolls mounted on rear axles at both ends. These rollers weigh 8 to 10 tons. This is the most used type applicable for a variety of works, *viz.*, earth work compaction, sub-base or base course rolling and for bituminous road rolling.

Static Tandom Compactors: Tandom compactors have two equal sized rolls and one centred in line-tandom. As these rollers have a smooth surface they are suitable for compacting bituminous layers. Improvements have been made on these types of compactors as tandom vibratory compactors. Large size tandom vibratory compactors are generally preferred now-a-days as they can be used either as static compactor or as a vibratory compactor as per the requirement.

2.4.2 Vibratory Compaction Equipments

Vibratory compactors may be categorized into the following groups (Sharma, 1988):

(*i*) Tandom vibratory compactors

(*ii*) Towed vibratory compactors

(*iii*) Towed sheeps-foot or Tamping-foot vibratory compactors

(*iv*) Self-propelled vibratory compactors

(*v*) Hand-guided vibratory compactors.

Tandom Vibratory Compactors: Two types of Tandom vibratory compactors are available, *viz.*, single-drum vibrating or double-drum vibrating. In the compactors with double-drum vibrating system two tandom wheels are provided with separate controlled vibrators in the front and rear rolls. Comparing single and double-drum vibratory compactors, the output of double-drum vibratory compactor is to be 80% more than the single-drum vibratory compactor. The double-drum vibratory type has an option to operate the single-drum or the double-drum.

Towed Vibratory Compactors: This type of compactor is especially used for compacting cohesive soils, fine and coarse grained mixed soil and rocky materials. The heavy type towed vibratory compactor is used in earth dam and embankment constructions. Because of the large amplitude, it shows more impact motion and therefore preferred for compacting cohesive soils, fine grained soils and mixed fine or coarse grained soils.

Towed Sheeps-Foot or Tamping-foot Vibratory Compactors: Towed sheeps-foot or tamping-foot vibratory compactor is useful in highly cohesive soils and soft rocks. The kneading and crushing effect of the feet improves the compaction performance. The compactors are provided with sheeps-foot or tamping-foot vibratory equipment. Tamping-foot are larger than the sheeps-foot and hence has a more contact area.

Self-Propelled Vibratory Compactors: These compactors are available with a weight of 8 to 10 tons dead weight. In one type large vibratory steel roll in front and two rubber tyres at the rear. The rubber tyres may be smooth or with treads. The smooth one can be used for bituminous work and the treads one for earth compaction work.

In the other type, vibratory steel roll is in the front and two static steel rolls at the rear for multi purpose work.

Hand-Guided Vibratory Compactors: Compactors of this type have duplex or double vibratory compactors with dual roll vibration and dual roll drive. These vibrators are small in size and thereby enable cross country mobility and excellent gradeability. They have provision for dual roll drive. These are especially used for compacting trenches, slopes, packing lots, small repair jobs and sports grounds, *etc.*

2.4.3 Rubber-tyred Compaction Equipments

Pneumatic-tyred or Rubber-tyred compactors are the most versatile equipment for general compaction as they are effective for low-cohesive soils and cohesionless soils including gravels, sands, clayey sands, silty sands and even sandy clays. The compaction produced is primarily by kneading action and some designs provide a "wobble-wheel" effect to increase kneading. These rollers are outfitted with a weight box or ballast box between two axles so that the total compaction load can be easily varied.

These compactors are provided with the tyre-configuration of odd numbers in front, *viz*, 5, 7, 9, *etc.*, with certain degree of overlap. Even numbers of tyres are provided at the rear. The wheels are mounted in two rows and spaced so that the wheels of the rear row track in the space between those of the front row. Such an arrangement does not create ruts and leave uncompleted areas. In the light rubber-tried rollers 7 to 13 wheels are mounted. The rollers are out-fitted with a weight box or ballast box between two axles so that the total compaction load can easily be varied. The rollers are available in a wide range of load sizes.

Wheel loads, the sizes, and infiltration pressures, number of passes and loose lift thickness are varied depending upon the soil type and the equipment available.

These rollers have the following advantages compared to steel rollers:

(*i*) Surface of layer is not bridged but uniform.

(*ii*) Bituminous layers compacted by rubber-tyre rollers show better sealed to keep away dirt and moisture.

(*iii*) Post-compaction by traffic is negligible when compacted by rubber-typed rollers.

(*iv*) Provides a kneading effect compared to smooth wheel rollers and thereby provide a better surface compaction.

(*v*) In compaction of asphalt or bituminous pavements no cracks are formed as that of pavements rolled by smooth wheel rollers.

2.4.4 Grid-rollers

These rollers are intermediate between smooth wheel and sheeps foot rollers, with their rotating wheels made of a network of steel bars forming a grid with square holes. These rollers provide less kneading action but high contact pressure. These are more suitable for course–grand soils.

Table 2.2 shows the type of roller suited for a particular type of soil and the corresponding rating of soil with reference to volume change, drainage, permeability and value as subgrade and base course.

Table 2.2. Soil Characteristics and Rating for Construction
(*Source: UGWES, 1953*)

Group Symbol (USC)	*Soil Type*	*Compaction Characteristics and Type of rollers suitable*	*Volume Change*	*Drainage and Permeability*	*Value as Subgrade not subject to frost*	*Value as Base for Pavement*
1	2	3	4	5	6	7
GW	Well-graded clean, gravels-sand mixtures	Good : tractor, rubber-tired, steel wheel or vibratory roller	Almost none	Good drainage, pervious	Excellent	Good
GP	Poorly graded clean gravels, gravel-sand mixtures	Good : tractor, rubber-tired, steel wheel or vibratory roller	Almost none	Good drainage, pervious	Excellent to good	Poor to fair
GM	Silty gravel, poorly-graded gravel-sand silts	Good : rubber-tired or light sheepsfoot roller	Slight	Poor drainage, semi-pervious	Excellent to good	Fair to poor
GC	Clayey gravels, poorly graded gravel-sand clays	Good to fair: rubber-tired or sheepsfoot roller	Slight	Poor drainage, impervious	Good	Good to fair
SW	Well-graded clean sands, gravelly sands	Good : tractor, rubber-tired or vibratory roller	Almost none	Good drainage, pervious	Good	Fair to poor
SP	Poorly graded clean sands, sand-gravel mixtures	Good: tractor, rubber-tired or vibratory roller	Almost none	Good drainage, pervious	Good to fair	Poor

(*Table Contd...*)

Table 2.2. (*Contd...*)

1	2	3	4	5	6	7
SM	Silty sands, poorly graded sand-silt mixtures	Good : rubber-tired or sheepsfoot roller	Slight	Poor drainage, impervious	Good to fair	Poor
SC	Clayey sands, poorly graded sand-clay mixtures	Good to fair : rubber-tired, or sheepsfoot roller	Slight to medium	Poor drainage, impervious	Good to fair	Fair to poor
ML	Inorganic silts and clayey silts	Good to poor : rubber-tired or sheepsfoot roller	Slight to medium	Poor drainage, impervious	Fair to poor	Not suitable
CL	Inorganic clays of low to medium plasticity	Good to fair : sheepsfoot or rubber-tired roller	Medium	No drainage, impervious	Fair to poor	Not suitable
OL	Organic silts and silty clays, low plasticity	Fair to poor : sheepsfoot or rubber-tired roller	Medium to high	Poor drainage, impervious	Poor	Not suitable
MH	Inorganic clayey silts, elastic silts	Fair to poor : sheepsfoot or rubber-tired roller	High	Poor drainage, impervious	Poor	Not suitable
CH	Inorganic clays of high plasticity	Fair to poor : sheepsfoot roller	Very high	No drainage, impervious	Poor to very poor	Not suitable
OH	Organic clays and silty clays	Fair to poor: sheepsfoot roller	High	No drainage, impervious	Very poor	Not suitable
Pt	Peat	Not suitable	Very high	Fair to poor drainage	Not suitable	Not suitable

2.4.5 Other Surface Compaction Devices

Heavy rubber-tired trucks and scrapers used in hauling operations can be used as compactors. Although these equipments do have equal weight as that of rubber-tired rollers they are not that efficient. Crawler-type tractors are capable of producing a moderate degree of compaction in cohesive soils and fairly effective in cohesionless soils.

Compaction equipment of smaller size should be used in the areas where large equipment can not be operated. Air tamps, gasoline-driven tamps, leaping-frog-type machines and small vibrating plate devices are available. Particularly hand-operated tamping and vibratory compactors are used for working in limited spaces and for compacting soils close to structures where care is required to prevent damage.

2.4.6 Dynamic Compaction

Dynamic compaction is a technique which was used long ago (Schultze and Muhs, 1967) has taken rebirth recently (Menard and Broise, 1975). In this method a very heavy weight (upto 45,000 kg) is dropped from a height of 15 to 40 m to fall freely back down to ground surface. This heavy impact on the ground leaves its mark behind. This impact energy on cohesionless soils cause liquefaction followed by settlement due to rapid drainage. In some soils radial fissures are formed around the impact facilitating rapid drainage.

This process is then repeated either at same location or subsequently over other parts of the area intended for stabilization at 5 to 10 m spacing. It is essential to avoid the formation of a thick surface crust at too early a stage in the operations. After allowing the required number of drops, the area shall be compacted at depth. However, the soil at the shallow depth from the surface will be in a great disturbed condition. This upper layer up to a depth of 2 m is then compacted using a smaller weight with greater area (and thus a lower stress) and is dropped from a small height (and hence imparts a lower energy). The usual energy per blow is in the range of 135×10^3 to 450×10^3 kg m, with values as high as 900×10^3 to 1800×10^3 kg m sometimes used. Generally a total tamping energy of 2 to 3 blows per sq. m. is used. Efficiency could be increased if the impact velocity exceeds the wave velocity in a liquefying soil.

Rigorous analytical model of the process is still not available. Only a preliminary beginnings of an understanding of this dynamic compaction process are available. When a weight strikes the ground surface it produces *P*, *S* and *R* waves. In nearly saturated soils the *P* waves cause direct compression and induce pore water pressure.

Densification is also brought in by *S* waves by mobilising shear forces in the particles. But the *R* waves and other surface waves are attempting to loose the surface soil and harmful in damaging the adjacent structure. Only empirical information is available to assess the depth of penetration of the compaction. Considering effective depth as a function of impact energy, the depth of penetration is within the following range (Dobson and Slocombe, 1982)

$$1.26\sqrt{Wh} < D < 3.16\sqrt{Wh}$$

where D = effective depth (m)

W = weight being dropped (kN)

h = height of drop (m)

As the analytical understanding is in the formative stage, it is recommended to have a small test-section at the site under consideration for necessary field evaluation (Koerner, 1985). Proper instrumentation can also be made and necessary realistic goals may be evaluated for the final densified soil mass.

This technique derives its merits for the following reasons:

(*i*) It is one of the simplest and most basic methods of compacting loose soil.

(*ii*) Depth of compaction can reach up to 20 m.

(*vi*) All types of soils can be compacted.

(*iv*) Produces equal settlements more quickly than surcharge type loading.

(*v*) It can be used to treat soils both above and below water table.

As soft soils are susceptible for disturbance, due to heavy impact, a careful evaluation is required in clayey soils and concrete or steel block, of about 45,000 kg weight with necessary arrangements for free fall and lifting and movement of crane to different locations is sufficient. One crane and tamper can treat from 300 to 600 m^2 per day. The major variables involve in the process are:

(*i*) magnitude of the weight, (*ii*) size of the weight, (*iii*) height of free fall of the weight, (*iv*) number of drops per location, (*v*) distribution of drop locations over the site, (*vi*) non-homogenity of the soil, (*vii*) strength, permeability and anisotropy of the soil, and (*viii*) degree of saturation of the soil.

The application of this technique is widely used and reported in literature (Leonards, *et. al.*, 1980 : Charles, *et. al.*, 1981; Ramaswamy *et. al.*, 1979, 1981; Welsh, 1982). From the above field applications it is evident that there is significant improvement attained in soil properties due to this technique. The improved properties of the soil after treatment have been evaluated in the above applications, (*i*) by conducting cone or standard penetration test, (*ii*) by monitoring surface settlement and pore water pressure and (*iii*) by conducting tests on the treated undisturbed samples for physical, hydraulic and mechanical properties. This method has been under estimated in the past but it is evidently a cost-effective method which could be used for improvement of all soil types in a varied field conditions.

2.5 SELECTION OF FIELD COMPACTION PROCEDURES

2.5.1 Choice of Equipment

Rollers: The selection of rollers should be primarily based on economic, that is, the chosen roller should obtain satisfactory compaction at the least cost. However, the selection of rollers is the sole responsibility of the designer in case of work-type specification is used or the contractor where the performance type specification is used. Further, the choice of a roller for a given job depends on the type of soil and the percentage of compaction or relative density. The soil characteristics and the choice of rollers along with the rating of soil for construction are presented in Table 2.2. Sheepsfoot and rubber-tired rollers are generally, preferred for cohesive soils. For cohesionless soils, rubber-tired rollers, vibrating compactors and crawler-tread tractors are usually preferred. Rubber-tired rollers, smooth-wheeled steel rollers, and vibratory compactors are preferred for base courses. Larger rubber-tired and high-tire pressure rollers are needed for jobs wherein high densities are required.

Auxiliary Equipment: In addition to rollers it is necessary to have earth work equipments. These are excavators, shovels, bull-dozers, tractors, motor graders, scrapers and loaders.

Excavator machines consist of the following components:

(*i*) An undercarriage which gives mobility to the excavator. This may be mounted with crawler track or wheel.

(*ii*) A superstructure with operator's cabin which could traverse through 360° or fitted on a rigid frame.

(*iii*) Hydraulically articulated booms and dipper arms with bucket.

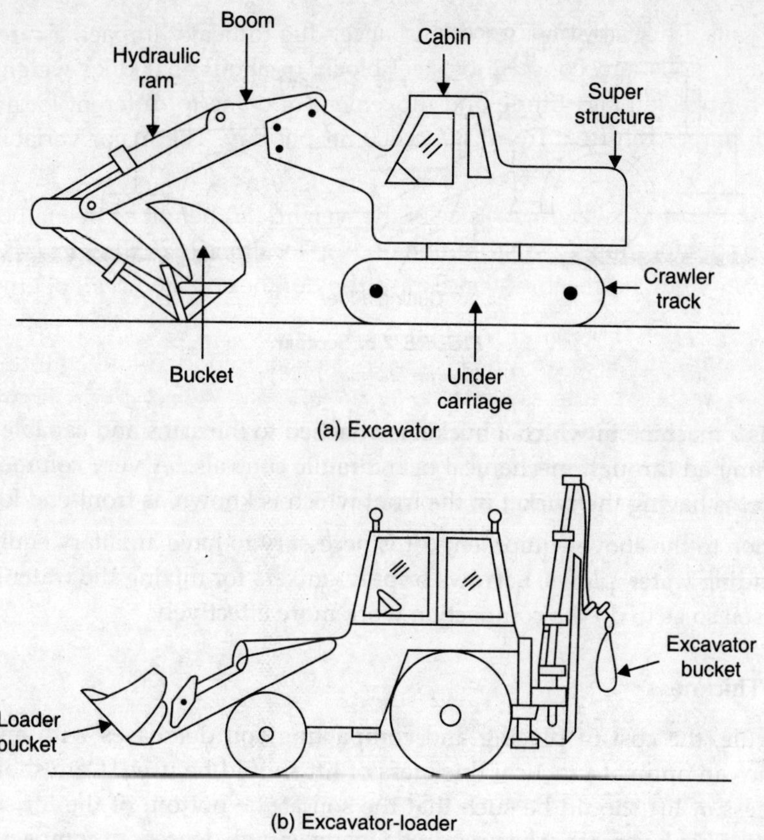

FIGURE 2.5. **Excavators.**
(*Source: Sharma, 1988*)

Excavators include dipper or power shovel, dragline, clamshell and drag shovel. Shovel is often used for a specific type of excavating machine with a short length boom and working with forward strokes.

Bull-dozer is an important equipment on a construction project. It is basically a scraping and pushing unit. Further it is a multi-purpose equipment which can be used for different purposes with some modifications.

A tractor is a multi-purpose machine which comes in varied types as light model to heavy model. The light model is used for agricultural or small haulage purposes. Heavy model equipped with several special rigs are used for earth moving work. Two principle applications of tractors are: (*i*) Cleaning and excavating machinery and (*ii*) Hauling and conveying machinery.

A grader is primarily a device for loading or finishing earthwork. Sometimes it is used also for trimming slopes and mixing gravel.

Scrapers are the devices to scrap the ground to load the material, to transport to the required distance, to dump at the intended site, to spread the dumped material over the required area, to attain the desired thickness and to return back to do the next cycle. In simple terms scrapers are designed to dig, load, haul, dump and spread.

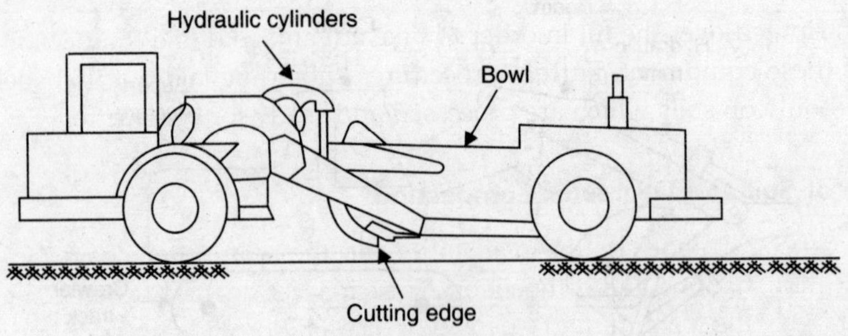

FIGURE 2.6. **Scraper.**
(*Source: Sharma, 1988*)

Loader is a machine in which a bucket is attached to the arms and capable of being raised, lowered and dumped through mechanical or hydraulic controls. A very common type is one in which the loader is having the bucket in the front which is known as front-end loader.

In addition to the above equipments, it is necessary to have auxiliary equipments, such as sprinkler for adding water, plows, harrows or pulvi-mixers for mixing the water into the soil and for drying the soil so as to do the compaction work more effectively.

2.5.2 Layer Thickness

As a general rule, the cost of placing and compacting soil decreases with an increase in lift thickness. Hence an optimal practical thickness of lift should be used. Care should be exercised that the thickness of lift should be such that the soil at the bottom of the lift should attain the required density. Thin layers must be preferred wherever high degrees of compaction are required. It is the common practice to use 15 cm thick lifts of cohesive soils of fills, subgrades and embankments. Thicker lifts may be preferred in embankments wherein the density requirement is only moderate. For base course and sub bases where light rubber-tired rollers are used, the lift thicknesses are limited to 75 to 100 mm. In cohesionless soils thicker lifts may be used compared to cohesive soils.

2.5.3 Wetting and Drying Methods

Wetting or Drying of soils can be done either in the pit or on the fill. As it is difficult to add more than 1 or 2% moisture on the fill and to maintain a uniform moisture and satisfactory working conditions, wetting soils in the pit has a definite advantage. Drying is usually attained by aerating the soil with the help of plows, disks, spring-tooth harrows and rotary tiller type mixers.

2.5.4 Construction Traffic

The effectiveness of construction traffic in compacting soils is similar to that of pneumatic-tired rollers. As the tyres do not give complete coverage with each pass and the tendency of drivers to follow only existing tracks, usually the soil is well compacted under the wheel tracks, leaving uncompacted materials between them. When the regular compaction is taken subsequently by the rollers, they are unable to compact the soft areas since they ride on the well compacted tracks. Thus the construction plan should be made such that any movement of equipment should be systematically distributed over the fill in order to take advantage of the compaction produced by the movement of these equipment. Further procedures should be laid out that such movements should be allowed only on soils which are processed and ready for compaction.

2.5.5 Selection of Soil and Degree of Compaction

The final selection of soil depends on the availability and the compaction characteristics, and cost involved for excavation, hauling and compaction. The choice of degree of compaction signifies the minimum percentage of the maximum dry density which will provide the necessary strength and incompressibility under worst condition. Table 2.3 gives a guide line for preliminary estimates when soil test data are not yet available (Sowers, 1979).

Table 2.3. Tentative Requirement for Compaction
(*Source: Sowers, 1979*)

Unified Soil Classification:	*Required Compaction-Percentage of Standard Proctor Maximum*		
Soil Symbol	*Class 1*	*Class 2*	*Class 3*
GW	97	94	90
GP	97	94	90
GM	98	94	90
GC	98	94	90
SW	97	95	91
SP	98	95	91
SM	98	95	91
SC	99	96	92
ML	100	96	92
CL	100	96	92
OL	—	96	93
MH	—	97	93
CH	—	—	93
OH	—	97	93

Class 1 = Upper 3 m of fills supporting 1 or 2 storey buildings
Upper 1 m of subgrade under pavements
Upper 0.3 m of subgrade under floors.

Class 2 = Deeper parts of fills under buildings
Deeper parts (up to 10 m) of fills under pavements and floors
Earth dams.

Class 3 = All other fills requiring some degree of strength or incompressibility.

2.5.6 Trial Embankments

In order to establish the number of passes required for a particular job, trial embankments may be used. This arrangement is necessary especially where the work-type specification is used. Trial embankments may be included in the job in the temporary structures such as cofferdams. If it is intended to include the trial embankment in the finished work, it should be located in a non-critical area. The trial embankments should be constructed with a soil with adequate strength and other properties which is going to be used in the final structure. The use of a trial embankment permits modification of the design or even a change in the compaction procedure were needed before the job is started. In a performance type of specification, the preliminary work is essentially a trial embankment since the contractor will be determining the number of passes required to obtain the specified density.

2.6 COMPACTION QUALITY CONTROL

It is essential to obtain the density and moisture content as required in the design. It could be achieved by taking compacted materials and testing them for moisture and density. Procuring samples and testing them is more important during the initial stages of the job while procedures are being worked out, it is absolutely essential to decide the supervisory staff (engineers and/or technicians) and required equipment for the testing be planned well in advance such that they will be available at the start of the construction. As the conventional methods of measuring density and moisture content are time consuming, rapid methods of determining these parameters should be used.

2.6.1 Moisture Content Measurements

Two types of moisture content samples are taken. One type called the construction control is intended to check whether or not a layer has the proper moisture content for compaction. The other type is taken in connection with the density tests to permit calculation of the dry weight. Usually the density tests are made before any appreciable drying of the sample and accordingly this moisture content corresponding to this is recorded as the moisture content at which the soil was compacted and termed as record samples.

In the conventional oven drying method several hours of drying is needed which may cause a construction delay and hence more rapid methods of determining the moisture and density should be used. Several rapid methods of measuring moisture content are available including heating the soil in an open pan over a hot plate or gas flame, the alcohol burning method, the alcohol method with a hydrometer and a moisture meter using calcium carbide to generate acetylene in a closed container connected to a pressure gauge. All the rapid methods, except the quick ovens, require individual attention for each sample. Necessarily the moisture content determined by rapid methods require a correction to make them agree with the conventional methods. Further, these methods are suitable for cohesionless materials as they yield results similar to oven drying, but give only approximate results for many types of cohesive soils (Hunt, 1986).

Moisture content can also be obtained from penetration readings taken with a Proctor needle in a sample compacted in a test mould, by comparing with a laboratory curve.

Many a times the engineer or the technician on the job develops a feel for the soil being compacted and can tell by rubbing the soil in the hands and by observing the soil in the process of compaction whether the moisture is adequate or not. This ability to estimate moisture content is developed by the engineer or the technician by observing the early periods of job by making a large number of moisture content determination and compares with the conventional ones.

As it is highly impossible to process at a given specific mositure accurately, it is necessary that the field inspection have a range within which the moisture content is acceptable. The range of moisture content to be adopted should be given by the designer, as this moulding moisture content affects the strength and other properties of the compacted material.

2.6.2 In-situ Density Measurements

In-situ measurements of densities frequently are made on a grid basis, requiring one test per so many square metres or cubic metres of compacted soil area or volume respectively. As per the norms of USBR one test per 2000 m^3 for embankment construction. Industrial buildings supported on load bearing fills should have a minimum of one test for every 1000 m^3 (Hunt, 1986). Further tests should be made wherever under or over compactions were doubted.

In a cohesive soil chunk samples or driven samples are obtained to determine the density (IS : 2720—Part 29, 1979). In a cohesionless soils where a sample could not be cut the sand cone density test (IS : 2720—Part 28, 1974) or the rubber balloon method (IS : 2720—Part 34, 1972) is used.

After the dry density is determined, the observed value is compared with standard maximum dry density and the percentage of density is evaluated. Layers which are not compacted to the required level should be given additional compaction. Where several soil types are encountered laboratory curves should be available for each type for necessary comparison of densities.

2.6.3 Nuclear Moisture-density Method

Surface-type nuclear moisture-density equipment are used nowadays because of the rapid results that can be obtained. The principal components in this type of equipment are the nuclear source which emits gamma rays, detector to pickup the gamma rays or photons passing through the tested soil, and a counter or scalar for determining the rate at which the gamma rays reach the detector. The nuclear sources are radium-beryllium and cesium-americium- beryllium combinations.

During operations of the equipment the gamma rays penetrate into the soil, where some are absorbed, but some reach the detector by direct transmission or after hiting soil mineral electrons. The amount of gamma radiation reaching the detector is inversely proportional to the soil density and determined by keeping a nuclear count rate received at the detector. The density is obtained using the observed count rate from a calibration curve provided by the manufacturer of the equipment.

Moisture determination are obtained from a "thermal neutron count". Alpha particles emitted from the americium or radium source bombard a beryllium target and causes the beryllium to emit fast neutrons. These fast neutrons lose velocity if they strike the hydrogen atoms in water. Based on this moisture results are provided as weight of water per unit of volume. After deducting the water determining from wet density determination, the dry density is obtained. Determination of moisture content by this method may have significant error if the soil contains iron, boron or cadmium.

2.6.4 Hilf's Method

The rapid compaction control procedure recommended by Indian Standard (IS : 2720—Part 38, 1976) gives the exact percentage of laboratory maximum dry density and a close approximation of the difference between optimum moisture content and fill moisture content of a field density sample, without any determination of moisture contents. By this moisture procedure it is possible to effect compaction control within one hour after conducting the field test.

Although this method was developed by Hilf (1966) primarily in controlling the placement of earth-fill in dam construction, the method is also applicable to control earth-fill placement in the construction of highway embankments, canal embankments and similar structures built of cohesive soils. The reader may refer to the Indian Standard (IS : 2720—Part 38, 1976) for detailed procedure.

Chapter 3

DRAINAGE METHODS

3.1 INTRODUCTION

Groundwater is usually considered as one of the most difficult problems that has to be handled in any civil engineering construction. Under many field situations during construction operation it may be necessary to eliminate seepage pressure to increase the shearing resistance or to reduce the danger of frost damage. These could be done by reducing the neutral stresses effectively by adopting any one of the drainage methods. Dewatering systems and drains are the two methods which are used to improve the ground condition before, during and after construction.

Dewatering systems essentially consist of lowering the water table to a required elevation and to establish below this level a system of collectors located in wells, galleries or ditches. The collected water at the collectors are pumped. Continuous pumping from excavations or from natural ground is a costly affair and the continued flow from the surrounding ground may endanger the stability of adjacent structures. Further heavy inflow may lead to erosion or collapse of the sides in an open excavation. In certain situations there may be instability of the base of excavation due to upward seepage.

On the other hand drains are provided to serve to control the flow, in some cases for lowering the water table, in others for reducing pore pressure and seepage forces.

In order to ensure in selecting a proper drainage method, it is important to obtain all the necessary information before commencing work, and this aspect should not be neglected at the site investigation stage.

3.2 SEEPAGE

A material is said to be porous if it contains interstices, or void spaces; within which the material is absent. The void spaces may be discontinuous or continuous. For a medium to be permeable at least a portion of the void spaces in the medium should be continuous. Eventually all soil deposits are permeable media.

Permeability of a soil is its capacity to transmit a fluid to pass through its interconnected void spaces. The flow process could be under the action of an externally applied force or due to a diffusion process. The externally applied load may be a pressure difference due to heads of water between two points and the diffusion process is the dispersion of molecules activated by differences in their concentration from point to point. Flow through the medium is quantified by a material characteristics termed the coefficient of permeability, k, (where $k = v/i$ and v is the discharge velocity and i the hydraulic gradient) after Darcy.

Permeability depends on the type of soil and the permeant. The characteristics related to permeant are viscosity, unit weight and the polarity of pore fluid. The characteristics related to soil are particle size, void ratio, composition, fabric and degree of saturation. The coefficient of permeability may have a range of variations. Typical ranges of permeability for different soil types and the drainage characteristics are listed in Table 3.1.

Table 3.1. Typical Permeability Ranges for Different Soils
(*Source: McCarthy, 1982*)

Soil type	*Relative degree of permeability*	*Coefficient of permeability k (cm / sec)*	*Drainage properties*
Clean gravel	High	1 to 10	Good
Clean sand, sand and gravel mixtures	Medium	1 to 10^{-3}	Good
Fine sands, silts	Low	10^{-3} to 10^{-5}	Fair through poor
Sand-silt-clay mixtures, glacial tills	Very low	10^{-4} to 10^{-7}	Poor through practically impervious
Homogeneous clays	Very low to practically impermeable	less than 10^{-7}	Practically impervious

Water can rise in a soil through connected voids to elevations above the ground water table due to capillary action. Such an action provides the moisture that results in heaving of buildings and pavements from the volume increase of expansive soils or from freezing. The potential for the detrimental effects of capillary rise is a function of soil type and the depth to the water table.

The velocity of flow through the soil voids is referred to as seepage velocity (v_s) which is always greater than discharge velocity (v). The flow of water in the soil causes stresses called seepage pressures (j) which is equal to the hydraulic gradient times the unit weight of water ($j = i\gamma_w$). In a vertical flow condition in a soil if the seepage pressure exceeds the submerged weight of the soil and the soil column is uplifted. This phenomenon is referred to as *quick condition* (boiling or liquefaction). Thus the most dangerous areas for liquefaction are those where the upward gradient is large and the counter balancing weight is small. Such susceptible areas are along the toe of a slope or bottom of an excavation.

The analysis of flow through soils is performed using a flow net. A flow net is a graphical presentation of flow consisting of a net of flow lines and equipotential lines which is an useful tool to determine the rate of flow, seepage pressures, pore water pressures and gradients. In general two flow conditions are encountered, *viz.*, confined flow and unconfined flow. Confined flow refers to the case where the phreatic surface is known whereas the unconfined flow refers to a condition in which the location of phreatic surface is not known.

3.3 FILTER REQUIREMENTS

In any drainage system perforated pipes and conduits of perforated pipes or pipe lines with open joints are usually provided. The space between the natural soil and the pipe is filled with a coarse-grained material known as a filler. If the voids of filler are larger than the finest grains of the adjoining soil, there is a possibility of these fine particles to fill the voids and accumulate and block the flow. On the other hand if the voids in the filler are as small as those in the soil, then there is a possibility of the filler material washed into the conduits and pipes and thus leading to erosion of the natural soil. Both these are undesirable conditions. A filler material that overcomes the above two conditions is referred to as a filter.

Filters are also used in earthdams, cofferdams and sheet pile structures as a transmission medium or to prevent piping. To prevent piping, the filter material should have adequate weight. If a filter extends across a boundary between coarse and fine soils, different materials have to be used.

Thus in general a filter or a drain material should satisfy the two requirements apart from adding weight, *viz.*,

(*i*) the gradation of filter material should be capable of forming small size pores such that the migration of adjacent particles through the pores is prevented.

(*ii*) the gradation of the filter material should be such that it allows a rapid drainage without developing large seepage forces.

The above requirements are satisfied by adopting a suitable grain-size distribution for the filter material based on the material to be protected. The following filter criteria are to be adopted (Bertram, 1940).

$$\frac{D_{15}\,(\text{filter})}{D_{85}\,(\text{protected soil})} < 4 \text{ to } 5 \qquad \ldots(3.1)$$

This criterion emphasizes that the D_{15} size of the filter soil should not be more than four or five times the D_{85} size of the protected soil. The second criterion is:

$$\frac{D_{15}\,(\text{filter})}{D_{15}\,(\text{protected soil})} > 4 \text{ to } 5 \qquad \ldots(3.2)$$

This criterion emphasizes that the D_{15} size of the filter soil should be more than four or five times the D_{15} size of the protected soil.

The additional requirement which has to be satisfied as regards to openings in the mesh screen or perforated pipe is that the maximum size of the filter material should be at least twice that of the openings in the mesh screen (or perforated pipe if no mesh is provided).

As the loss of head due to percolation through the filter should be smallest possible, large filters are usually made up of several layers. Each of these layers should satisfy the above requirements with respect to the preceding layer. Such composite filters are said to be graded filters. As a rough guideline, the grain-size distribution curves of the fine and coarse-grained soils should be roughly parallel.

3.4 GROUNDWATER

Groundwater can be encountered in various modes of occurrence. The term water level, water table or phreatic surface has been defined as the level within a body of subsurface water at which water pressures are equal to atmospheric pressure. The neutral stress on this surface is zero. The water that occupies the voids of the soil above water table, called upper zone, constitutes soil moisture. This water may be found as gravity water, hygroscopic moisture, perched water or in the capillary fringe. This zone may be fully or partially saturated. The zone below the water table is saturated and present in natural formations such as *aquifers, aquitards* and *aquicludes.*

Aquifers are relatively pervious soil and rock formation through which groundwater flows induced by gravity. In nature the most familiar situation is a stratum of coarse sand in which the groundwater surface rises and falls with changes in weather, discharge and groundwater use. Such an aquifer is termed a free or *unconfined aquifer* wherein the groundwater table is likely to slope in the same direction as the ground surface. Further the slope of water table is generally flatter and more uniform (Fig. 3.1). A relatively impervious soil or rock is termed as an aquiclude. For

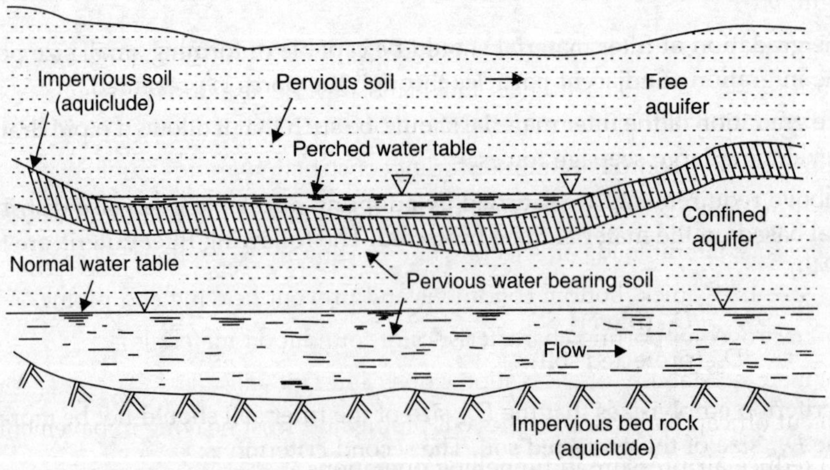

FIGURE 3.1. **Water Table and Aquifers.**

example, the ratio of permeabilities of an aquifer to an aquiclude may exceed 100 : 1. The groundwater pattern is more complex, in formations that consist of alternate aquifers, and a aquicludes. A sagging aquiclude below an aquifer creates a basin that holds groundwater perched above a general water table (Fig. 3.1). When an aquifer is confined between two aquicludes, it is called *confined aquifer* and is capable of carrying water under pressure. In such a situation, the elevation of zero pressure is above the upper surface of the water and the groundwater is said to be under artesian pressure. Artesian pressures are usually developed by sloping aquifers. Artesian aquifers may extend over several hundred square metres of areas.

Engineers are primarily concerned with aquifers, aquiclude and the capillary zone. For example, water from aquifers flowing into excavations has to be controlled to maintain dry working condition or to reduce pressures on structures. If aquifers are under artesian conditions could cause boiling or piping and uplift the bottom of excavations. And aquicludes, represented by saturated silts, can allow seepage into excavations and cause weakness zones in excavation.

Thus the engineers should be aware that groundwater conditions are transient and that those encountered during investigation will not necessarily be those existing during construction or at sometimes latter as it undergoes a large seasonal and periodic variations.

The water table position has a significant impact in civil engineering constructions in a number of ways. For example, the following water table positions may be of concern to civil engineers:

(*i*) excavation below the water table in granular soils require dewatering,
(*ii*) a sudden water table rise can cause uplift of basement floors or of a structure itself,
(*iii*) fluctuation in water table position affects the active zone in swelling soils and
(*iv*) excess groundwater pumping may result in ground subsidence or even ground collapse.

3.5 GROUNDWATER AND SEEPAGE CONTROL

Groundwater and seepage control needs a most significant consideration in many of the civil engineering works such as in the stability of natural slopes and cuts, dams and levees, excavations for structures, open-pit mines, tunnels and shafts, burried structures, pavements and side-hill fills. Further, the control should be ensured during construction period and as well after construction.

The necessary controls required during construction are to (Hunt, 1986):

(*i*) provide a dry excavation and permit construction to proceed efficiently
(*ii*) reduce lateral loads on sheeting and bracing in excavations
(*iii*) stabilize 'quick' bottom conditions and prevent heaving and piping
(*iv*) improve supporting characteristics of foundation materials
(*v*) increase stability of excavation slopes and side-hill fills
(*vi*) cut off capillary rise and prevent piping and frost heaving in pavements
(*vii*) reduce air pressure in tunnelling operations.

The following controls are required after construction to (Hunt, 1986):

(*i*) reduce or eliminate uplift pressures on bottom slabs and permit economics from the reduction of slab thicknesses for basements, burried structures, canal linings, spillways, dry docks, *etc*.

(*ii*) provide for dry basements.

(*iii*) reduce lateral pressures on retaining structures.

(*iv*) control embankment seepage in all dams.

(*v*) control seepage and pore pressures beneath pavements, side-hill fills, and cut slopes.

(*vi*) prevent surface and groundwater contamination from pollutants.

The major control methods are:

(*i*) Dewatering systems,

(*ii*) Drains and

(*iii*) Cut-offs and barriers.

The dewatering systems and drains are dealt subsequently and cut-offs and barriers are dealt in Chapter on Grouting methods.

3.6 METHODS OF DEWATERING SYSTEMS

Groundwater can be controlled by adopting one or more types of dewatering systems or drains appropriate to the size and depth of excavation, geological conditions, and characteristics of the soil. In earlier days the control of groundwater in construction work was accompanied by one or a combination of the following methods: (*i*) constructing ditches and dikes to intercept seepage from the slopes combined with a controlled rate of excavation, (*ii*) sheeting combined with pumping from sumps, and (*iii*) pumping from deep sheeted sumps dug outside of the working area.

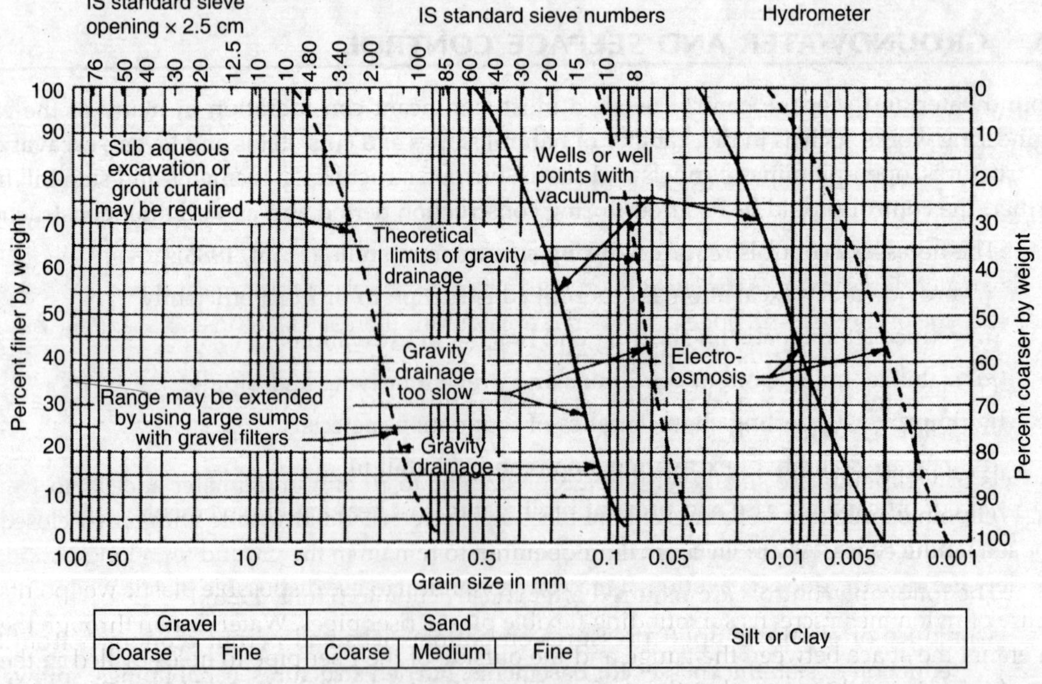

FIGURE 3.2. **Dewatering Systems Applicable to Different Soils.**

(*Source: IS: 9759–1981*)

After the advent of suitable well-installation equipment, techniques and pumping machines, the control of groundwater has much improved. Various types of drainage systems which may be considered most suitable based on the grain-size distribution of soils is presented (IS : 9759–1981) in Fig. 3.2. Different dewatering systems and drains are subsequently described.

3.6.1 Open Sumps and Ditches

The essential feature of the method is a sump below the ground level of the excavation at one or more corners or sides (Fig. 3.3). In order to prevent standing water on the floor of excavation, a small grip or ditch is cut around the bottom of the excavation, falling towards the sump. In the case of large excavations which have to remain open over long periods, it is advisible to design these drainage ditches with much more care.

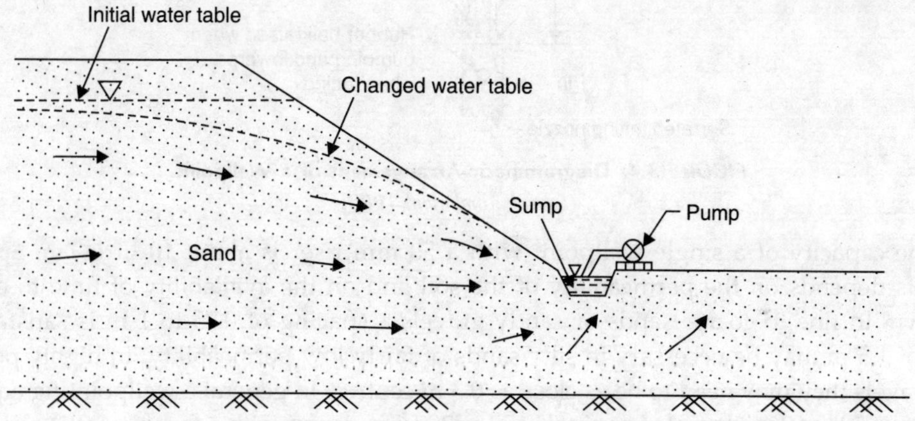

FIGURE 3.3. **Dewatering Through Sumps.**

This is the most widely used of all methods of groundwater lowering. It is the most economical method for installation and maintenance which could be applied to most soil and rock conditions. This method is also more appropriate in situations where boulders or other massive obstructions are met with in the ground. The greatest depth to which the water table can be lowered by this method is about 8 m below the pump level. This method has a serious disadvantage that the groundwater flows towards the excavation with a high head or a steep slope and hence there is a risk or collapse of the sides. In open or timbered excavations there is risk of instability of the base due to upward seepage towards pumping sump.

3.6.2 Wellpoint Systems

Filter wells or wellpoints are small well-screens of sizes 50 to 80 mm in diameter and 0.3 to 1 m length. Wellpoints are either with brass or stainless-steel screens and are made with either closed ends or self jetting types. Where wellpoints are required to remain in the ground for a long period, *e.g.*, for dewatering a drydock excavation, it may be economical to use disposable plastic wellpoints which are of nylon mesh screens surrounding flexible plastic riser pipes. Water drawn through the screen enters the space between the gauge and the outside of the riser pipe to holes drilled in the bottom of this pipe and then reaches the surface. The wellpoints are installed by jetting them into the ground. A diagrammatic arrangement of a wellpoint is shown in Fig. 3.4. A typical layout of a wellpoint system is shown in Fig. 3.5.

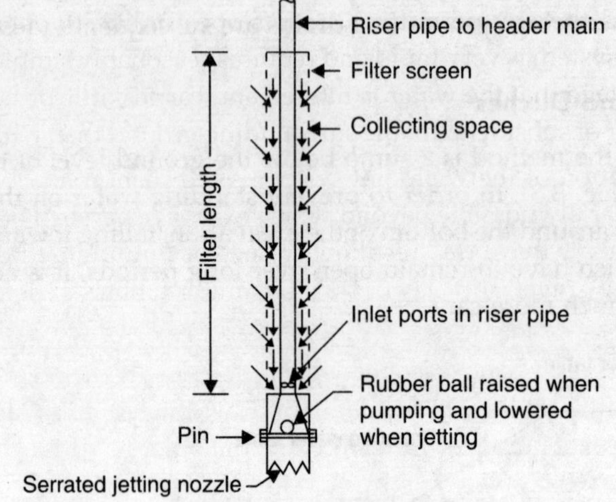

***FIGURE 3.4.* Diagrammatic Arrangement of a Wellpoint.**

(*Source: Tomlinson, 1980*)

The capacity of a single wellpoint with a 50 mm riser is about 10 litres/m. Spacing of wellpoints depends on the permeability of the soil and on the availability of time to effect the drawdown. In fine to coarse sands or sandy gravels a spacing of 0.75 to 1 m is satisfactory. A spacing of 1.5 m may be necessary in silty sands of fairly low permeability. In highly permeable coarse gravels they may need to be as close as 0.3 m centres. In general a wellpointing equipment comprises of 50 to 60 wellpoints to a single 150 or 200 mm pump with a separate jetting pump. The wellpoint pump has an air/water separator and a vacuum pump as well as the normal centrifugal pump.

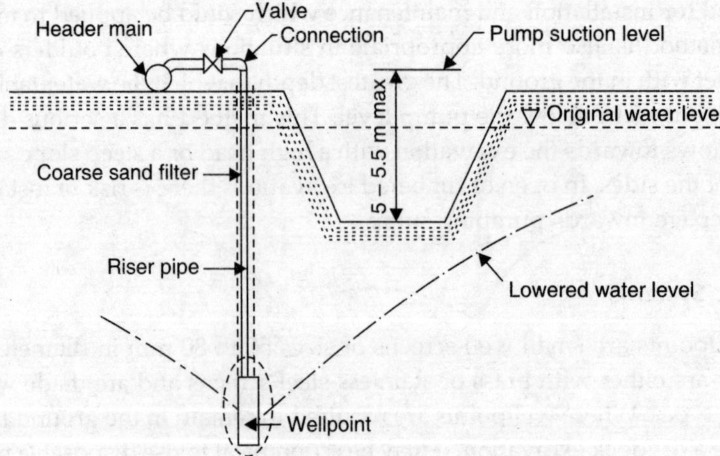

***FIGURE 3.5.* Single-stage Wellpoint Installation.**

Wellpoint system is the most commonly used method for construction purposes. A wellpoint system is suitable when the site is accessible, and where water bearing stratum to be drained is not too deep. Wellpoint system acts most effectively in sands and sandy gravels of moderate permeability. Further, in the wellpoint system the water is drawn away from the excavation, thus stabilizing the sides and thereby permitting steep slopes unlike in open-sump pumping. The installation of wellpoint system is very rapid and requires reasonably simple and cheap equipment. In this there is an advantage that the water is filtered and carries little or no soil particles. Because of this there is less danger of subsidence of the surrounding ground than with open-sump pumping.

The serious limitation of the wellpoint system is the suction lift. A lowering of about 6 m below pump level is generally possible beyond which excessive air shall be drawn into the system through joint in the pipes, valves, *etc.*, resulting in loss of pumping efficiency. If the ground is consisting mainly of large gravel, stiff clay or soil containing cobbles or bolders it is not possible to install wellpoints.

For dewatering deeper excavations the wellpoints must be installed in two or more stages as shown in Fig. 3.6. There is no limit to the depth of drawdown in this way, but the overall width of excavation at ground level becomes very large. On the other hand it is possible to avoid multi-wellpoint stages by excavating down to water level before installing the pump and header.

When wellpoints are used in braced excavations (Fig. 3.7) they are placed close to the toes of the sheet piles. This is done in order to ensure lowering the water level between the sheet pile rows. Wellpoints are provided in conjunction with the sheet piles under the following conditions: (*i*) to prevent quick condition of the bottom when the sheet piles are of limited penetration and (*ii*) to eliminate hydrostatic pressure on the back of a sheetpile cofferdam thus allowing higher bracing to be used.

As an alternate to the conventional wellpoint system with surface pumps, one can use a jet-eductor wellpoint system. A jet-eductor wellpoint system consists of a wellpoint attached to the bottom of jet-eductor pump, with one pressure pipe and a slightly larger return pipe. These two pipes along with the wellpoint and jet-eductor pump are installed in a cased hole and surrounded

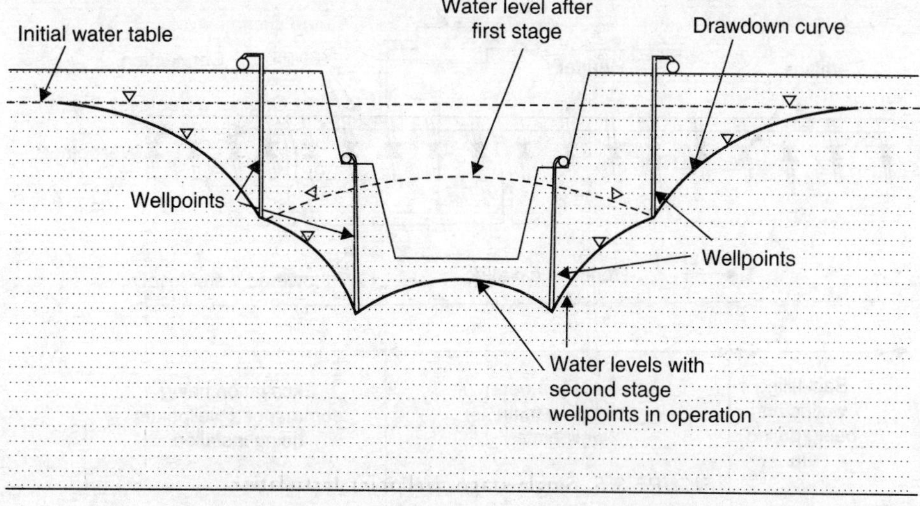

FIGURE 3.6. **Multistage Wellpoint Operation.**

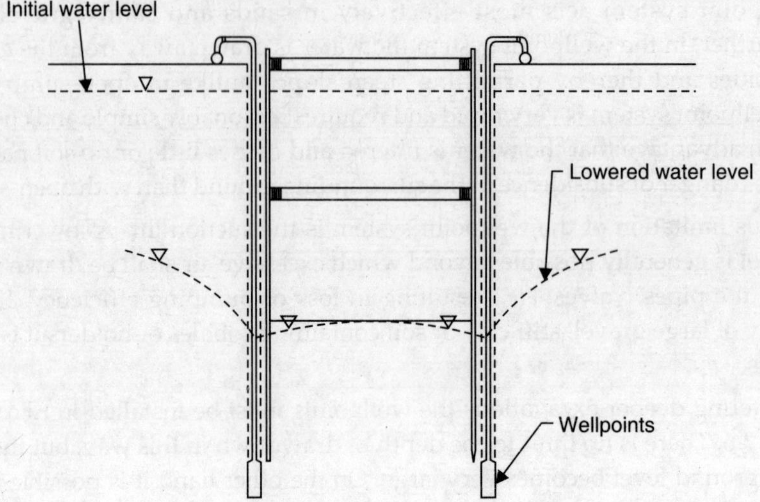

FIGURE 3.7. **Wellpoints in Braced Excavation.**

with a filter sand, if necessary. This method has the advantage that the drawdown of the groundwater is not limited in depth by the suction lift of a pump at ground level. The serious disadvantage with this system is that the jet-eductor pumps usually have a low efficiency.

For large excavations or where the depth of excavation below the water table is more than 10 to 15 m or there is a necessity to reduce the artesian pressure from a deep aquifer it is desirable to use deep-wells (dealt subsequently) with or without wellpoints.

Wellpoints are installed either by the progressive system or the ring system. In the progressive system the header is laid out along the sides of the excavation (Fig. 3.8).

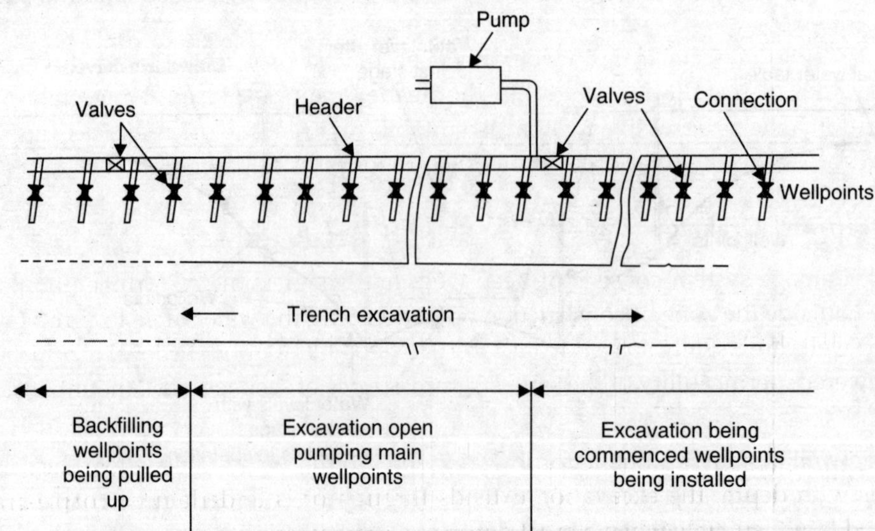

FIGURE 3.8. **Single-stage Wellpoint Installation by Progressive System.**

(*Source: Tomlinson, 1980*)

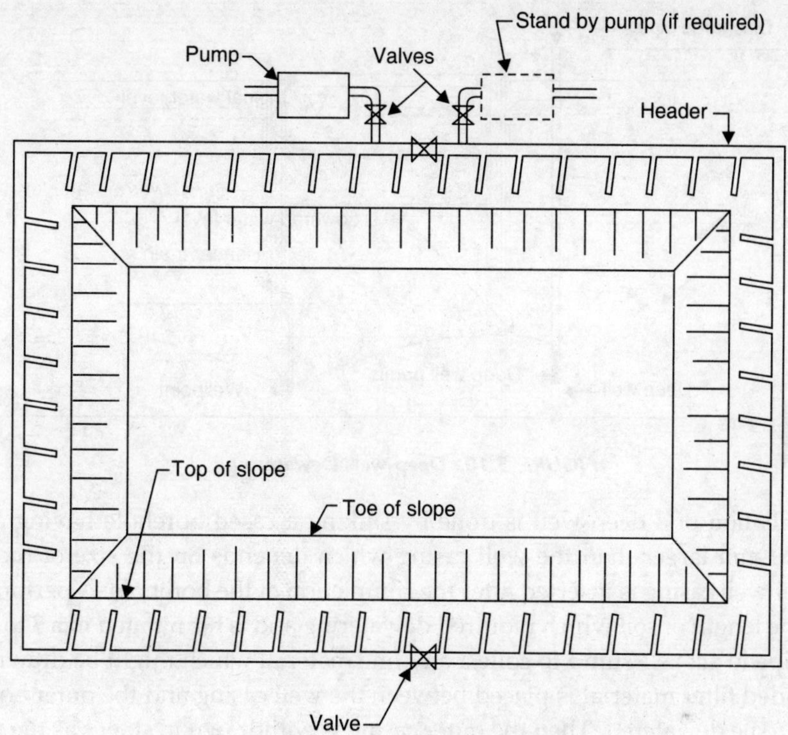

FIGURE 3.9. **Wellpoint Installation by the Ring System.**
(*Source: Tomlinson, 1980*)

Pumping is done continuously in one length as further points are jetted ahead of the pumped down section and pulled out from the completed and backfilled lengths. Only one header is sufficient for narrow excavations whereas for wide excavations or in soils containing layers of less permeable materials, the header must be placed on both sides of the excavation. In the ring system of installation (Fig. 3.9) the header main is placed completely to surround the excavation. The progressive system is suitable for trench work whereas the ring system is most suitable for rectangular excavations such as for piers or basements and for long trenches.

3.6.3 Deep-well Drainage

Deep-well drainage system consists of deep-wells and submersible or turbine pumps which can be installed outside the zone of construction operations and the water table lowered to the desired level. Deep-wells are usually spaced from 8 to 80 m depending upon the level to which water table must be lowered, permeability of the sand stratum, source of seepage and amount of submergence available.

Deep-well system is suitable for lowering the ground water table where the soil formation is pervious with depth, the excavation extends through or is underlain by coarse-grained soils. This method is also suitable when a great depth of water lowering is required or where a head due to artesian pressure has to be lowered in a permeable strata at a considerable depth below the excavation level. Deep-wells may be combined with the wellpoint system on certain field conditions for lowering the groundwater table (Fig. 3.10).

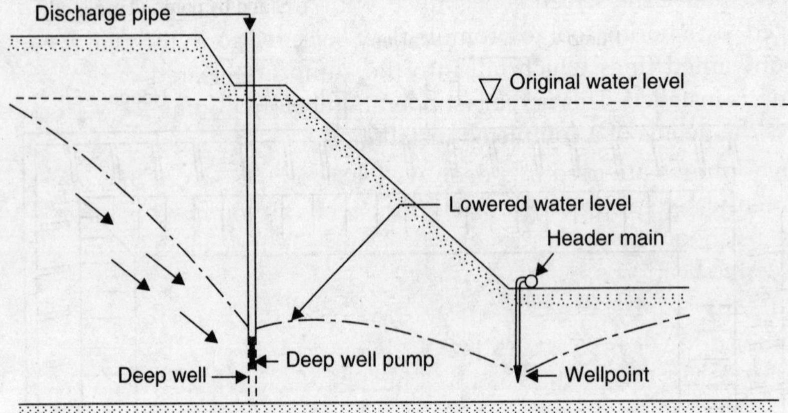

FIGURE 3.10. **Deep-well Dewatering.**

The installation of a deep-well is done by sinking a cased borehole having a diameter of about 200 to 300 mm larger than the well casing which depends on the size of the submersible pump. The inner well casing is inserted after the completion of the borehole. A perforated screen is installed over the length of soil which required dewatering and is terminated in a 3 to 5 m length of unperforated pipe to act as a sump to collect any fine material which might be drawn through the filter mesh. Graded filter material is placed between the well casing and the outer borehole casing over the length to be dewatered. Then the outer casing is withdrawn in stages as the filter material

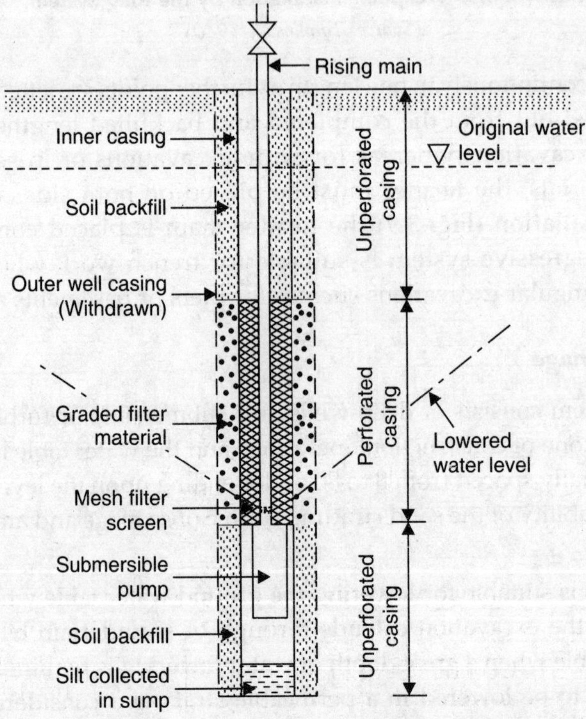

FIGURE 3.11. **Details of Deep-well Installation.**

(*Source: Tomlinson, 1980*)

is placed. The space above the screen is backfilled with any available material. The water in the well is then 'surged' by a boring tool to promote flow back and forth through the filter, and at the same time any unwanted fines which fall into the sump are cleared out by bailer before the submersible pump is installed. This is the last operation before putting the well into commission. Figure 3.11 shows the details of a completed installation.

If centrifugal pumps are used in a deep-well system, the tops of the screens should be set below the computed water surface in the well. If the wells are pumped by deep-well pumps, the bottoms of the wells should be set to provide sufficient length of submerged screen to admit the flow without excessive head loss.

As heavy boring plant is used to sink the well in very adverse formations like boulders, rocks or under other difficult field environment, the cost of deep-well system is relatively high. Thus it is advised to restrict this method to jobs which have a long construction period such as dry docks or access shafts for long sub-aqueous tunnels.

3.6.4 Vacuum Dewatering Systems

Gravity methods, such as wellpoints and deep-wells, are not much effective in the fine-grained soils with permeability in the range of 0.1 to 10 × 10^{-3} mm/s. Such soils can be dewatered satisfactorily by applying a vacuum to the piping system. A vacuum dewatering system requires that the well or wellpoint screens, and riser pipe be surrounded with filter sand extending to within a few metres of the ground surface. The top few portion of the hole is sealed or capped with an impervious soil or other suitable material. By having the pumping main a vacuum pressure, the hydraulic gradient for flow to the wellpoints is increased. This method is most suitable in layered or stratified soils with coefficient of permeability of the range 0.1 to 10 × 10^{-4} cm/sec. A typical vacuum dewatering system in a stratified soil is shown in Fig. 3.12. In this system the wellpoints should be placed closer than the conventional system. It is common to use suction pump in this system and the practical maximum height of lift is about 3 to 6 m.

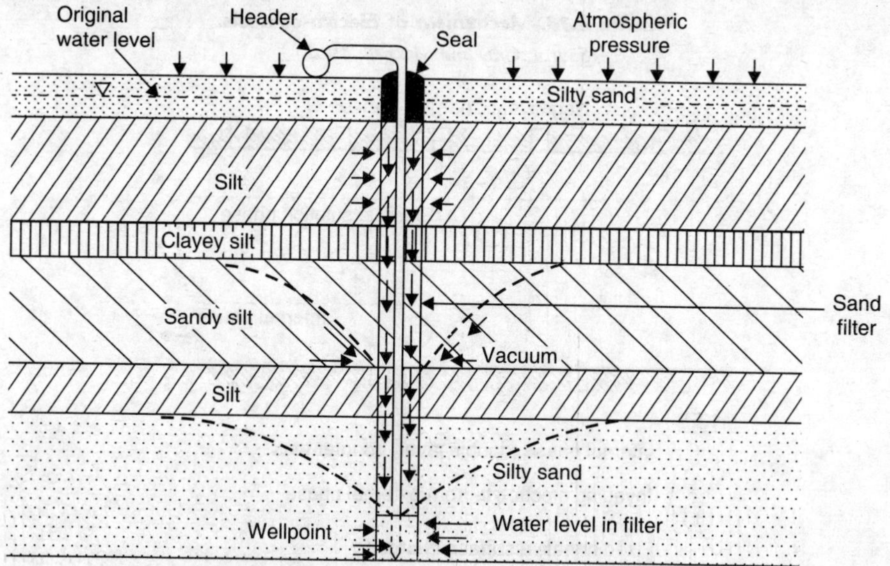

FIGURE 3.12. **Vacuum Dewatering System.**
(Source: Tomlinson, 1980)

3.6.5 Dewatering by Electro-osmosis

When an external electromotive force is applied across a solid–liquid interface the movable diffuse double layer is displaced tangentially with respect to the fixed layer. This is electro-osmosis. As the surface of fine-grained soil particles causes a net negative charge, positive ions (cations) in solution are attracted towards the soil particles and concentrate near their surfaces. Upon application of an electromotive force between two electrodes in a soil medium the positive ions adjacent to the soil particles and the water molecules attached to the ions are attracted to the cathode and are repelled by the anode (Fig. 3.13). The free water in the interior of the void spaces is carried along to the cathode by viscous flow. By making the cathode a well, water can be collected in the well and then pumped out. Hypothetical distribution of ions between external and internal phases in a clay pore is shown in Fig. 3.14. A comparison of electro-osmotic flow with hydraulic flow through a

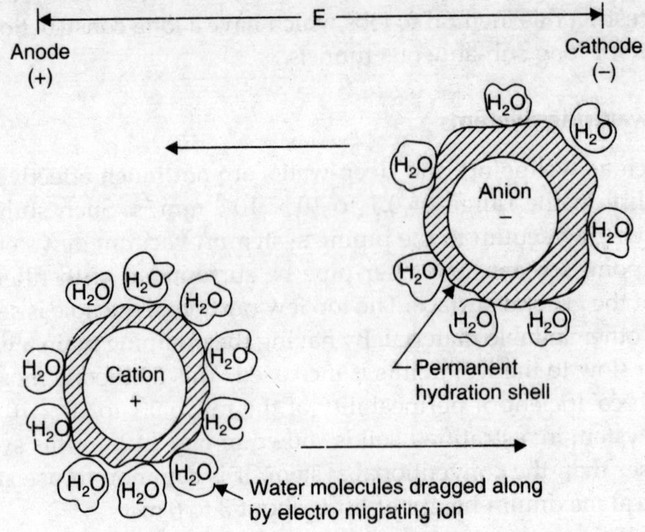

***FIGURE 3.13.* Mechanism of Electro-osmosis.**

(Source: Gray and Mitchell, 1967)

Negatively Charged Surface

External solution phase

Internal phase

KEY

+ Internal phase (double layer) counter ions

⊖ ⊕ Invading electrolyte from external phase

⊟ Fixed negative surface charges

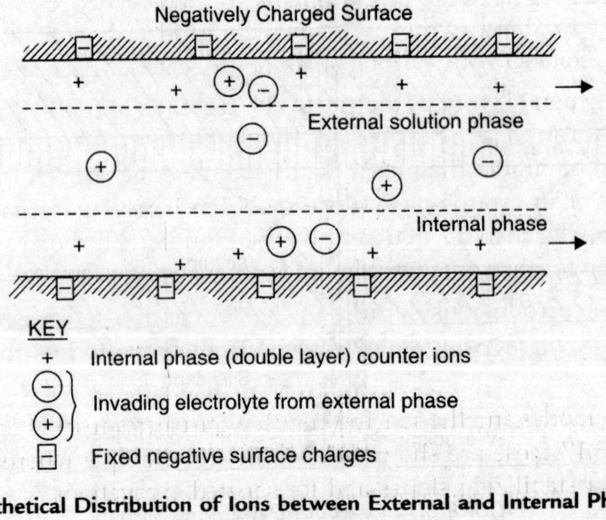

***FIGURE 3.14.* Hypothetical Distribution of Ions between External and Internal Phases in a Clay Pore.**

(Source: Gray and Mitchell, 1967)

single capillary is shown in Fig. 3.15. The electro-osmotic flow (Q_e) produced by an applied electric field is given by an expression similar to hydraulic flow (Q_h) as

$$Q_e = k_e i_e A \quad \text{...(3.3)}$$

where k_e is called the *electro-osmotic permeability*, the unit being cm/sec for the constant potential gradient of one volt/cm. Different theories (Helmholtz, 1879; Schmid, 1950–52; Tadashi, 1961; Gray and Mitchell, 1967) have been postulated to explain the electro-osmotic flow. While the hydraulic coefficient of permeability (k) and the rate of hydraulic flow depend on the size of voids in the soil and consequently on the grain-size, whereas the coefficient of electro-osmotic permeability and the rate of electro-osmotic flow are almost independent of grain-size. For most soils it lies within the range of 0.4 to 0.6 × 10^{-4} cm/sec (Casagrande, 1949, 1952).

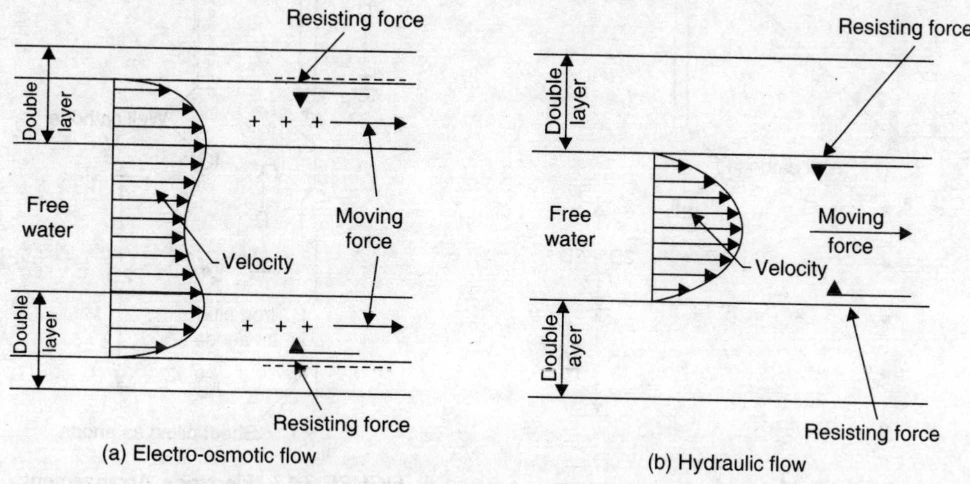

***FIGURE 3.15.* Comparison of Electro-osmotic and Hydraulic Flows.**

The potential gradient should not exceed 0.5 volt/cm in order to prevent considerable loss of energy due to heating of the ground. Apart from this limitation, there exists a fairly wide range of variations in installations to comply with equipment which the contractor has at hand. For example, equal gradients can be achieved with various potentials by changing the spacing of the electrodes.

Electrodes. The general layout of the electrodes depends upon the purpose for which they are intended. Sheet piles of any shape can be used. The simplest type of anode for normal application are old pipes, of 25 mm or 50 mm diameters, which can be easily driven into the soil. These can either be placed singly or in groups of two or three. The individual pipes forming one anode should not be closer together than 0.7 m otherwise the group acts like one single electrode with a diameter only slightly larger than that of the individual pipes.

Since the anodes corrode considerably in the course of a few weeks of electro-osmotic treatment they should be replaced as soon as the current drops to less than 30% of the initial consumption.

The length of the anodes and the spacing between the electrodes will differ in every practical case. In Figs. 3.16 and 3.17 these are shown for a common example where it is the intention to apply electro-osmosis for stabilizing slopes and for sheeted excavations.

Where excavations are to be stabilized during construction it is advisable in general to install cathodes in the form of wells from which the accumulating water can be pumped according to need. In contrast to normal wellpoints these cathode-wells should extend from the surface to the bottom so as to allow the water to discharge into them over the whole of their height. Such cathode-wells have the advantage of serving also as drainage wells for layers and pockets of sand which are frequently present in silty soils and which otherwise would hardly be affected by electro-osmosis. Two basic arrangement of cathode-wells are shown in Fig. 3.18. Either the well is arranged separately but in close proximity to the cathode (Fig. 3.18 *a*) or the cathode and the well combined as one body (Fig. 3.18 *b*).

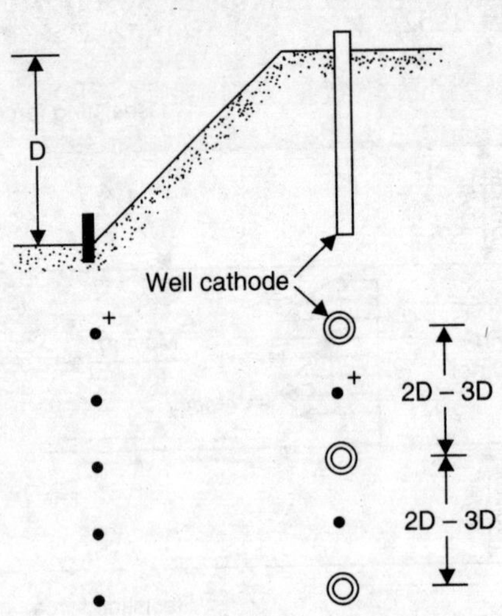

FIGURE 3.16. **Electrode Arrangements for Slopes.**

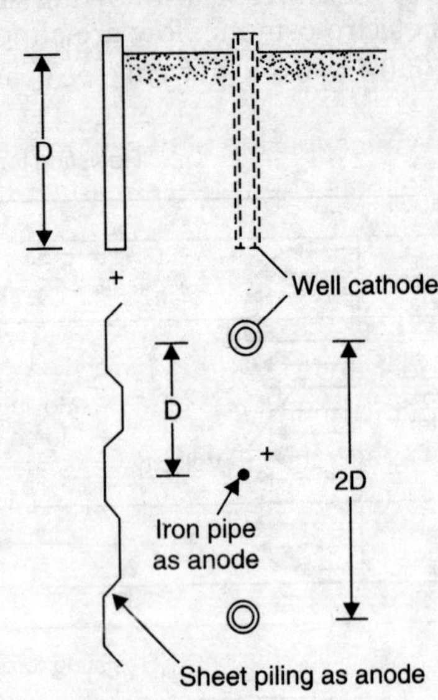

FIGURE 3.17. **Electrode Arrangements for Sheated Excavation.**

In case of cathodes embedded in pervious filter, as indicated in Fig. 3.18 (*a*), the pore water entering into the filter well lose practically all of its electrical contact with the electrode and thus reducing the electro-osmotic transport of water to a fraction of the full capacity of this method.

The well and electrode combined into one body (Fig. 3.18 *b*) consists of a metal grid or slotted pipe in the inside, a copper gauze surrounding it and a jute or cotton fabric in between. The

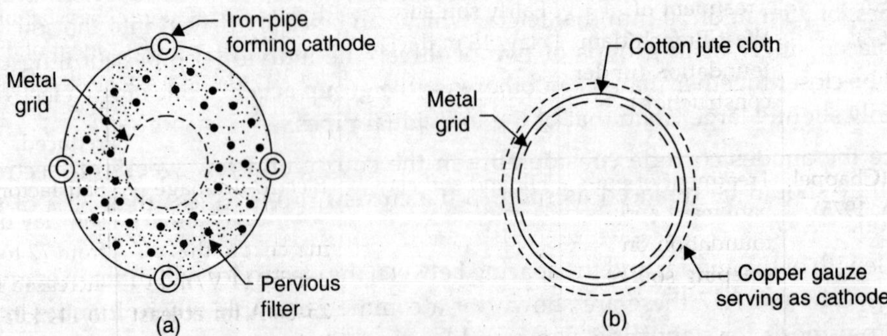

FIGURE 3.18. **Combination of Cathode and Well.**

copper gauze serves as cathode and it is to be placed into immediate contact with the soil which is being subjected to electro-osmosis. As a rule, the anodes and cathodes should reach to equal depths. Anodes and cathodes should extend in depth at least 1.5 m below the bottom of the slope or excavation (IS : 9759–1981). If the electrodes are arranged in a row, Indian Standard (IS : 9759–1981) recommends that the cathodes to be placed in one or more lines and spaced on 7.5 to 10.5 m centres, with anodes installed midway between the cathodes. The cathode-wells should have a diameter not less than 100 mm.

Safety Measures. The following simple rules should be made known to everybody on the site:

(i) Only persons wearing rubber boots should be admitted into the neighbourhood of the electrodes.

(ii) While working between anode and cathode neither the electrode nor the wiring should be touched in such a manner that while one hand is on the electrode, the other gets into contact with the ground or with the other electrode. Serious harm could be the consequence, even with less than 100 volts.

(iii) Where the excavation is carried out by machinery special attention has to be paid by the operator to avoid the occurrence of short circuits. This may endanger the operator himself in the first place. No particular precautions are necessary if the machine operates within a space enclosed by anodes such as sheeted excavations of the type shown in Fig. 3.16.

Case histories of electro-osmosis treatment of soft clays are presented in Table 3.2 (Pilot, 1981). A comparative study of dewatering systems discussed above is presented in Table 3.3 (IS : 9759–1981).

Table 3.2. Case Histories of Electro-osmosis
(*Source: Pilot, 1981*)

Location and reference	*Object of treatment*	*Soil treated*	*Procedure*	*Results and comments*
Norway. (Bjerrum et al., 1967).	To increase strength prior to earthworks.	Sensitive clay, about 12 m thick.	Electro-drainage 65 V/m at 250 A for 4 months.	Satisfactory treatment of 2700 m^2 soil; w dropped from 31 to 17%; s_u increased from 9 to 38 kPa.
Marren, USA. (Fetzer, 1967).	Treatment of West-Branch dam foundation (under construction).	Fairly soft silty clay, 18 m thick.	Electro-drainage: 8–25 V/m at 40 A for 10 to 12 months.	Satisfactory treatment of 580,000 m^2 soil; excess pore pressure greatly reduced.
Singapore (Chappel and Burton, 1975).	Treatment of embankment and its foundation on creeping soil.	Fill and soft clay, 5 m thick.	Electro-drainage with local hardening next to the anode: 14 V/m at 25–30 A for at least 9 days.	Satisfactory; w of soft clay decreased from 72 to 35%; s_u increased form 50 to 144 kPa.

(Table Contd...)

Table 3.2. *(Contd...)*

Bordeaux, France, (Caron, 1968).	Treatment prior to excavation.	Soft clay, 9 m thick.	Electro-drainage : 16–25 V/m at 20 A for 3 months.	Satisfactory treatment of 1000 m^2 soil.
Tunnis, Tunisia, (Caron, 1968).	Foundation soil treatment prior to construction of reservoirs.	Very soft clay, 15 m thick.	Electro-injection: ammonium chloride.	Satisfactory; s_u increased from 6 to 15 kPa.
Seattle, USA. (Dearstyne and Newman, 1963).	Treatment of foundation soil under existing runway.	Soft clay (through pumping) 1.5 m deep.	Electro-injection : H. 226, with 6 V/m at 4 A for 15 days.	Satisfactory.
Antwerp, Belgium, (De Beer and Wallays, 1965).	Treatment prior to excavation.	Very soft clay (w_n = 120–220%), 7 m thick.	Electro-drainage: 10 V/m at 45 A for one month.	Satisfactory.
Angers, France. (Caron, 1971; Peignaud, 1977).	To increase soil strength under bridge abutment.	Soft to stiff clay, 10 m thick.	Electro-injection : sodium silicate, with 28 V/m at 30 A for maximum 3 months.	Unsatisfactory—major implementation problems.
Dighi, India. (Sridharan, 1972).	To harden soil around a pile.	Soft clay around a test pile 150 mm as diameter.	Electro-drainage : 90 V/m at 8 A for 72 hours.	Satisfactory.

Note: w = water contact, s_u = undrained shear strength.

Table 3.3. Comparative Studies of Dewatering Systems
(*Source: IS: 9759–1981*)

Sl. No.	*Method*	*Soils Suitable for treatment*	*Uses*	*Advantages*	*Disadvantages*
1.	Sump pumping	Clean gravels and coarse sands	Open shallow excavations	Simplest pumping equipment	(*a*) Fines easily removed from ground (*b*) Encourages in sability of formation.
2.	Wellpoint systems with pumps	Sandy gravels down to fine sands (with proper control can also be used in silty sands)	Open excavations including rolling pipe trench excavations	(*a*) Quick and easy to install in suitable soils	(*a*) Difficult to instal in open gravels and grounds containing cobbles and boulders

(*Table Contd...*)

Table 3.3. *(Contd...)*

				(*b*) Economical for short pumping periods of a few weeks	(*b*) Pumping must be continuous and noise of pump may be a problem in a built up area (*c*) Suction lift is limited to 4.5 to 6.0 m. (*d*) If greater lowering is needed multistage installation is necessary.
3.	Deep bored filter wells with electric submersible pumps	Gravels to silty fine sand, and water bearing rocks	Deep excavation in through or above water bearing formations	(*a*) No limitation on amount of drawdown as there is for suction pumping (*b*) A well can be constructed to draw water from several layers throughout its depth (*c*) Wells can be sited clear of working area (*d*) No noise problem if mains electricity supply is available.	
4.	Electro-osmosis	Silts, silty clays and some peats	Open excavation in appropriate soils or to speed dissipation of construction pore pressures	Any appropriate soil can be used when no other water lowering method is applicable	Installations and running costs are usually high.

(Table Contd...)

Table 3.3. (Contd...)

5.	Jet-eductor system	Sands (with proper control can also be used in silty sands and sandy silts)	(a) Deep excavation (in space of confined that multistage well pointing cannot be used) (b) Usually more appropriate to low permeability soils	(a) No limitation on account of draw-down (b) Raking holes are possible	(a) Initial installation is fairly costly (b) Risk of flooding excavation if high pressure water main is ruptured (c) Optimum operation difficult to control.

3.7 DESIGN STEPS FOR DEWATERING SYSTEMS

A complete dewatering system should be capable of lowering the groundwater table up to a required depth, should intercept seepage into the excavation and at the same time should remove the surface water which would otherwise affect the operation of the dewatering system or construction. Thus the selection of an appropriate dewatering system and its design depends on geological, soil and water-table conditions at the site, the magnitude of the excavation, surface-water runoff and construction requirements. The essential steps in designing a dewatering system is presented below.

3.7.1 Subsoil Investigation

Paramount importance should be given to know the characteristics of soils adjacent and beneath an excavation. The depth and spacing of borings have to be properly planned to ascertain the thickness of stratified soils and soils with significant variation. Samples should be taken at frequent intervals so as to identify each soil type, permeability characteristics of the soil, and layers of clay or any other impervious material. The position of water table and substratum pressure should be carefully recorded. Grain-size distribution and permeability are the two important factors to be determined.

Table 3.4. Permeability Values of Pervious Stratum
(*Source: IS: 9759–1981*)

Sl. No.	*Types of Sand*	*Coefficient of Permeability (k) cm/sec*
1.	Very fine sand	1 to 50
2.	Fine sand	51 to 200
3.	Fine to medium sand	201 to 500
4.	Medium sand	501 to 1000
5.	Medium to coarse sand	1001 to 1500
6.	Gravel and coarse sand	1501 to 3000

Permeability of the sand strata to be dewatered or in which the hydrostatic pressure is to be reduced should be determined. Indian Standard (IS : 9759–1981) recommends a field pumping test for this case. However, various methods may be adopted to determine the permeability. A rough approximation of permeability may be obtained from Table 3.4 (IS : 9759–1981). A better estimate could be from a comparison of grain-size distribution curves or from the Hazen's empirical relationship, *viz.*,

$$k = C_1 D_{10}^2 \qquad ...(3.4)$$

where, D_{10} = Diameter of particle (cm),

k = Coefficient of permeability (cm/sec),

C_1 = Constant varying between 100 and 150 in case of pumping test is not conducted.

This approximate expression will be valid for fairly uniform sands in loose state with uniformity coefficient not greater than 2. Laboratory determination of k yield reliable results for uniform sands but the results will be misleading in case of well graded or stratified or gravelly sands. It is warranted to assess the permeability by a suitable field pumping test in case of large dewatering projects.

3.7.2 Source and Water Table Details

Source of seepage and knowledge of the water table and any substratum pressure at a site are most important factors to be considered while designing a dewatering or a pressure-relief system. The source of seepage depends on the geological features of the area, nearby streams or other bodies of water, on the degree of perviousness of the stratum, and amount of drawdown. Streams close to the wells may act as line source of seepage depending on the distance of the wells from the effective source of seepage. A flow may be from an aquifer being drained, the distance to which is commonly known as the radius of influence. The radius of influence may be estimated from a drawdown curve established from a field pumping test. An empirical equation (Eq. 3.5) based on Sichordt's equation (as recommended by IS : 9759–1981) may be used to estimate the radius of influence.

$$R = C'(H - h_w)\sqrt{k} \qquad ...(3.5)$$

where, R = radius of influence (m)

C' = a constant = 0.9 (for gravity flows)

H = depth of natural water table (m)

h_w = head at the well (m)

k = coefficient of permeability (10^{-4} cm/sec).

If the well is located close to a river, the source of seepage may be considered as the river, provided the distance L from the well to the river is less than $R/2$.

Generally the initial elevation of the water table is considered in deciding the elevation at which the first stage of wellpoints/wells would have to be installed, as the water table or artesian pressure may be affected frequently due to seasonal variations or with the stage of adjacent water source. In order to ensure dry working conditions during construction it is preferable in most soils to maintain the water table 1 to 2 m below the bottom of the excavation. Some groundwaters may contain iron or dissolved salts which over an extended period of time may partially or fully clog the screens, filters or surrounding soil and corrode the screens or metallic parts resulting in the reduction of efficiency of dewatering system. Thus due consideration should be shown on the chemical properties of groundwater in the design, operation and maintenance of dewatering system.

3.7.3 Distance of Wellpoints/Wells from Source of Seepage

Based on the field condition the value of R or L should be ascertained. In this regard, the following considerations should be taken care of (IS : 9759–1981):

(*a*) If the actual radius of influence is large compared with the radius of the well, only an approximate estimation of R may be sufficient since the discharge is not much sensitive to the value of R.

(*b*) An accurate estimation of L should be made for a particular dewatering system, since the discharge is inversely proportional to the value of L.

3.7.4 Effective Well Radius

The effective well radius, r_w, for a wellpoint/well is decided based on the installation of wells with or without filter. If a wellpoint/well is installed without a gravel or sand filter the effective well radius can be taken as one half the outside diameter of the well screen. When a gravel or sand filter is used around the well screen the well radius is taken as one-half the outside diameter of the filter. If a well screen is installed without a filter in the pervious stratum but a natural filter formed around the screen due to surging, the effective well radius will exceed one-half of the outside diameter of the well screen. As the extent of the developed filter due to surging is indefinite it is generally adequate and conservative to use a value of r_w, corresponding to the outside diameter of the well screen.

3.7.5 Discharge Computations

Before designing a dewatering and/or pressure relief system, it is necessary to know the rate at which water must be removed from the pervious strata in order to achieve the required groundwater lowering or pressure relief. Based on this, the design should decide the number, size, spacing, and penetration of the wellpoints/wells and the size and capacity of collectors and pumps. Thus it is mandatory to establish the fundamental relationships between discharge from wells and wellpoints and the corresponding drawdown effected in the pervious strata. This is done by following the design principles as applicable to laminar and continuous flow, through a porous medium.

In a dewatering system with a single line of closely spaced wells, an approximate solution for the drawdown produced by these wells can be obtained by considering the line of wells equivalent to a drainage slot. The validity of this assumption depends upon the spacing of the wells; as wells are more closely spaced, they tend to approach a continuous line sink or slot. Their spacing is so determined that the head along the line of wellpoints is essentially the same as would exist at a slot. The following procedure is recommended by Indian Standard (IS : 9759–1981) to be followed in this respect:

(*i*) If H is the head corresponding to the natural water table and h_o is the head at the slot, then the head reduction $(H - h_o)$ at a slot required to produce the desired residual head h_D should be computed from equations given in Tables 3.5 to 3.8 along with the factors in Figs. 3.19 and 3.20.

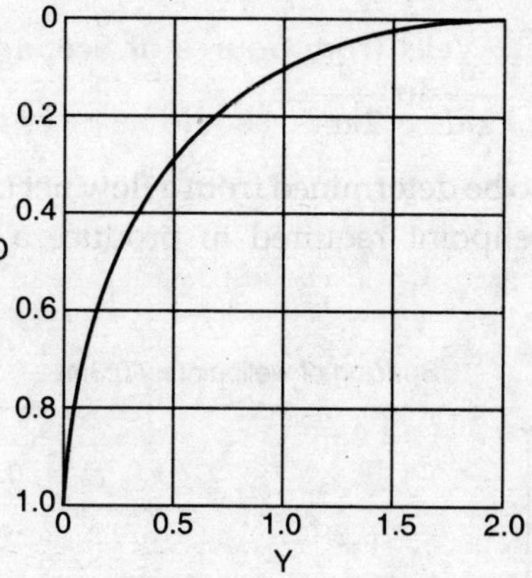

***FIGURE 3.19.* Factor Y versus Ratio W/D.**
(*Source: IS: 9759–1981*)

(*ii*) Assuming that $h_o = h_D$, and that $(h_o - h_w)$, the head difference (h_w being the head at the well) is small (assumed as 0.001 H), the wellpoint spacing can then be computed from the following equations:

(*a*) For artesian case :

$$\frac{h_D - h_w}{H - h_D} = \frac{a}{2\pi L} \ln \frac{a}{2\pi r_w} \qquad \text{...(3.6)}$$

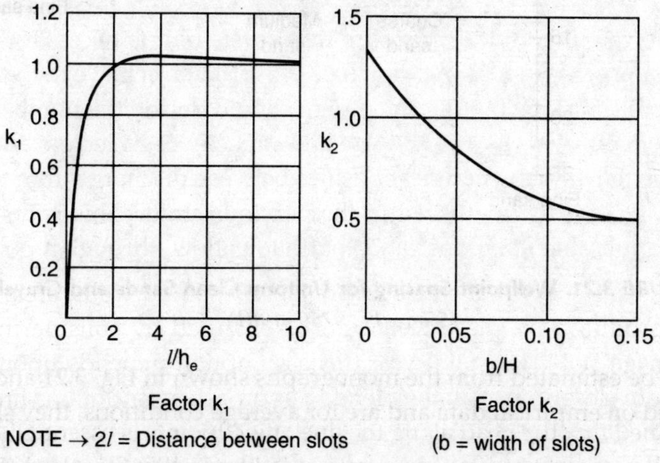

***FIGURE 3.20.* Computation of Factors k_1 and k_2.**
(*Source: IS: 9759–1981*)

where, L = distance of wellpoints from the line source

a = spacing of wellpoints, and

r_w = radius of wellpoints

(*b*) For gravity case:

$$\frac{h_D^2 - h_w^2}{H^2 - h_D^2} = \frac{a}{2\pi L} \ln \frac{a}{2\pi r_w} \qquad ...(3.7)$$

Wellpoint spacing can also be determined from a flow net (reader may refer to IS : 9759-1981). An approximate spacing of wellpoint required to produce a given groundwater lowering in

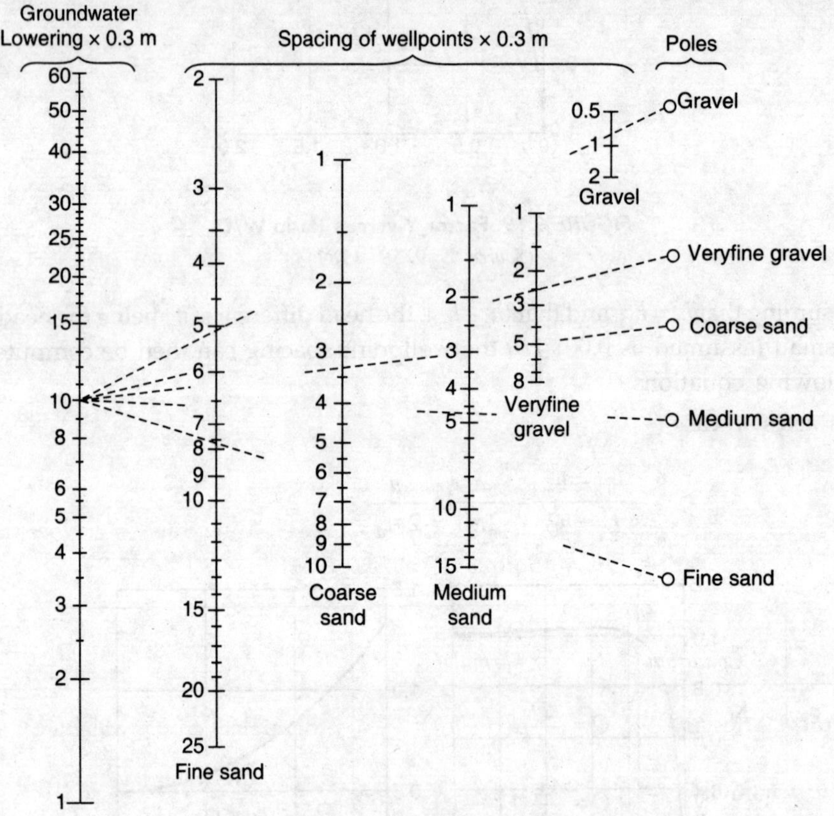

FIGURE 3.21. **Wellpoint Spacing for Uniform Clean Sands and Gravels.**
(*Source: IS : 9759–1981*)

various soils can also be estimated from the monographs shown in Fig. 3.21 and Fig. 3.22. As these monographs are based on empirical data and are for average conditions, they should be used with caution.

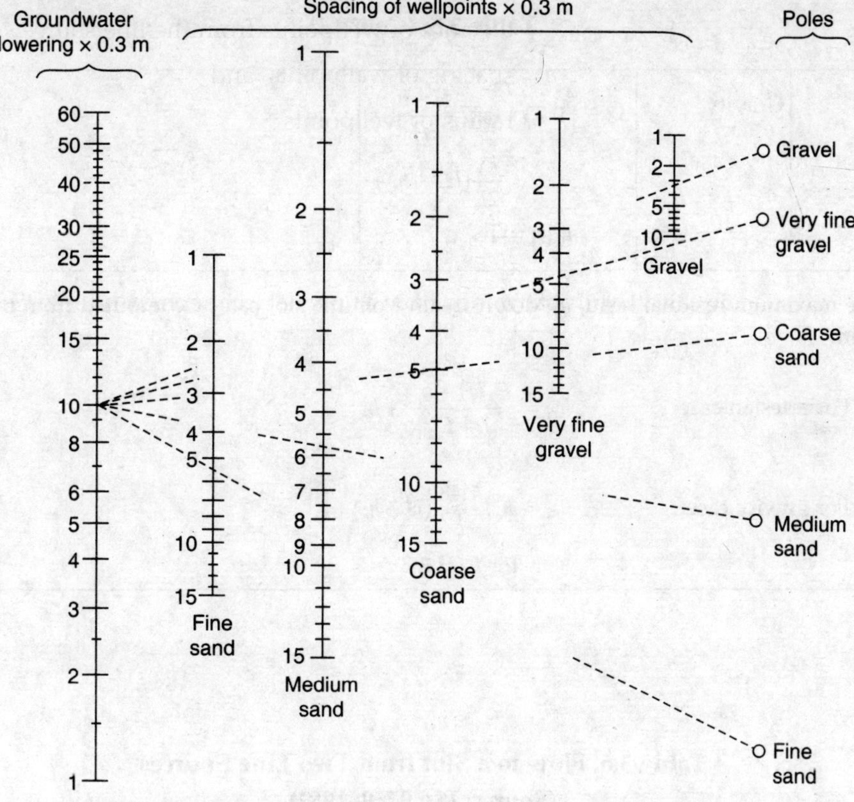

FIGURE 3.22. **Wellpoint Spacing for Stratified Clean Sands and Gravels.**

(*Source: IS : 9759–1981*)

Table 3.5. Flow to a Slot from a Single Line Source
(*Source: IS : 9759–1981*)

Penetration	*Flow Condition*	*Discharge Formulae*	*Remarks*
Fully penetrating slot	Artesian	$Q = \frac{kDx}{L}(H - h_e)$	H = original groundwater level
	Gravity	$Q = \frac{kx}{2L}(H^2 - h_e^2)$	h_e = groundwater level at the use Q = flow rate L = distance of the slot from the line source D = depth of pervious stratum x = distance perpendicular to the direction of flow
Partially penetrating slot	Artesian	$Q = \frac{kDx(H - h_e)}{L + E_A}$	E_A = extra-length factor as described by various authors

(*Table Contd...*)

Table 3.5. (*Contd...*)

	Gravity	$Q = \left(0.73 + 0.27\dfrac{H - h_e}{H}\right)$ $\dfrac{kx}{2L}(H^2 - h_D^2)$ for $L/H \geq 3$	

Note. The maximum residual head, h_D downstream from the slot can be computed from the following expression:

(*i*) For artesian case: $$h_D = \frac{E_A(H - h_e)}{L + E_A} + h_e$$

(*ii*) For gravity case: $$h_D = h_o\left[\frac{1.48}{L}(H - h_e) + 1\right]$$

for $L/H \geq 3$

Table 3.6. Flow to a Slot from Two Line Sources
(*Source: IS : 9759–1981*)

Penetration	*Flow Condition*	*Discharge Formulae*	*Remarks*
Partially penetrating slot	Artesian	$Q = \dfrac{2kDx(H - h_e)}{L + yD}$	Q = flow to the slot; y = a factor which depends on the ratio W/D
	Gravity	$Q = \dfrac{0.73 + 0.27\,H - h_e}{H}\,\dfrac{kx}{L}(H^2 - h_e^2)$	W = penetration of the slot into the pervious stratum (To be determined from Fig. 3.19).
Fully penetrating slot		The flow is twice the one computed from Table 3.4 for the respective cases	**Note.** The slot is midway between the line sources.

Note. At distance y from the slot, in excess of about 1.3 D the head h increases linearly as y increases, and can be computed as follows:

$$h = h_o + (H - h_e)\,\frac{y + D}{L + yD}.$$

Table 3.7. Flow to Two Partially Penetrating Slots Midway Between and Parallel to Two Line Sources

(*Source: IS: 9759–1981*)

Flow Condition	*Discharge Formulae*	*Remarks*
Artesian	Flow from one source to the closest of the two slots is obtained from equations of Table 3.5. Head h_D midway between slots is obtained from Table 3.5.	
Gravity	Flow from either slot is determined from Table 3.5. The head h_D midway between the slots is given by $h_D = h_e\left[\frac{k_1 k_2}{L}(H - h_e) + 1\right]$	k_1 and k_2 can be obtained from Fig. 3.20

Table 3.8. Head Reduction for Finite Length of Slot

(*Source: IS : 9759-1981*)

Penetration	*Flow Condition*	*Head Reduction at the Wells*
Fully penetrating well	Artesian	$H - h_w = \frac{Q_w}{2\pi k_D}\ln\frac{a}{2\pi r_w} + \frac{Q_w L}{kD_a}$
	Gravity	$H^2 - h_w^2 = \frac{2Q_w L}{ka} + \frac{Q_w}{\pi k}\ln\frac{a}{2\pi r_w}$
Partially penetrating	Artesian	$H - h_w = \frac{Q_w}{kD}\left(\frac{L}{a} + \theta_a\right)$
	Gravity	The formula for fully penetrating case can be used provided Q_w is computed from an appropriate equation.

Note 1. Q_w = discharge per well.

Note 2. θ_a = uplift factor (for details as given in various text books).

Note 3. h_w = head at the well.

(*iii*) After the wellpoint spacing and the head h_w at the wellpoint have been computed, the flow Q_w per wellpoint can be computed from the equation given in Table 3.8.

(*iv*) The above value of h_w should be equal to or greater than the value of h_w computed from equation (3.7). The total head loss in a wellpoint connection should be estimated afresh:

$$h_w = M - V + H_e + H_w \qquad \ldots(3.8)$$

where, M = distance from base of pervious stratum,

V = vacuum at pump intake,

H_e = average head loss in header pipe up to pump intake,

H_e = (head loss due to screen entrance (H_e))
+ (friction loss due to flow through the well screen (H_s))
+ (friction loss due to flow in the riser pipe (H_r))
+ (velocity head (H_v).)

3.7.6 Design of Filters

Based on the filter criteria, discussed in Section 3.3, the filter requirements for filter material shall be as under:

Character of Filter Materials	*Ratio R_{50}*	*Ratio R_{15}*
Uniform grain-size distribution (C_u = 3 to 4)	5 to 10	—
Well graded to poorly graded (non-uniform); subrounded grains	12 to 58	12 to 40
Well graded to poorly graded (non-uniform);	9 to 30	6 to 18

where,

$$R_{50} = \frac{D_{50} \text{ of filter material}}{D_{50} \text{ of material to be protected}}$$

$$R_{15} = \frac{D_{15} \text{ of filter material}}{D_{15} \text{ of material to be protected}}.$$

3.7.7 Design and Selection of Well Screens

The design of wells and wellpoints must ensure that there will be little resistance to water flowing through the screen and riser pipe, prevent infiltration of sand during pumping, and resist corrosion by water and soil. Commercially available wellpoints are of brass or stainless-steel screens mounted over galvanized, tin-dipped, or stainless-steel suction pipe. A high-capacity type of wellpoints should be used when large flows are foreseen in the field. The mesh or slot size of a screen should be smaller than the 80% size (D_{80}) or 70% size (D_{70}), respectively, of the sand in which the wellpoint will be installed, so as to avoid infiltration of sand. While draining silty soils, the wellpoint should be provided with a graded filter.

Screens commonly used for deep large diameter wells are slotted steel or wood stave pipe or perforated steel pipe wrapped with galvanized, trapezoidal-shaped wire. Slotted steel or wooden screens are commercially available with various width of slots. In order to enhance the efficiency of the well and at the same time to prevent infiltration of sand, it is desirable to use a graded filter around the screens. The riser pipe usually consists of either steel or wood.

As per Indian Standard (IS : 9759–1981) the following criteria should be observed in designing and selecting well screens or wellpoints:

Slot width ≤ D_{70} (filter or aquifer sand)

Hole diameter or width < D_{80} (filter or aquifer sand).

3.7.8 Selection of Pumps and Accessories

Pumps, headers, discharge lines and power unit must be of sufficient capacity to remove the required flow from the wells or wellpoints and conduct it away from the dewatered area. The selection of a pump and power unit depends on various factors such as required discharge, suction lift plus positive head including hydraulic head losses, air-handling capacity, power available, fuel economy and durability of units. When selecting a pump and motor unit, it is necessary to adopt a conservative approach as the estimated flow is not exact and the efficiency may be reduced in long time pumping.

Centrifugal pumps are used to pump water in collector pipes connected to wells or wellpoint system and removing surface water from sumps. The selected pumps should have sufficient air handling capacity, and they should be able to produce a high-vacuum. A wellpoint pump consisting of a self-priming centrifugal pump with attached vacuum pump is generally adequate and develops 6 to 7.5 m of vacuum. If the depth of water table lowering is large (> 4.5 m) but the rate of pumpage for each wellpoint is relatively small (< 10 to 15 gpm), installation of a single stage wellpoint system at the top of the excavation or water table, with attached jet-eductor pump may prove to be advantageous than a multistage wellpoint system. Water table up to 30 m can be lowered using a jet-eductor wellpoint system.

In selecting the power unit to be used for driving the pumps, consideration should be given to the initial cost of the unit, and the cost of operation, including maintenance and fuel. The capacity of a pump required for a dewatering system is decided by the horse power as obtained from the formula

$$\text{Horse power} = \frac{(\text{Total discharge in gpm}) \times (\text{Total dynamic load})}{3960 \times (\text{Efficiency of the pump and engine})} \qquad \ldots(3.9)$$

where total dynamic load = (operating vacuum at the pump in take) + (discharge friction losses.)

Header pipes are of 15 to 30 cm diameter are relatively light weight steel or plastic pipes. Headers for wellpoint system are provided with inlets for connection of wellpoints at short intervals. The connection of collector lines to the header pipes should be through non-return valves. Discharge line may be of steel, aluminium or plastic pipes which should conduct the flow with relatively small head loss.

3.7.9 Wellpointing in Deep Excavations

If the water table could not be lowered more than 6 m or if a deep excavation has to be made, it is advantages to use multistage wellpoints as discussed earlier. The lowest header of the multistage system should be located not more than about 4.5 m above subgrade to ensure that proper drawdown of the groundwater level can be achieved with the vacuum available in the line. Observations should be made at every stage of operation for discharge and groundwater lowering.

3.7.10 Deep Bored Wells

The procedure for designing a deep-well system is similar to that of wellpoints. The well should be sufficiently large to accommodate pump and to keep the head loss low. Deep-wells are usually of 15 to 45 cm diameter with screens of 6 to 22.5 m length. Pumping from wells can be done using surface pumps with their suction pipes installed in bored wells. This could be effectively used only up to a depth of 7.5 m. If centrifugal pumps are used in a deep-well system, the pumps can be located on the excavation slopes and connected to a common header pipe. For deep excavations submersible pumps should be installed, with a rising main to the surface. Sufficient depth of pervious material should be there below the level to which the water table is to be lowered, for adequate submergence of well screen and pump. For large diameter pipes (> 150 mm) turbine or submersible pumps are used.

3.7.11 Control of Surface Water

In laying out a dewatering system, proper steps should be taken to control surface water so as to prevent flooding of pump resulting in failure of the system. Further, uncontrolled run-off may cause serious erosion of slopes. The required measures are providing dikes, ditches, sumps, pumps and mulching and seeding to minimise slope erosion. The following factors should be considered while designing and selecting measures to control water (IS : 9759–1981):

(*a*) duration of construction
(*b*) frequency of rainfall occurrence
(*c*) intensity of rainfall and the resulting run-off
(*d*) size of area to be protected, and
(*e*) available sump storage.

A dike can be built around the top of the excavation to eliminate run-off into the excavation from the surrounding area. Dikes should be large enough to prevent water from overtopping them. The top of the dike should be at least 30 cm above the level of surface water to be impounded. Generally dikes are of width 40 to 150 cm and with side slopes of 1 on 2 or 2.5. Water retained by the dikes can be pumped out or conducted to sumps in the bottom of the excavation by pipes or line channels and then pumped out of the excavation.

Ditches should be designed with adequate allowance for silting, free board and storage. Velocities of flow should be low enough to reduce the extent of maintenance.

3.8 DRAINS

A complete drain consists of three components, *viz.*, filter, conduit or collector, and disposal system.

As discussed earlier a filter is essential for continued efficiency of the drain and to prevent seepage erosion during high hydraulic gradients. The water is collected in the drain conduits from the filter and is carried away. Ordinarily, the conduit is 5 to 10 times larger than its hydraulic dictate to allow for variation in soil permeability and to accommodate some silting. Commercial pipes have perforations of 8 to 9 mm in diameter and require a gravel filter with a maximum size of 12 to 15 mm. The permanent and simple disposal system is gravity. During adverse conditions such as wet weather, high water table and difficult topography, gravity system of disposal can not function. In such situations pumping has to be resorted to which will be costlier over a long period.

3.8.1 Open Drains

The oldest method of draining excavations, roads, *etc.* is by open drains, *viz.*, a ditch or a sump. A sump is merely a shot ditch which could be constructed easily with unskilled labour. Details of this method is discussed in Section 3.6.1.

3.8.2 Closed Drains

When seepage erosion or piping is troublesome or where a permanent drain is desired, perforated pipe can be laid at a required depth in ditches and the ditch is back filled with a suitable filter material (Fig. 3.23). As far as possible pipes should be laid in straight lines. Openings should be provided for every 30 to 50 m to flush out the pipe occasionally. Also manholes should be provided at changes in direction and at intervals of 100 to 150 m along straight sections.

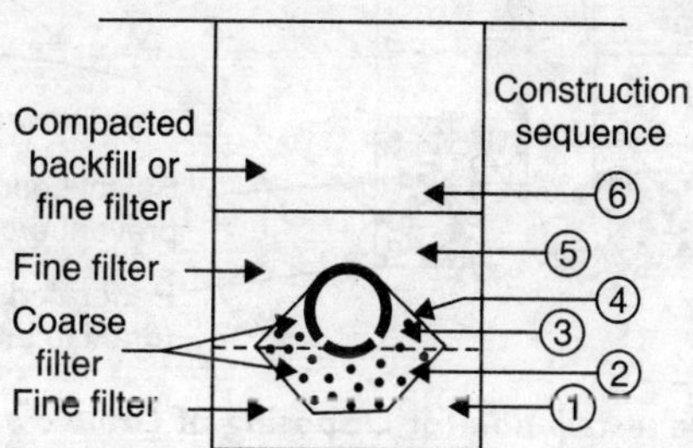

FIGURE 3.23. **Pipe Drain With Double Filter Layers.**

3.8.3 Horizontal Drains

If field situations warrant to avoid open-cut work or non-availability of adequate submergence, the groundwater can be lowered by means of a Ranny drainage system. This system consists of a reinforced concrete shafts or wells from which a number of horizontal perforated pipes are fixed. These pipes may be extended to a required length in any direction. Water collected in the well is pumped out by means of a turbine pump. This system may not be effective in lowering water in stratified soils.

3.8.4 Drainage after Construction

Preventing groundwater from seeping into a structure or through a finished structure may be necessary in order to obtain proper use of the structure or to protect it from drainage. In some situations, the dewatering systems utilized during construction may be further used to protect the structure.

In a particular field condition if it is necessary that an usable part of a structure has to be located below the groundwater table, it is essential to build the facility utilizing waterproof design and construction techniques. One of the desirable features for a subaqueous structure is to have all seams and joints with water stops or to avoid completely seams and joints.

Foundation Drains. Where groundwater is present in the vicinity of a structure, provision should be made to quickly carry away the water from the building. In the worst condition the effect of groundwater may be allowed only on the exterior side of a building. When the depth below the water table is not too great, it is feasible to control the water by foundation drains. An arrangement of such a drain is shown in Fig. 3.24. However, such drains should not be placed lower than the bottom of the footing. Such drains consist of perforated pipes or pipes with open joints so that the groundwater can enter into the pipe. A suitable filter should surround the pipes. The collected water in the pipes is disposed off by gravity flow to a storm drain system or other drainage facility such as a ditch, dry well, *etc*. If disposal by gravity flow is not possible, the drainage water has to be directed to sump pit or other collector and pumped to a disposal.

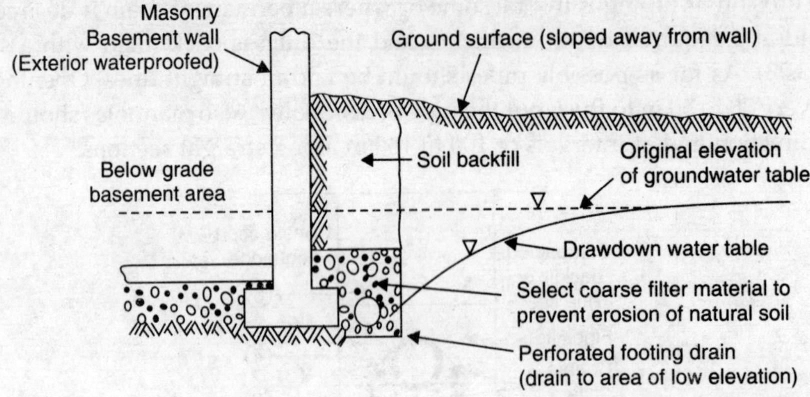

FIGURE 3.24. **Footing Drain Tile Installation for Disposing of Groundwater Against a Basement Wall.**

Blanket Drains. Continuous drainage blankets are sometimes constructed beneath dams and basement floor slabs to provide a highly permeable drainage path for removal of groundwater acting against the bottom of the slab. If an escape path is provided the uplift pressure can be reduced and the possibility of seepage through the floor arrested. The blanket consists of a fine filter layer in contact with the soil followed by a coarse filter cum collector layer (Fig. 3.25) and the latter is in contact with the underside of a masonry dam or basement floor. The blanket is connected by conduits to a sump where the collector water is pumped out or to drainage pipe where disposal is by gravity. Such blankets are also provided beneath pavements to prevent capillary flow upward.

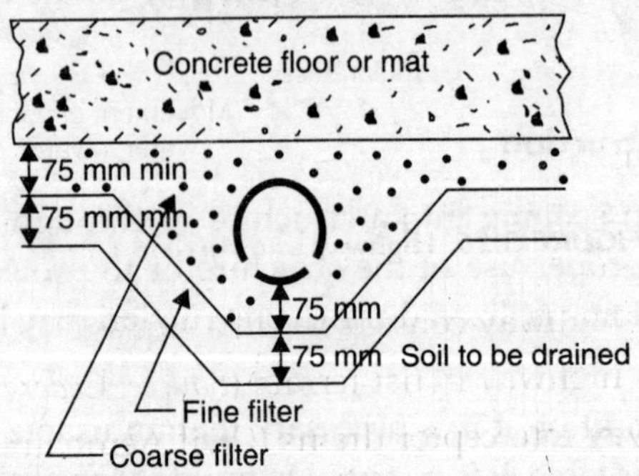

FIGURE 3.25. **Blanket Under-drain with Pipe Disposal—Minimum.**

Interceptor Drains. In paved highways and airfield runways trench drains are provided parallel to the shoulder. Such drains are termed as interceptor drains which are provided to lower the groundwater table to a level beneath the pavement and to permit easy lateral drainage for water finding its way into the coarse base material. The purpose of such drainage facility is to keep the base and subgrade soils dry so as to maintain adequate strength and stability. The drains also provide a means for disposal of surface and near surface water and also shall help to intercept underground flow trying to enter the pavement from the side areas. Similarly, open drainage ditches located adjacent to shoulder area may help to intercept surface and near surface water flowing towards the roadway area from the sides also prevent development of excess pore water pressure. Figure 3.26 shows different application of interceptor drains.

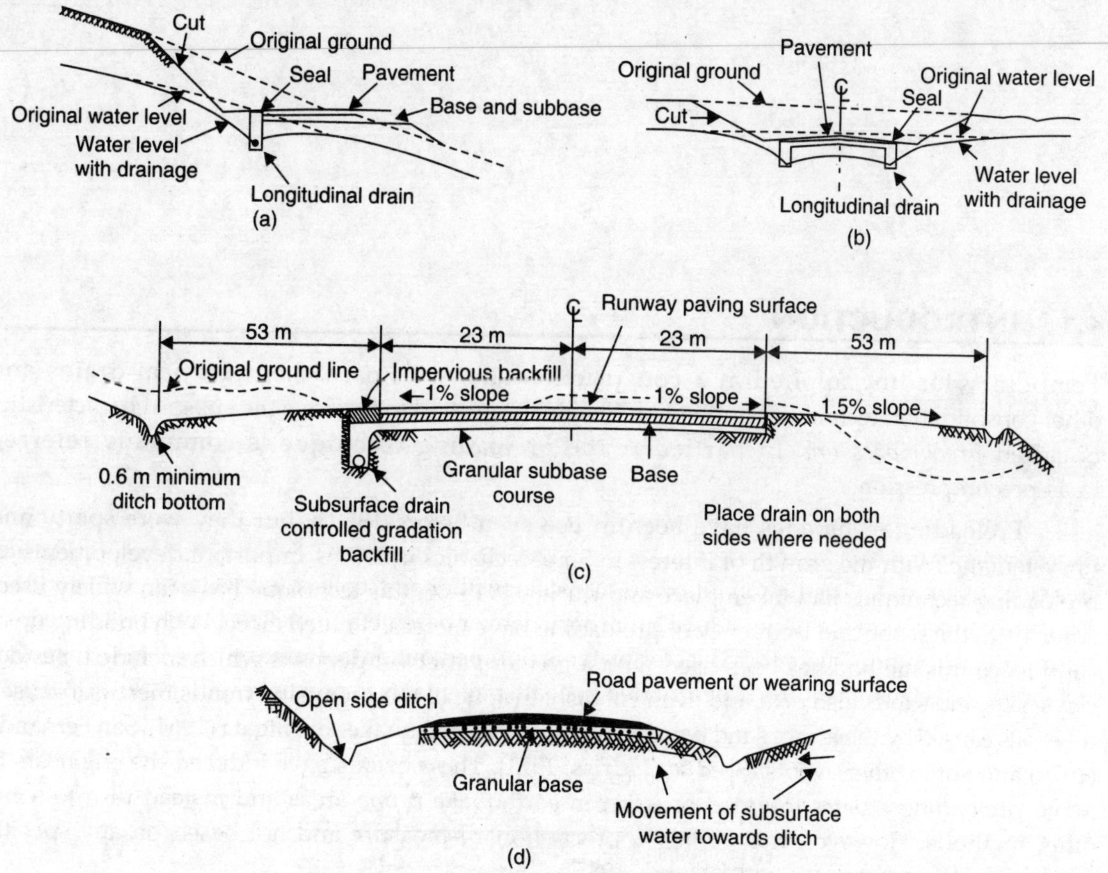

FIGURE 3.26. **Highway and Airfield Drains**

(*a*) Interceptor drain for highway constructed in a side hill (*Source: Cedergren, 1967*);
(*b*) Interceptor drain for highway in flat terrain (*Source: Cedergren, 1967*);
(c) Typical airport runway interceptor drains (Civil Aeronautics Administration);
(*d*) Typical open-shoulder ditch for roads.

Chapter 4

Precompression and Vertical Drains

4.1 INTRODUCTION

Temporary loading applied at a construction site, with or without vertical drains and other consolidation techniques applied on/inside the ground to improve the subsoil characteristics is called *precompression*. In particular the preloading technique is commonly referred to as precompression.

Preloading techniques have been in use even before 1920's, but they were spotty and unsystematic. With the growth of interest in Soil Mechanics in 1930's, important developments on preloading techniques had taken place and during 1945–65, this technique has been widely used. After 1965, the scope and frequency of preloading have increased much more. With building up of confidence this method has been tried with a greater variety of deposits which included, besides clays, silts and sands also peat and even rubbish. In spite of the promising trends there is localised reversals caused by the fear of sand liquefaction after the earthquakes of Nligata (1964), San Fernando (1970) and some other events (Seed and Idriss, 1971). These events have induced the engineers to avoid preloading at sites founded on sand in earthquake-prone areas and instead turn to some other methods. However, this is only a precautionary measure and not based on any specific experience (Stamatopoulos and Kotzins, 1985).

The preloading technique along with construction of different types of vertical drains for accelerating consolidation have been discussed in this Chapter. Also consolidation based on electro-osmotic and chemico-osmotic principles have also been dealt.

4.2 COMPRESSIBILITY OF SOILS AND CONSOLIDATION

A soil layer undergoes compression due to compressive stress caused by construction activities or otherwise. The compression may be attributed to rearrangement of particles, seepage of water, crushing of particles and elastic distortions. The compression may be progressive and cumulative, dependent on the type, magnitude and duration of load, and on the properties of the material. The civil engineer pays equal importance to compression along with shear failure in designing structures. The magnitude of compression may be detrimental for some structures or may hamper the normal functioning of conventional structures.

4.2.1 Compressibility

A change in the stress system acting on a soil mass causes a change in the volume of the mass. Such changes in volume, result in change in the permeability characteristics of the soil, change in the interparticle forces, change in shearing resistance and displacement of the soil mass. Considering the condition that a change in stress system results in a volume decrease, will cause compression or settlement of the boundaries of the mass.

All soils undergo elastic distortion almost immediately after the application of load including saturated soils under no drainage condition due to induced shear strains (Leonards, 1962). Compression caused due to these process is termed *initial or immediate compression* (Fig. 4.1).

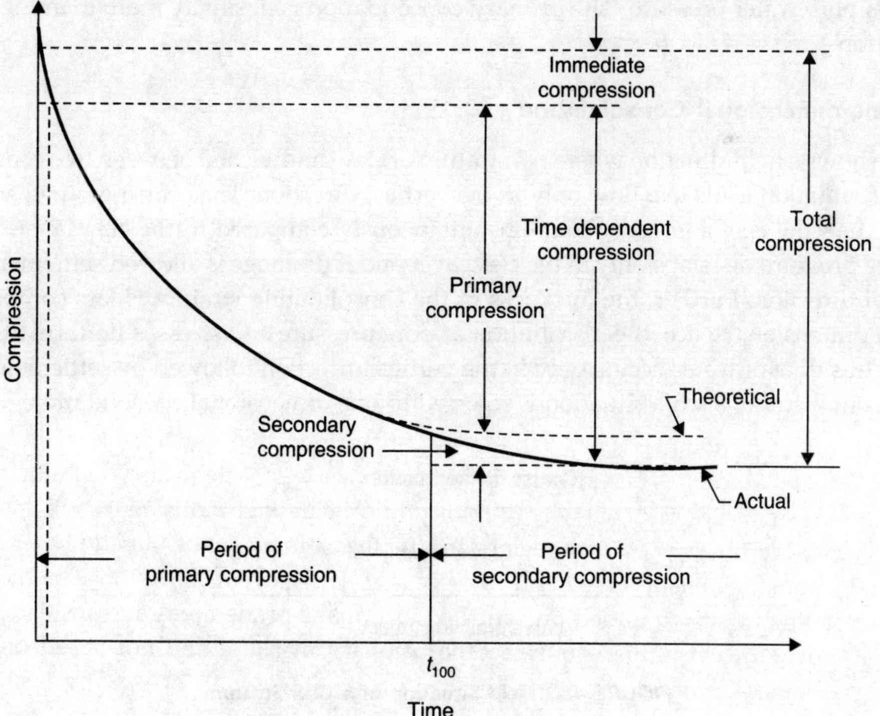

FIGURE 4.1. **Components of Total Compression.**

When a soil mass is subjected to stress in such a manner that its volume is decreased, there are three possible factors to which this decrease might be attributed: (*i*) compression of solid matter, (*ii*) compression of water and air within the voids, and (*iii*) escape of water and air from the voids. The solid matter and the pore water, being relatively incompressible, do not undergo appreciable volume change when subjected to compressive load usually encountered in soil masses. If the soil is saturated, the decrease in volume of the soil may be justifiably considered as entirely due to an escape of water from the voids. In a partially saturated soil mass the situation is much more complex since presence of a compressible gas within the voids may undergo appreciable compression, even though there may be no escape of water from the voids. The time-dependent compression in a partially saturated soil is beyond the scope of this book. The time-dependent compression in a saturated soil is discussed further.

As majority of the natural deposits are saturated, the compressibility of a soil mass may be attributed fully to the rigidity of soil skeleton which in turn depends on the structural arrangement of particles and the bonding forces between the particles. Thus in a saturated soil-water system the compressibility of mineral skeleton is generally so large compared with the compressibility of water. If such a system is loaded, the applied pressure is initially transferred to pore water as excess pressure and once drainage is permitted, a portion of the applied pressure is transferred to the mineral skeleton causing a reduction in excess pore water pressure followed by a compression. This process of gradual load transfer from pore water to soil skeleton and the corresponding gradual compression is termed *consolidation*. That part of consolidation which is completely controlled by the resistance to flow of water under the induced hydraulic gradient, is called *primary consolidation*. The other part, called *secondary consolidation* (*creep*), is due to the plastic deformation of the soil at zero excess pore water pressure. The primary consolidation is normally more than the secondary consolidation.

4.2.2 One-dimensional Consolidation

Figure 4.2 shows a field situation wherein a stratum of clay sandwiched between two coarse-grained strata. This situation leads to a flow only in the vertical direction. The coarse-grained strata which are stiffer than the clay and consolidate instantaneously compared to the clay layer. Hence, the excess pore pressure develops only in the clay layer and if drainage is allowed settlement occurs in the vertical direction. Further, the thickness of the consolidating stratum is less compared to the horizontal dimension. Hence, the distribution of pore pressure and stress is uniform at all vertical sections. Thus flow of water occurs only in the vertical direction followed by settlement and there is no horizontal strain. Such a situation is referred to one dimensional consolidation.

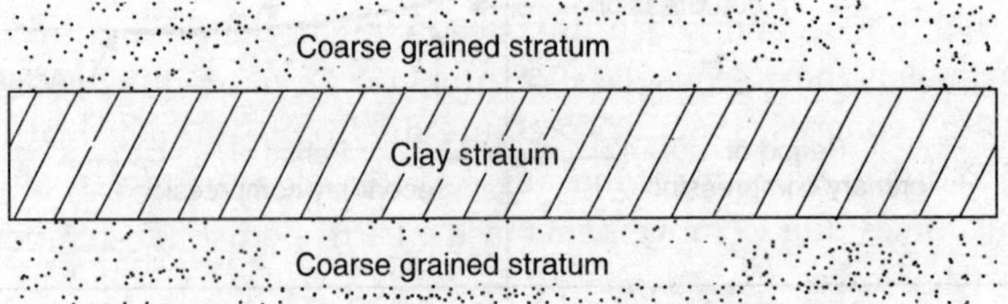

FIGURE 4.2. **Field Situation of a Clay Stratum.**

In some field problems the consolidation may be in two or three directions. For example, the consolidation may be in horizontal and vertical directions, when an oil tank is placed on a clay stratum having a thickness greater than the diameter of the tank, and under a long embankment. In axisymmetric loading such as under a circular tank the flow will be in vertical and radial directions and accordingly the consolidation will be in the vertical and radial directions. Solutions to vertical and radial consolidations are available to calculate the time required for consolidation. Exact solutions to two and three-dimensional consolidation problems are not available. However, most available two and three-dimensional solutions are based on the assumption that the total stress remains constant.

4.2.3 Compressibility Characteristics of Soil Deposits

In the natural process of deposition, fine grained soils undergo the process of consolidation under their own weight of overburden pressure. After an elapse of several years a state of equilibrium is reached and the compression ceases. This equilibrium may be established only in horizontal ground surfaces whereas in sloping ground surfaces creep may be taking place. Season after season, the process continues followed sometimes by erosion or removal of overburden pressures. Sometimes due to frequent deposition the process of consolidation may be continuously taking place. So fine grained soil deposits exist in the field under different conditions and their stress history should be known.

Normally-Consolidated Soils: Soils of low to moderate plasticity include clays, silts, organic and micaceous silts with plasticity indexes up to 30 (as well as sands, gravels and porous rocks). Such a soil deposit is said to be normally-consolidated if it has never been acted on in its history by effective overburden pressure greater than those existing at present. The characteristic *e*-log *p* curve for normally-consolidated soil is shown in Fig. 4.3 (*a*). The greater part of the curve is approximately straight and the slope is expressed as *compression index*. Experience has revealed that the natural water content of these soils is commonly close to the liquid limit. If the water content is lower than liquid limit, the sensitivity (a measure of the consistency of a soil) of the soil is low and if greater than liquid limit, the soil is said to be highly sensitive. In any event these soils are always soft to a considerable depth.

Over-Consolidated Soils: A soil deposit, of low to medium plasticity, which has been fully consolidated under a pressure, p_c, greater than the present overburden pressure, p_o, is said to be *overconsolidated* (preconsolidated or precompressed). Most undisturbed soils are over consolidated to some degree and exhibit the characteristic *e*-log *p* curve of Fig. 4.3 (*b*). The ratio $(p_c - p_0)/p_o$ is referred to as the *overconsolidation ratio* (OCR) and p_c the *preconsolidation pressure.*

The preconsolidation pressure can be estimated from the *e*-log *p* curve [Fig. 4.3 (*b*)]. Tangents are drawn to the initial, flat section and to the steep straight-line portion of the curve and the intersection is approximately the preconsolidation pressure. Overconsolidation may also be caused by a variety of factors (Leonards, 1962): (*i*) pressure due to overburden that was subsequently eroded, (*ii*) desiccation due to exposure of the surface, often accompanied by alteration of the clay minerals, (*iii*) tectonic forces due to movements in the earth's crust, (*iv*) temporary overloading, and (*v*) sustained seepage forces.

Some of the other factors (Sowers, 1979) such as (*i*) changing the physico-chemical bonds between the clay particles or by introducing stresses by the expansion/contraction of the grains during the alteration process may show effects of preconsolidation. For example, residual soils, weathered rocks and some partially indurated rock exhibit preconsolidation from this source. Removal of high concentrations of cations due to leaching may have the same effect in some clays deposited under salt water. All these are referred to as *pseudo-preconsolidation* (Sowers, 1979).

Preconsolidated clays have water contents much less than their liquid limits. Heavily overconsolidated clays exhibit water contents less than plastic limits. A soil will not settle appreciably until the stress imposed by the structure is greater than the preconsolidation pressure.

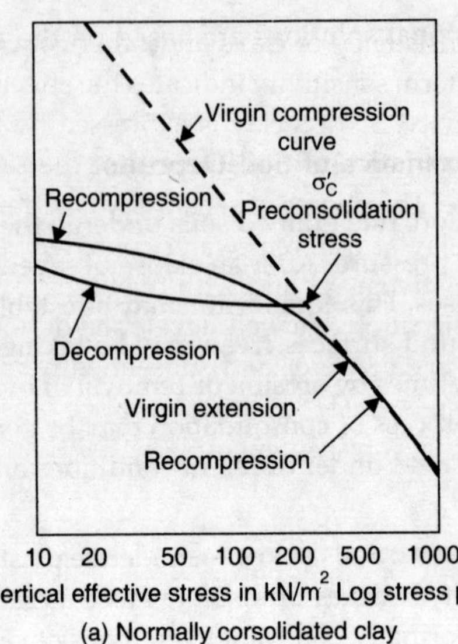

(a) Normally corsolidated clay

(b) Low to moderate plasticity preconsolidated clay

(c) Highly sensitive clay

(d) Cemented soil partially saturated

(e) Highly plastic clay partially saturated

Note. σ'_z = Effective vertical stress; $\sigma'_o = p_o$; $\sigma'_c = p_c$

***FIGURE 4.3.* Stress Void Ratio Curves of Typical Natural Soils.**

(*Source: Sowers, 1979*)

Sensitive Clays. The consistency of clays and other cohesive soils is usually described as soft, medium, stiff or hard. The term sensitivity indicates the effect of remoulding of the consistency of a clay. The degree of sensitivity, S_t, of a clay is expressed by the ratio between the undrained strength of an undisturbed specimen and the strength of the same specimen at the same water content but in a remoulded state. The degree of sensitivity is different for different clays and it may also be different for the same clay at different water contents. Sensitivity, S_t, varies from 2 to 16 with the lowest value for insensitive clays and with the highest value for the extrasensitive clays. In some locations, because of formation of a well-developed or honeycomb structure or to leaching of soft glacial clays deposited in salt water and subsequently uplifted, the S_t value may be still higher and they are known as *quick clays*. Sensitive clays are preconsolidated to some degree. The water contents of highly sensitive clays is generally higher than their liquid limits. Figure 4.3 (*c*) shows the *e*-log *p* curve for sensitive clays and other soils with a flocculent or well developed skeletal structure. The curve is flat up to the preconsolidation pressure, drops sharply and then gradually flattens to approximate a straight line. The sharp drop in void ratio may be attributed to a structural break down in which the bonds between the particles are broken and the particles rearrange themselves into a more dense orientation (Sowers, 1979).

Cemented Soils. Cemented soils and similar soft rocks depict *e*-log *p* curves similar to preconsolidated soils. The preconsolidation effect may be attributed to the bonding between the particles. If the bonds break suddenly due to increased pressure, the *e*-log *p* curve will be similar to that of sensitive clays. For such materials with water-sensitive bonding two pressure-void ratio curves may be visualised (Sowers, 1979): undisturbed (usually moist but not saturated) and inundated, as shown in Fig. 4.3 (*d*). The drop from the higher value (undisturbed value) may occur due to inundation under small pressure (may be due to self weight of soil). Due to this behaviour serious sudden settlements may occur in dry regions where cemented soils have been wet due to construction activities (Sowers, 1979).

Expansive Soils. Partially saturated highly plastic expansive clays and some highly micaceous soils may be subjected to high preconsolidation due to desiccation. These soils have high affinity to water and absorb water if they have access to water and expands [Fig. 4.3 (*e*)]. Two *e*-long *p* curves may be developed for such soils: undisturbed but partially saturated, and inundated (Sowers, 1979).

Organic and Reclaimed Soils. Peats and highly organic soils are extremely compressible depending on the initial void ratio. The *e*-log *p* curves will be similar to that of clay deposits but with less preconsolidation. The virgin curve in *e*-log *p* plot is seldom a straight line. The compressibility of reclaimed soils can not be measured in the laboratory because of the nonhomogenity of the material. However, the compressibility characteristics is similar to that of organic soils. Much of the compression is crushing of hollow bodies and distortion of resilient materials (Sowers, 1979).

Sand. Sandy soils compress less under static loads, than under dynamic loads. A static load induces a compression in a sand mass. In general, the compression in non-cohesive soils occurs immediately after the application of load. The compression is due to volume changes caused by lateral yielding or shear strain that takes place in the soil.

If the loaded area is large, the compression is governed by the contact pressure, which is a function of the modulus of subgrade reaction at different points. In such situations, compression versus pressure curves may be used, the curve being obtained by field testing or by use of a simple compression test in the laboratory in which the lateral deformations are prevented. However, the laboratory curve should be used with caution. In non-cohesive soils the most common method to predict the settlement is by adopting semi-empirical rules relating compression to the results of field penetration tests.

4.2.4 Methods of Evaluating Compressibility

As discussed earlier a soil mass is said-theoretically-to be fully consolidated under a given pressure when the excess pore water pressure is zero, depicting the termination of primary consolidation. The slow continued compression that continues after the excess pore water pressures have substantially dissipated is due to secondary compression (Fig. 4.4). As the secondary compression is time dependent, the longer the clay remains under a constant effective stress the denser it becomes. Practical evidence show that creep ceases or becomes so small that it is not measurable. The relative importance of secondary and primary compression varies with the type of soil and also with the ratio of stress increment to initial stress. Both primary and secondary compressions are time-dependent compression which are dealt below. Immediate compression which is generally less for cohesive soils may be computed from elastic properties of soils.

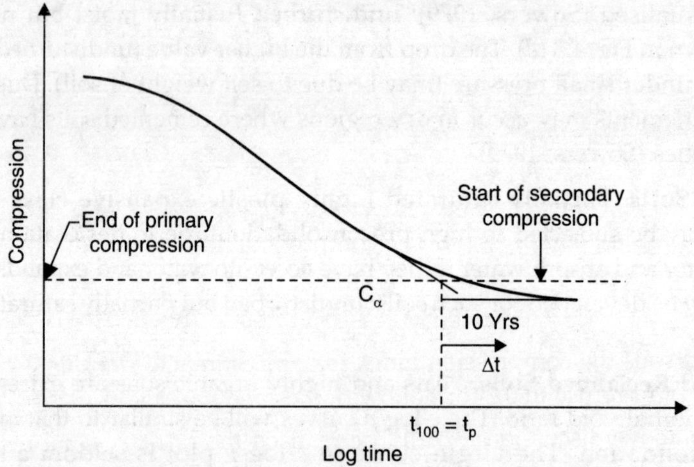

FIGURE 4.4. **Secondary Compression and Definition of its Rate.**

Primary Compression: If the soil beneath a structure contains layers of sand or stiff clay alternatively with layers of soft clay, the compressibility of the sand and stiff-clay strata can be disregarded (Terzaghi and Peck, 1967).

The compressibility of layers of clay depends on the magnitude of the greatest pressure experienced by the deposit and the liquid limit of the clay. The primary compression, S_c, of the normally consolidated clay of thickness *H* with initial void ratio, e_o, due to a change in pressure *p* is computed from Eq. (4.1)

$$S_c = H \frac{C_c}{1+e_o} \log_{10} \frac{p_o + \Delta p}{p_o} \qquad ...(4.1)$$

where C_c is the compression index from a field *e*-log *p* curve. The field *e*-log *p* curve has to be extrapolated from the results of the laboratory *e*-log *p* curve by a suitable procedure (Schmertmann, 1955). C_c can be estimated roughly from a known liquid limit (Terzaghi and Peck, 1967) as

$$C_c = 0.009\,(w_L - 10\%) \qquad ...(4.2)$$

For soils of very low plasticity and porous rock, Sowers (1953) has found the compression index to be related to the undisturbed void ratio, e_o,

$$C_c = 0.75\,(e_o - a) \qquad ...(4.3)$$

where *a* is a constant whose value is from 0.2 for porous rock to 0.8 for highly micaceous soils. Both the expressions are approximate at the best and considerable variations should be expected. Equation (4.2) may be erring in the unsafe side if the liquid limit is greater than 100%, if the natural water content at a depth of 7 to 10 m is greater than the liquid limit or if it contains a high percentage of organic material (Terzaghi and Peck, 1967).

The compressibility of a precompressed clay depends on liquid limit and on the stress ratio $\Delta p/(p'_o - p_o)$ where Δp is the pressure added by structure to the present overburden pressure p_o and p'_o is the maximum pressure that has ever acted on the clay. If the above stress ratio is less than 50%, then the primary compression is about 10 to 25% of that a similar normally consolidated clay. For stress ratio greater than 100% the influence of the precompression on the settlement of the structure can be disregarded (Terzaghi and Peck, 1967).

The compressibility of heavily precompressed beds of clay is usually irrelevant. Large and heavy structures on such stiff clay may be damaged due to moderate differential settlement. However, a settlement computation may be made from a carefully conducted consolidation test results on an undisturbed sample.

For sensitive clays, cemented soils and expansive soils consolidation test results may be used for evaluating primary compression. The compression index for peat can be estimated from the initial void ratio e_o (Sowers, 1979) as

$$C_c = (0.5 \text{ to } 0.7)\, e_o \qquad ...(4.4)$$

For reclaimed soils the compression index may be estimated from Eq. (4.4), the value of e_o varies from 0.15 for rigid refuse to 0.7 for hollow bodies and garbage.

Based on the above discussion it could be realised the primary compression is evaluated based on a justifiable value of C_c. It is quite reasonable to expect C_c to depend on in-situ void ratio, soil structure and the amount and type of clay minerals and the index properties. Based on these factors, different equations are suggested by various authors and reported (Table 4.1) by Bowles (1988).

Equations incorporating w_n and e_o and/or w_L applicable to preconsolidated soils also. It is reasonable to take an average value of C_c based on the above equations and if the scatter is large C_c should be used from the consolidation test results (Bowles, 1988).

Table 4.1. Equations for the Compression Index from Different Sources
(*Source: Bowles, 1988*)

Compression Index, C_c	*Comments*	*Source*
$0.009\,(w_L - 10\%)$	± 30% clays of moderate S_t	Terzaghi and Peck (1967)
$0.37\,(e_o + 0.003\,w_L + 0.0004\,w_n - 0.34)$	Statistical analysis	Azzouz et. al. (1976)
$0.5\left(\frac{1+e_o}{G_s}\right)^{2.4}$	$e_o \le 0.8$	Bowels (1988) and Rendonherrero (1980)
$-0.0997 + 0.0009\,w_L + 0.0014\,I_p + 0.0036\,w_n + 0.1165\,e_o + 0.0025\,C_p$	134 soils analysed	Koppula(1981)
$0.2343\,e_o$	—	Nagaraj and Murthy (1985, 1986)
$0.009\,w_n + 0.005\,w_L$	—	Koppula(1986)

Note. Use w_L, w_n, I_p as percentage and not decimal

$$e_o = w_n\,G_s$$

G_s = Specific gravity of soil solids

Equations that use e_o (or w_n) and w_n are for both normally and over consolidated soils.

Secondary Compression: It is difficult to evaluate secondary compression. However, by maintaining a constant pressure on a clay long enough past the point of primary consolidation, a relationship between secondary compression and time may be obtained, provided temperature control and equipment corrosion are taken care of. The magnitude of the secondary compression is referred to as a coefficient of secondary compression and expressed as the slope C_α of the final portion of the time compression curve (Fig. 4.4).

$$C_\alpha = \frac{\Delta H/H}{\log(t_p + \Delta t/t_p)} \quad \ldots(4.5)$$

where t_p = time when primary consolidation is complete

Δt = time increment producing ΔH

H = thickness of laboratory sample

Therefore, the secondary compression, $S_s = \Delta H$, for the time increment Δt for a stratum of thickness H is

$$S_s = HC_\alpha \log \frac{t_p + \Delta t}{t_p} \quad \ldots(4.6)$$

Secondary compression may be significant in highly plastic soils and especially in organic soils. Typical values for the coefficient of secondary compression for some soils are given in Table 4.2 (Ladd, 1967).

Table 4.2. Typical Values of C_α

(*Source: Ladd, 1967*)

Type of soils	*Value of C_α*
Normally consolidated clays	0.005 to 0.02
Very plastic soils, Organic soils	$\geq$ 0.03
Precompressed clays with OCR > 2	< 0.001

4.2.5 Rate of Consolidation

The consolidation of clay under a load in a consolidation test does not take place instantaneously. As the clays have less permeability, all the entrapped water in the voids can not escape immediately when a load is applied. However, the flow does take place rapidly at first, as it continues, the pressure drops and the rate of flow decreases. As the water is forced out, the particles come closer causing settlement at the surface of the specimen. The rate of settlement is rapid at first, and then decreases to a small fairly constant value.

The progress of consolidation occurs at different rates in different parts of the specimen. As the drainage facilities are better at the upper and lower boundaries the progress of consolidation is rapid than at the middle of the specimen.

A theory, known as theory of consolidation, (Terzaghi, 1925, 1943), has been developed for computing the rate of consolidation based on the assumption that the laws of hydraulics govern the decrease of porewater pressure. The following relationship between the degree of consolidation U and time as

$$U = f(T_v) \qquad \ldots(4.7)$$

where

$$T_v = \frac{c_v}{d^2} t \qquad \ldots(4.8)$$

and

$$c_v = \frac{k}{m_v \gamma_w} = \frac{\Delta k}{\gamma_w} \qquad \ldots(4.9)$$

T_v is a dimensionless number called the *time factor*; c_v is known as *coefficient of consolidation*; d is the drainage path (= half the thickness of sample), t is the time corresponding to U, m_v is *coefficient of volume compressibility*, k is coefficient of permeability and D is the constrained modulus. Figure 4.5 is a graphical representation of Eq. (4.7).

In order to estimate the degree of consolidation or the rate of settlement, the prime factor the coefficient of consolidation, c_v has to be evaluated properly. The coefficient of consolidation can be evaluated by means of a consolidation test, by fitting the experimental deformation-time curve to the theoretical curve for Eq. (4.7). Two popular methods are A. Casagrande's (1936) log-time fitting method and Taylor's (1948) $\sqrt{\text{time}}$ fitting method. These methods yield some what different values. The $\sqrt{\text{time}}$ method usually gives a larger value of c_v than the log-time method, and this method is usually preferred.

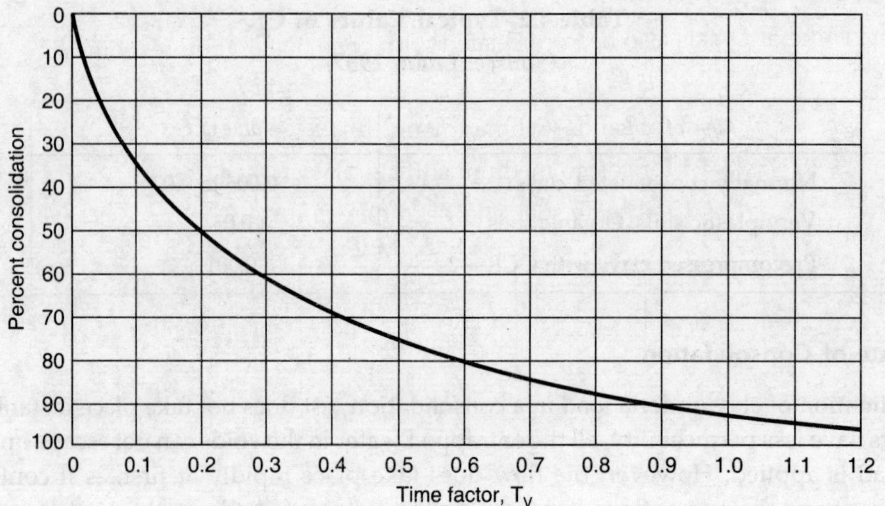

FIGURE 4.5. **Time Rate of Consolidation for a Stratum Drained on both Surfaces and any Distribution of Stress or for a Stratum Drained on one Surface and a Uniform Distribution of Stress.**

An estimation of c_v is feasible if the constrained modulus D is known for the load level from the expression

$$c_v = \frac{Dk}{\gamma_w} \qquad \text{...(4.10)}$$

The best way to determine D $(= 1/m_v)$ is from a consolidation test results. The parameter D also can be estimated by indirect methods from simple tests such as water content, cone penetration test, standard penetration test, *etc.* Values from indirect methods could be used for preliminary rough estimates of settlement or rate of settlement.

The value of D which is important in the evaluation of settlement calculation is the value applicable in the stress range between the natural overburden stress (NOS) and the vertical stress due to natural overburden stress and the stress after surface loading ($\Delta\sigma'_v$).

The NOS can be easily evaluated. But $\Delta\sigma'_v$ is generally unknown. However, in most problems it can be assumed to be on the order of 0.1 MN/m^2 and hence the average stress will be roughly equal to NOS + 0.05 MN/m^2. The value of D at this particular stress level is of interest and has been called as *specific constrained modulus,* D_s. The parameter D_s is not greatly influenced by moderate variations in $\Delta\sigma'_v$, and the error involved in accepting 0.05 MN/m^2 is small (Stamatopoulos and Kotzians, 1985).

The parameter D_s can be estimated from the empirical calculations from Fig. 4.6 for sands, silts, clays and moderately organic soils. For soils below the water table D_s may be estimated from water content and for soils above water table, D_s may be estimated either from the natural dry density or the natural void ratio.

For cohesionless soils D_s has been correlated with the standard penetration resistance (N), Parry (1972), as:

$$D_s = 50\,N \qquad \text{...(4.11)}$$

where D_s is in units of kgf/cm^2.

The value of D_s has also been correlated with cone penetration resistance q_c as

$$D_s = aq_c \quad \ldots(4.12)$$

where a = 2 for sands (Schmertmann, 1970)
= 4 for soft to medium stiff silt-clays (OECD, 1979)
= 1 for peaty soils and peat (OECD, 1979).

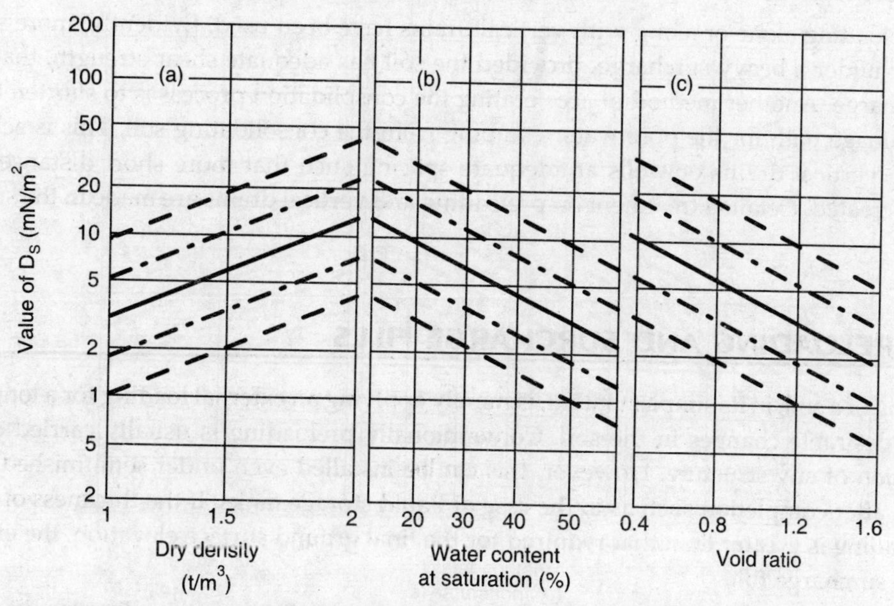

FIGURE 4.6. **Estimation of D_s: *(a)* From Dry Density, *(b)* From Water Content at Saturation, and *(c)* From Void Ratio. Valid for all Ordinary Soils Within the Limits of Water Content, Dry Density and Void Ratio Coefficients of Correlation = 0.63 – 0.64 Outer and; Intermediate Lines Bracket 95 and 67% of Results Respectively.**

(Source: Stamatopoulos and Kotzias, 1973)

The parameter D_s has also been correlated with C_c as (Coumoulos, 1979):

$$D_s = \frac{1 + e_o}{0.434\, C_c}\, \sigma_v \quad \ldots(4.13)$$

where σ_v is the effective consolidation stress and under in-situ stress conditions this will be equal to the normal overburden stress.

As c_v varies from load increment, loading and unloading and different samples of the same soil, it is extremely difficult to select a value of c_v for use in a particular engineering problem. Thus the observed rate of settlement or heave of a structure is several time faster than that estimated from a selected c_v value from a consolidation test (Branwell and Lambe, 1968). This anomaly may be due to various assumptions made in the one-dimensional consolidation theory. However, predictions of rate of consolidation are useful only to know in advance the approximate time required for consolidation.

4.2.6 Accelerating Consolidation

Accelerating the rate of consolidation can be useful in certain civil engineering works, such as highway and airfield embankments on compressible soils, wherein it is essential that the major portion of the settlement should have occurred during construction and not after construction (Rutledge and Johnson, 1958). Also in order to increase the shear strength of a dam foundation to permit more rapid construction acceleration has been resorted to (Lane, 1948).

Preloading alone or along with vertical drains have been used. Evidently, more settlement will occur under a heavy surcharge, provided the soil has adequate shear strength, than under a light surcharge. Another method of accelerating the consolidation process is to shorten the length of the drainage path for the pore water escaping from the consolidating soil. This is achieved by providing vertical drains or wells at adequate spacing such that more short distance drainage paths are created. Detailed treatment on preloading and vertical drains are made in the subsequent sections.

4.3 PRELOADING AND SURCHARGE FILLS

Preloading of a soil, in its simplest intent, is merely applying an external loading for a long duration to cause desirable changes in the soil. Conventionally preloading is usually carried out before construction of any structure. However, this can be installed even under semifinished condition and even after completion such as in the case of liquid storage tanks. If the thickness of fill placed for preloading is greater than that required for the final ground surface elevation, the excess fill is termed a surcharge fill.

It has been reported (Stamatopoulos and Kotzias, 1985) that preloading has been used in many civil engineering works such as road embankments, bridge abutments and box culverts, warehouses, gravity quay walls, housing complexes, runways, storage tanks, multistored structures, canals and industries.

Preloading methods have several advantages, in comparison with the other methods adopted for improving ground support, such as (*i*) the cost involved is comparatively less and vary between 10 to 20% without using vertical drains and 20 to 40% with the use of vertical drains, (*ii*) especially attractive when the fill material, after completion of preloading, is subsequently used in the same project as fill material or for site preparation, (*iii*) simple conventional construction equipment needed for earth moving job is sufficient for the preloading works, (*iv*) cost of monitoring instruments are cheap and the time needed for installation is only two to three weeks, (*v*) effect of preloading can be observed periodically from the field-instrumentation and makes possible future predictions about the behaviour, and (*vi*) provide uniform improved properties of the ground.

The decision to adopt preloading technique will be quite reasonable and successful only when the following conditions are satisfied (Stamatopoulos and Kotzias, 1985): (*i*) there should not be any base failure during preloading or during the operation of final structure, (*ii*) the duration of preloading will be within the time allotted by the construction schedule, (*iii*) there will be no damage to adjoining structures, (*iv*) there will be no undue disturbance to nearby communities, (*v*) settlements after construction will be within the range of tolerances, and (*vi*) the cost is less compared to other methods.

4.3.1 Precompression Principles

Preloading increases the pore water pressure in the soil and, as the consolidation process occurs, an increase in the effective stress takes place in the soil accompanied by surface settlement. The preloading material placed over the ground to be improved in amounts sufficient to produce a stress in the soil equal to that anticipated from the final structures. The time required for attainment of full consolidation varies directly as the square of the layer-thickness and inversely as the permeability. Thus preloading alone-can be sufficient only when the thickness of layer is less and more construction time is allowed in the development of the project; otherwise surcharge fill may be successfully used.

The ratio of the weight used for preloading to the weight of the final structure to be constructed on the improved soil is called coefficient of surcharge (m_s). The higher the value of m_s, the less time required for consolidation under full preload and the higher the proven post-preloading factor of safety against a base failure. On the other hand higher value of m_s leads to more construction time and expense for preload heap and more attention to be paid for stability. Generally m_s varies from 1.0 to 2.0 and the correct choice of m_s is a vital part of the decision making process. Although the surcharge fill is not going to improve the rate of consolidation, the pressure increase proportionately increase the amount of consolidation. Thus a permanent fill in combination with a surcharge fill can cause more amount of settlement in shorter time.

Using conventional theories of consolidation, the amount of surcharge fill required and the loading period can be estimated. Surcharge fills generally enable to cause both primary and most of the secondary compression. In highly organic deposits and in sanitary landfills, the secondary compression will be predominating. While computing the time required for full consolidation, it has to be remembered that the average degree of consolidation throughout the consolidating layer and the value at each depth vary. Hence, the analysis has to include the possible existence of residual excess pore water pressure after removal of surcharge. In case residual excess pore water exists, the settlement would occur under the permanent fill. To avoid this it is desirable to base the required surcharge loading duration on the consolidation time required for points most distant from drainage boundaries (Mitchell, 1976). The rate of preload and surcharge fill placement may be controlled depending on the rate at which the soil is gaining strength. These should be checked during the fill placement by means of piezometers and in-situ strength tests.

4.3.2 Preloading Methods

Heaping of fill materials is the most common method of preloading. In some cases the material is left completely or partly and the removed material may be reutilized in the same project for another preload or for embankment construction. There is danger of a base failure in heaping fill material, still this method is generally adopted as it is less costly for all types of structures and locations. Instead of heaping fill, the required weight for preloading can also be applied to the ground by constructing a peripheral dyke and filling the enclosed area with water.

In normal practice the most frequent method of preloading is embankment loading. In most cases three to eight months are required from the beginning of embankment placement to the end when the load is removed. The height of preload soil heap in most cases is 3 to 8 m above the original grade and the minimum and maximum height being 1.5 m and 18 m respectively. The usual range of settlement observed is from 0.3 to 1 m and in exceptional cases this may lead to 2 m. Table 4.3 illustrates some case histories of preloading by fill (Pilot, 1981).

In another method of preloading, the final structure is used as a vehicle for load application. This method has been extensively applied for liquid storage tanks. Prior to taking up soil improvement the tank is constructed and then filled with water incrementally. At every stage of additional water loading, the completion of consolidation is ensured. After the final increment of filling and allowing sufficient time for the rate of settlement to decrease almost completely, the tank is emptied and its base levelled by jacking. As stated above this method gained importance in liquid storage tanks as they are usually built of flexible steel plates. Further, there is no botheration of bringing, filling and removing fill material which tremendously reduces the cost and time of construction.

Table 4.3. Some Case Histories of Preloading by Fill

(Source: Pilot, 1981)

Location and reference	*Soil description*	*Performance data*					
		Permanent load (kPa)	*Surcharge load (kPa)*	*Time to overload removal (months)*	*Total settlement under over load (m)*	*Heave, on removal of overload (m)*	*Additional settlement after removal of overload load (m)*
Fukuroi, Japan (Ueda, 1971)	Recent, fine alluvium, 20 m thick; w_n = 30 to 60%, s_u = 50 to 200 kPa, c_v = 2.5 to 4.5 × 10^{-7} m^2/sec	42	84	7	0.39	0.04	0
Tampa, USA (Wheeless and Sowers, 1972)	Sandy clay, 6 m thick under 5 m of sand	41	174	3.5	0.07	0.02	0
La gunillas, Venezuela (Lambe, 1972)	Clay and silt, about 10 m thick; s_u = 25 to 100 kPa	20.0	22.5	36	1.10	0.05	0
Berthier ville, Canada (Samson and Garneau, 1973)	Silty clay, about 60 m thick	160	160	11	1.20	0.04	0
Arles, France (Simon and Brigando, 1976)	Clay and peat, 11 m thick; w_n = 20 to 120%, s_u = 10 to 20 kPa	150	176	20	1.20	0.08	0.06

Note. w_n = natural moisture content; s_u = undrained shear strength
c_v = coefficient of consolidation.

Another most effective method of preloading is by lowering the water table provided the soil conditions permit. In highly permeable soil the effect will be more. As the water table is lowered the effect of buoyancy is lost and the soil above water table gains unit weight by about 10 kN/m^3. As an approximation every meter lowering of water table will produce about the same loading as half a meter depth of fill. The lowering of water table is done by a suitable dewatering system. The rate of settlement may be increased substantially by combining the method of lowering the water table with that of heaping fill.

Another approach is inundating or preponding. Lowering of water table is applicable when the water table is high. When the water table is low, a load can be applied to some soils by opposite action, *i.e.*, by inundating or preponding the surface. The effect of preponding is breaking loose bonds between particles, increasing surface tension forces and the weight of water. The combined effect depending on the type of soil cause adequate densificaiton of soil. This technique is also referred to as hydrocompaction.

Kjellman (1952) proposed a vacuum preloading method. In this a 150 mm layer of sand is placed on the surface of a soft clay, the layer is covered with an impervious membrane. An application of a vacuum of 60 to 80 kPa is induced in the sand which acts as an equivalent overload. For better performance, it is the usual practice to use vertical drains in conjunction with this method. The method derives the advantages (Pilot, 1981) for the following reasons:

(*i*) there are no problems of embankment stability to consider.

(*ii*) it eliminates the need for backfill material which is usually expensive and often unavailable.

(*iii*) the installation and removal of the means of applying the preload are readily accomplished.

Table 4.4 shows four case histories of vacuum preloading (Pilot, 1981).

Jacking is another procedure adopted in preloading. This method is mostly applied to individual footings of either new buildings to which extra stories are to be added. Jacking is a standard method of preloading footings and piles in underpinning.

In large areas if only moderate improvement is needed on a specific area wherein heavy loading is planned, it is advisable to use preloading in combination with piling. In this approach, preloading near the piles must be completed first so as to prevent down-drag forces on piles are minimised. In general, mixed methods provide increased safety for structures that are sensitive for settlement, and low cost of improvement for pavements, utilities and small buildings.

Table 4.4. Case Histories of Vacuum Preloading
(*Source: Pilot, 1981*)

Location, structure, reference	*Soil description*	*Performance data*	
		Technique employed	*Settlement (m)*
Alfortville, France; gas-holder foundation. (Soletanche, 1960)	Power station soot, 3 m thick, over sand-gravel mixture.	Membrane method, applied in two stages: 1. 40 to 90 kPa for 40 days 2. 40 kPa, 3 years later, for 60 days concurrently with filling of reservoir.	 0.24 0.10

(*Table Contd...*)

Table 4.4. *(Contd...)*

Philadelphia, USA; runway (Halton. et al, 1965)	Silt and organic clay, 7 m thick, over sands and gravels.	Well method (without membrane) in conjunction with pumping from well-points and sand drains; maximum 40 kPa negative pressure applied for 40 days.	0.10
Inland Sea, Japan; reservoir foundation, (Teshima and Ogawa, 1967)	Clay and silt, 6 m thick, under layer of fine sand lying over sands and gravels.	Well method (without membrane) in conjunction with pumping from well-points and sand drains ; vacuum applied for 60 days.	0.3 to 0.5
Brest, France; hydraulic fill, (Paute, 1970)	Clayey silt fill, 6 m thick, over 2 m of slime lying over sands and gravels.	Well method (without membrane) 20 to 45 kPa negative pressure applied for about 30 days,	
		–wells with 10 m spacing	0.03 to 0.12
		–wells with 3.3 m spacing	0.19 to 0.42

4.3.3 Construction Requirements

In order to achieve the best performance by the preloading technique various factors have to be considered and the salient factor are explained below.

Ground Investigation: Ground investigation has to be planned with utmost care and the subsurface conditions should be investigated thoroughly. Two aspects have to be decided before selecting this method, *i.e.,* the ground should be able to support the weight of the preload for which the subsoil should have adequate shear strength and the time required for the settlement to complete, that is, for the ground improvement to be complete. Preloading requires a higher investigation effort than other methods as switching over to a new technique in the middle of operation is not feasible.

Preliminary investigation may be made by geophysical methods followed by boring and sampling. It is recommended to have atleast 2 to 3 borings for sites up to 1000 m^2. For sites 1000 to 5000 m^2 the minimum number of borings should be three plus one for every 1000 m^2 in excess of 1000 m^2. Undisturbed samples should be taken for about 3 m and standard penetration tests for every 1 m. Borings should be deep enough to reach the underlying formation of hard soil or up to a depth of 30 m from the ground surface whichever is less.

Soil Types: Practically preloading technique has been applied effectively on every type of naturally laid or man-made soil. Under natural soils, loose sands and silts, soft silty clays, organic silts or alluvial deposits and less frequently organic soils like peat have been treated successfully. However, thick deposits of plastic clay and sanitary fills are still doubtful to treat satisfactorily. Preloading can be applied to soils above or below the water table.

Space, Materials and Access: Adequate space is required for preloading. About 10 m or more of area outside the perimeter of the planned structure is needed to perform the preload operations. Evidently the space requirement may not be a problem in new projects but in extensions to existing installations space may be critical.

Another important requirement is the availability of a suitable fill material. Among the fill materials, granular materials are the most desirable since they do not turn to mud during rainy weather and has adequate strength and stability and can be reused after the completion of preloading. Ore and industrial products are also satisfactorily used. In some cases fine-grained materials, like clayey soils, have also been used in practice successfully with the constraint that it could be possible to construct only during dry weather.

The other aspect is the access of the fill material. The transport of large quantities of fill materials, through inhabited areas, may cause inconvenience to public by way of dust, noise and burden on traffic. Transport of fill material by sea, lake or river may be acceptable, especially if dredging is required for deepening the water. If the preloading is to be done by water, there should be adequate source of water near to the site.

Site Preparation: Before starting the preload operation the site must be cleared by surface vegetation, the topsoil up to a depth of about 0.5 m must be removed and then the surface must be covered by a base layer of free draining material. The necessity for removal of surface vegetation is to prevent further settlement due to a large-term decay and also to facilitate the placement of base layer. The base layer is of about 0.6 m thick and must be a non-cohesive admixture of gravel and sand. The base layer plays a two fold role: (*i*) to receive and discharge the water that reaches the surface of the compressible soils during the consolidation process, and (*ii*) to provide a comfortable working surface on which equipment can move in rainy weather without hinderance. In areas where the sand-gravel is expensive or not available, it is advisable to use, geosynthetics as they offer a high water conductivity and considerable strength.

4.4 MONITORING OF COMPRESSION

It has been discussed in Section 4.2, that the compression caused by surface load is divided into two categories, *viz.*, time-independent and time-dependent. The time-independent is the initial compression or the initial settlement. The time-dependent are primary and secondary compressions. Field time-settlement observations reveal that settlements could occur over periods of several months to many years. It is necessary to predict the time rate and the duration and amount of settlement for proper monitoring of the preloading operation. In preload problems settlement-time predictions are of the following types (Stamatopoulos and Kotzias, 1985):

(*i*) Prediction of the behaviour of large-scale preloading or test fill from the results of borings and tests.

(*ii*) Prediction of the behaviour of large-scale preloading from that of a test fill.

(*iii*) Prediction of the behaviour of the permanent structure from that of the preloading.

The first of these predictions are fully dependent on the results of borings and tests on laboratory samples which are used on theories of consolidation for necessary prediction. The other two predictions are based on empirical methods based on field observations rather than theoretical methods. However, the second and third method should be considered to be more reliable than the first, as they are based on field observations rather than from theory and small scale tests.

4.4.1 Predictions From Borings and Tests

The time prediction of initial compression is trivial since the determinations take place immediately after the load application. The time prediction of secondary compression could be determined approximately from Eq. (4.6). Thus only the time for primary compression has to be determined using the well established Terzaghi's theory of consolidation. Equation (4.8) is rewritten as

$$T_v = \frac{c_v}{d^2}t = \frac{Dk}{\gamma_w}\frac{1}{d^2}t \qquad \ldots(4.14)$$

A determination of T_v for a given time t from Eq. (4.14) will enable to determine the degree of consolidation U from Fig. 4.5. Orelse T_v can be obtained from Fig. 4.5 for a known degree of consolidation and the corresponding t can be obtained from Eq. (4.14). In either case c_v or D and k along with d should be known. The coefficient of consolidation can be obtained from a laboratory consolidation test data or the values of D and k have to be estimated. The former is simple as it requires only a compression versus time curve from a laboratory test. The latter is more complex because it must be based on both the laboratory consolidation test (to determine D) and the field permeability test (to determine k). However, it is more appropriate to predict the rate of settlement from field data. D can be reasonably estimated by taking it equal to specific constrained modulus, D_S (Stamatopoulos and Kotzias, 1985). Thus it remains to estimate k and D.

In the field the soil is found in a heterogeneous condition with sand seams, clay pockets, stratified layers, decayed roots and other relatively pervious organic remains. In such irregular and erratic medium, the value of k (or c_v) when measured in small laboratory sample may be misleading. Hence it is appropriate to use k from field measurements.

The field condition is not that simple to determine d as in a laboratory condition for a homogeneous soil with double drainage. If sand or highly permeable layers present within the soil, as in Fig. 4.7 (*b*), the drainage path may be much different from the conventional ideal condition [Fig. 4.7 (*a*)].

Thus the uncertainties in k (or c_v) and d can be overcome by any one of the following approaches (Stamatopoulos and Kotzias, 1985):

(*i*) d is known from borings, so k must also be measured in the field in order to include the effect of pervious material.

(*ii*) k is known from field tests, but d is not absolutely known.

(*iii*) the factor k/d^2 is not known reliably.

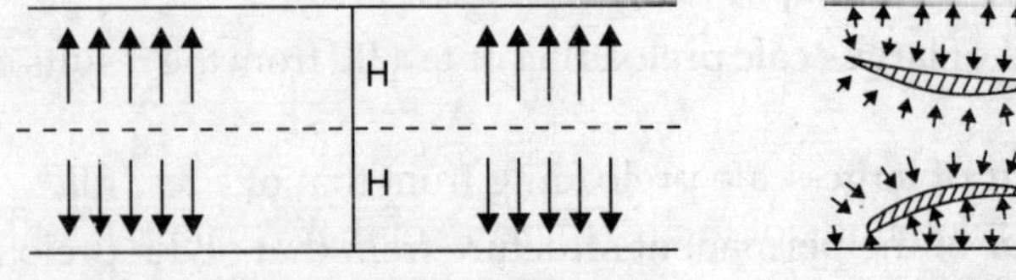

(a) As assumed in the derivation of Terzaghi's theory of consolidation

(b) As it could happen in a field problem

***FIGURE 4.7.* Arrows Showing Water Drainage Paths.**

Of these three approaches the first is the most convenient because by retaining the theory of consolidation, predictions may be made using $c_v = \dfrac{Dk}{\gamma_w}$. The second and third approaches summarily reject the theory or consolidation and fully based on empirical approach. Figure 4.8 shows a comparison from field curve and predictions, based on (*i*) c_v being determined from laboratory test data, and (*ii*) c_v being determined from field permeability and taking D equal to the laboratory determined D_s (Stamatopoulos and Kotzias, 1985). It could be realised that the second prediction is close to the true one.

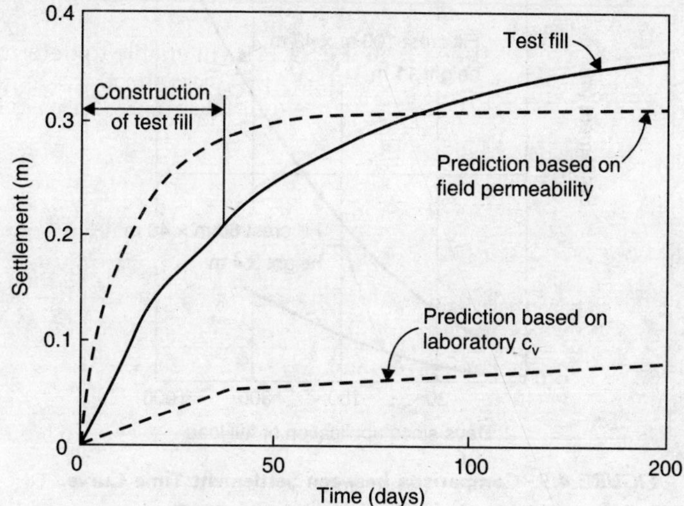

***FIGURE 4.8.* Comparison Between a Settlement-Time Curve Obtained from a Test Fill and Predictions Based, First, on Laboratory c_v and, Second on Field Permeability. The Second Prediction Affords a much Better Approximation to the Field Curve.**
(*Sourco: Stamatopoulos and Kotzius, 1985*)

The theory of consolidation is only applicable for fully saturated soils. In partially saturated soils, compression may occur due to surface loads even without expulsion of water and the theory becomes irrelevant. In the field partially saturated soils are encountered such as dumps and man-made fills, locations where the water table has been lowered, and loess soils after inundation. In such situations it is a good practice to take samples, wherever possible, and study the compression-time behaviour by laboratory tests. In situations where samples could not be taken due to oversize materials, field plate-bearing tests may be conducted to assess the rheological properties (Stamatopoulos and Kotzias, 1985).

4.4.2 Prediction From Test Fills

A reliable estimate of time rate and the magnitude of total settlement of the actual preloading can be made from a test fill. The width of test fill must be wide enough (or as wide as the large-scale preloading) such that the load is exerted to the deepest compressible soil. If the large-scale preloading is not going to experience a base failure or excessive creep, then the development of settlement will be in the same way as in the test fill. The large-scale preloading subjected to the fill load is expected to settle by an amount equal to the settlement of the test fill, corresponding to the same time, multiplied by the ratio of preloading to test pressures. Figure 4.9 shows a comparison between settlement-time curves for test fill and the large-scale porloading (Stamatopoulos and Kotzias, 1985). It could be observed both the curves are similar in shape and the settlement of the large-scale loading is about 3.5 times the settlement of the test fill and the settlements are very nearly proportional to the heights of fill, *i.e.*, 11 m : 3.4 m.

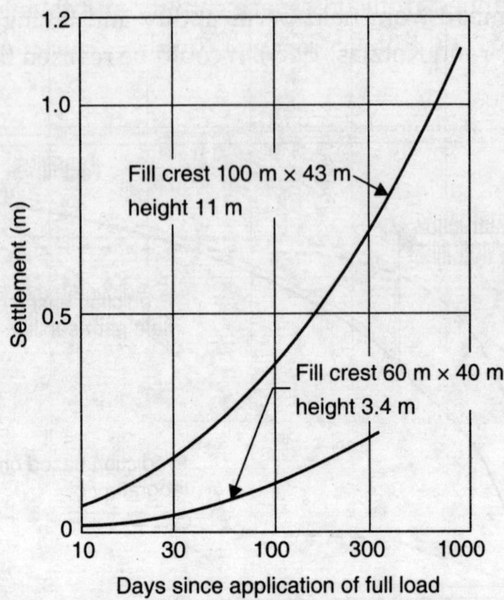

FIGURE 4.9. **Comparison between Settlement Time Curve.**
(*Source: Stamatopoulos and Kotzias, 1985*)

4.4.3 Monitoring Time Rate of Settlement

An effective tool for monitoring the danger of a base failure is to maintain plots of average daily rate of settlement and time during the load accumulation. The time rate of settlement during fill build up is predictably related to the rate of load application. One such correlation is shown in Fig. 4.10 for a coastal area underlaid by erratic deposits (Stamatopoulos and Kotzias, 1985). Similarly the time rate of settlement during fill build up can be related to pore pressure induced in the soil, both quantities reflecting the intensity of the process of consolidation. One such correlation is shown in Fig. 4.11 (Stamatopoulos and Kotzias, 1985). High rates of settlement may demand a fill discontinuation so as to avoid a failure.

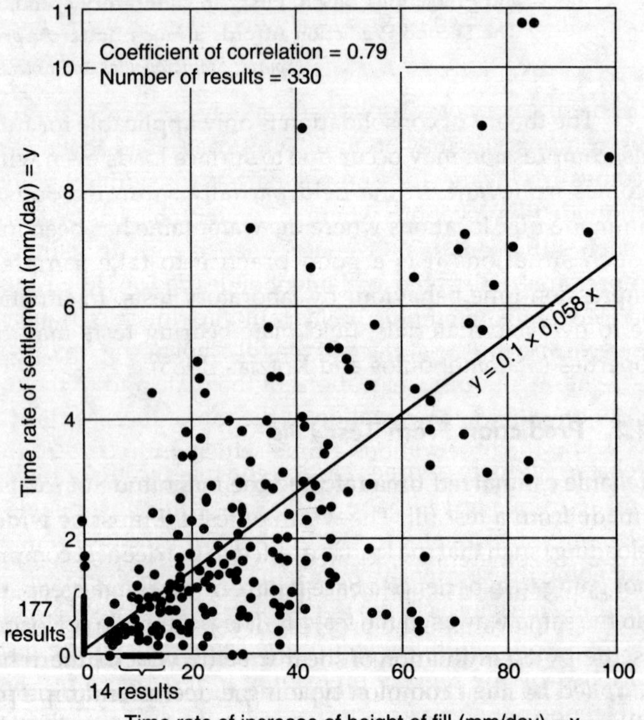

FIGURE 4.10. **The Time Rate of Settlement Correlates well with the Time Rate of Loading (Coastal Area Underlaid by Erratic Deposits).**
(*Source: Stamatopoulos and Kotzias, 1985*)

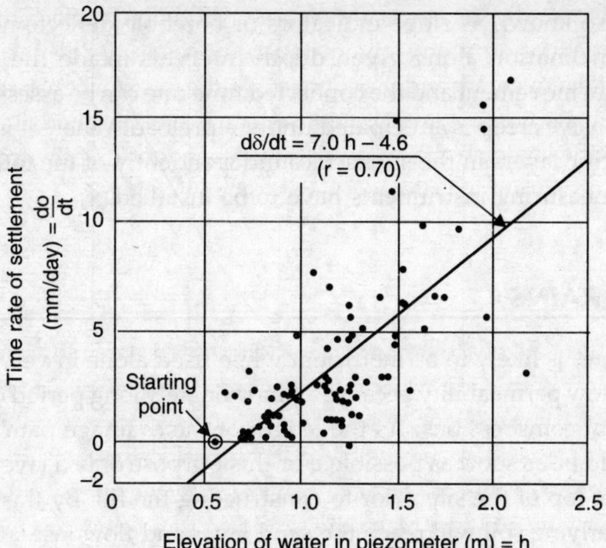

FIGURE 4.11. **The Time Rate of Settlement Correlates Well with the Piezometric Head, the Coefficient of Correlation Equals 0.70.**
(*Source: Stamatopoulos and Kotzias, 1985*)

4.4.4 Instrumentation

In order to monitor the progress of consolidation of a preloaded soil certain instruments have to be installed. Instruments are mainly of two types, *viz.*, settlement plates and piezometers and the former is used to measure the surface settlements and the latter is used to measure the pore water pressure. Sometimes to measure the side movements or the compression between specific elevations inclinometers are used.

Settlement plate consists of a steel plate which is installed on a prepared surface at an elevation at which the settlement is desired. At the centre of the plate a steel riser pipe or rod is connected. As the preload fill is build up, the riser is extended, with the top kept a few centimetres above or below the surface of the fill, depending on the work, so as to prevent damage to the riser by equipment. Initial elevations of the plate and of the top of the steel riser are obtained. For each addition of riser the elevation of the top of the riser is also obtained before and after the addition. Based on the observations a time-settlement record of the plate is obtained. Settlement plates are generally preferred because the observations obtained from them are the easiest to interpret and less in cost.

Piezometers are located at critical locations within the subsurface. These help at first the development and later the dissipation of pore water pressure. Measurements of pore water pressures are used to evaluate the progress of consolidation of embankments or of soft foundation soils and to help in the evaluation of shear strength which in turn helps to determine the possibility of a base failure. The most common type is the open stand pipe piezometer which is installed in a boring. Other types of piezometers may be useful under special conditions such as when there is (*i*) an excess time lag due to low permeability of the soil, (*ii*) a need to control from a central room, or (*iii*) a desirability to avoid borings. For details of different types of piezometers reference may be made to USBR (1974).

Inclinometers, also known as slope indicators or borehole deflectometers, are installed to measure the average inclination along given depth intervals inside the boreholes. From the inclinations the horizontal movement and the connected time rate can be assessed. They are installed only in cases where intensive creep is anticipated under a preload. Where it is essential to measure the compression of specific layers in the subsurface independently of the total settlement, vertical differential settlement measuring instruments have to be installed.

4.5 VERTICAL DRAINS

The preloading techniques is likely to be inefficient when used alone in very thick soft clays or in soils with exceptionally low permeability because an inordinately long period of time will be needed to bring about significant compressions. As the length of the drainage path controls the time for consolidation, this should be as short as possible. For this purpose only a layer of sand (called sand blanket) is placed on the top of the site prior to constructing the fill. By this provision any water squeezed from the underlying soil will reach the sand layer and flow fast laterally to the fill edge as the permeability of sand is much larger than the underlying soil. Based on the same concept, radial improvements in preloading time can be effected by the installation of vertical drains to shorten the drainage path under which the clay will consolidate.

4.5.1 General Principle

Vertical drains are continuous vertical columns of pervious (sand or fibrous) material installed in clayey soils. These drains provide the pathway for the pore water to escape from the consolidating soil by travelling a shorter distance than would be necessary without them. Further, they allow the flow inside the soil to take place along the horizontal which is the direction of least resistance. Thus it serves the purpose of collecting and discharging the expelled water faster during the process of consolidation.

Installation of vertical drains in conjuction with preloading brings about the rapid dissipation of excess pore water pressure and thereby accelerating the primary consolidation. Vertical drains have no direct effect on the rate of secondary compression. However, the early completion of primary compression leads to the earlier occurrence of secondary compression and thus amounting to a total reduction in the time for consolidation.

In order to signify the relevance or otherwise of the use of vertical drains a sequence of events that may lead to use of vertical drains is illustrated in Fig. 4.12 (Stamatopoulos and Kotzias, 1985). Figure 4.12 (*a*) represents a laboratory consolidation test result on a 20 mm thick specimen. From the time-compression curve, the time required for primary consolidation is 3.7 hrs. Figure 4.12 (*b*) represents the time prediction for real problem without vertical drains. Extrapolating from the laboratory results to the real problem of a clay stratum of 6.3 m thick, with double drainage, time required for 70 and 88% consolidation have been estimated as 16 and 43 years respectively. As such a long time can not be accommodated in a construction project, vertical drains have been planned. Figure 4.12 (*c*) represents the time prediction for the same real problem with vertical drains. Evidently the estimate shows an encouraging results and enhances support for the use of vertical drains.

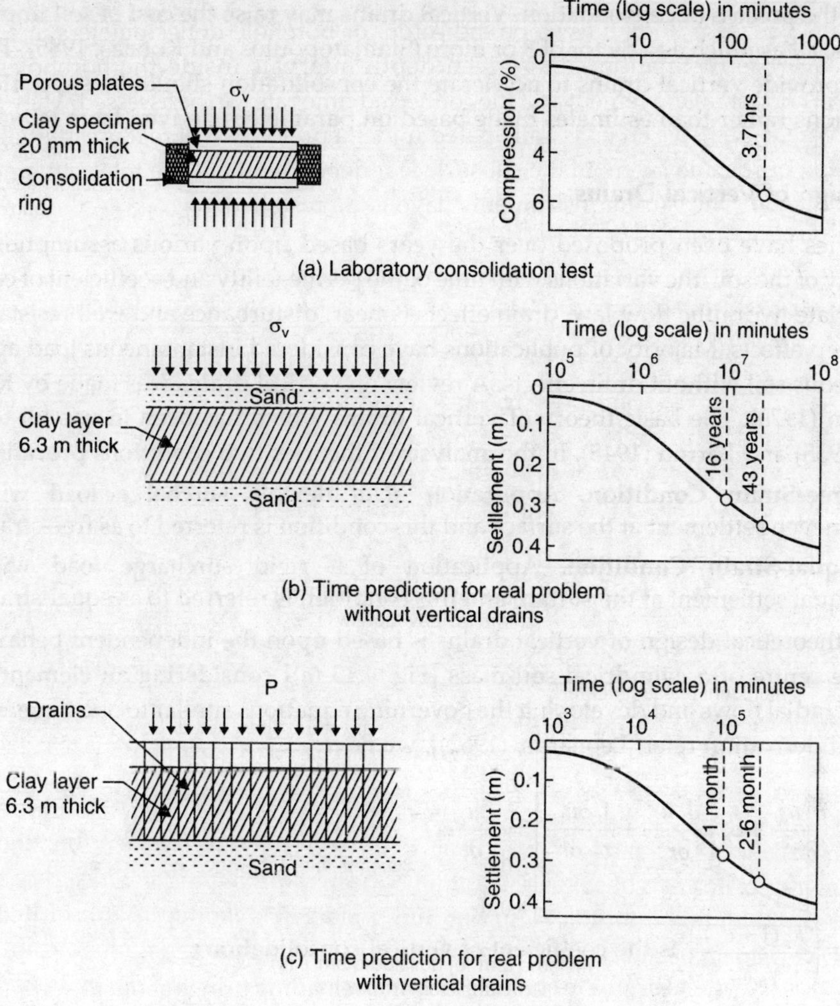

***FIGURE 4.12.* Comparisons between Time Curves.**

(Source: Stamatopoulos and Kotzias, 1985)

There is an important fallacy in this reasoning, namely that the time required for consolidation has been evaluated based on the laboratory results. The assumptions involved in the Terzaghi's theory may not be absolutely true in the field. The assumption of 20 mm specimens are made up of the same constituents and behave in exactly the same fashion as that of thick field layer may be true. Further, it has been pointed out that the time rate of consolidation can be 10, 100 or even 1000 times faster than that estimated from laboratory results (Simons, 1974; Stamatopoulos and Kotzias, 1983). Still, there is a growing suspicion that there are cases where vertical drains have been used without being needed (Lewis *et al.*, 1975). Vertical drains in certain field conditions may be unnecessary or even harmful. For example, in peaty soil the permeability is high enough so that primary consolidation is so rapid that vertical drains are unnecessary and vertical drains may be ineffective in clays with abundant pervious inclusions. Further in sensitive clays, soil disturbance may result in high initial pore water pressure and a zone of low permeability around the drain. Sometimes, the time required for the installation of vertical drains may be greater than the time saved from the process of consolidation. Vertical drains may raise the cost of soil improvement, in certain cases, by as much as a factor of 3 or more (Stamatopoulos and Kotzias, 1985). Therefore, the decision to provide vertical drains to accelerate the consolidation should be supported by strong field conditions rather than estimates made based on parameters derived from laboratory tests.

4.5.2 Design of Vertical Drains

Many theories have been proposed over the years based upon various assumptions about the homogeneity of the soil, the variations with time of the permeability and coefficient of consolidation, the appropriate hydraulic flow law, drain effects (smear, disturbance and well resistance) loading rate and creep effects. Majority of publications have considered instantaneous load application on a homogeneous soil without drain effects. A review on vertical drains was made by Richart (1959) and Johnson (1970). The basic theory of vertical drains which has been in use till today is after Rendulic (1935) and Barron (1948). In the analysis of theories two conditions prevail, *viz.*,

(*i*) **Free-Strain Condition.** Application of a flexible surcharge load will cause an uneven settlement at the surface and this condition is referred to as free-strain condition.

(*ii*) **Equal-Strain Condition.** Application of a rigid surcharge load will cause an equal settlement at the surface and this condition is referred to as equal-strain condition.

The theoretical design of vertical drains is based upon the independent behaviour of each drain in the centre of a cylindrical soil mass [Fig. 4.13 (*a*)] considering an element of soil with vertical and radial flows and developing the governing equations similar to one dimensional theory [for detailed derivation refer, Leonards, (1962)], we have

$$c_v \frac{\partial^2 u_w}{\partial z^2} + c_h \left[\frac{\partial^2 u_w}{\partial r^2} + \frac{1}{r} \frac{\partial u_w}{\partial r} \right] = \frac{\partial u_w}{\partial t} \qquad \text{...(4.15)}$$

where $c_v = \left(\dfrac{k_v (1 + e_o)}{\gamma_w a_v} \right)$ is the coefficient of vertical consolidation

and $c_h = \dfrac{k_h (1 + e_o)}{\gamma_w a_v}$ is the coefficient of radial or horizontal consolidation, where k_v and k_h are referred to a vertical and horizontal permeabilities.

Equation (4.15) is the governing differential equation for three -dimensional consolidation and may be considered to consist of two parts:

One dimensional flow:

$$c_v \frac{\partial^2 u_w}{\partial z^2} = \frac{\partial u_w}{\partial t} \qquad \text{...(4.16)}$$

Radial flow:

$$c_h \left[\frac{\partial^2 u_w}{\partial r^2} + \frac{1}{r} \frac{\partial u_w}{\partial r} \right] = \frac{\partial u_w}{\partial t} \qquad \text{...(4.17)}$$

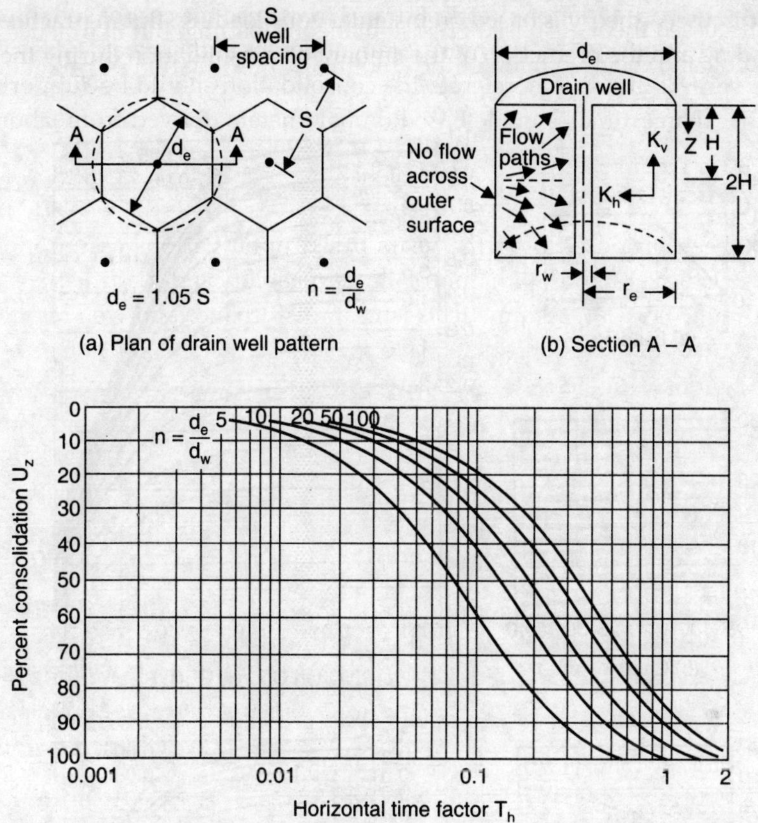

(c) Values of T_h as a function of U_z for various ratios of sand drain spacing to sand drain size

***FIGURE 4.13.* Theoretical Results for Radial Consolidation to Vertical Drain Wells.**
(*Source: Barron, 1948*)

A solution of Eq. (4.15) has been considered by Carilo (1942) as a combination of solutions of Eq. (4.16) and Eq. (4.17) and accordingly we have

$$(1 - U) = (1 - U_z)(1 - U_r) \qquad \ldots(4.18)$$

where U = degree of consolidation for three dimensional flow

U_z = degree of consolidation for one dimensional flow

U_r = degree of consolidation for radial flow

The time factor for radial flow, $T_r = \dfrac{c_h t}{(2r_e)^2}$ ...(4.19)

where r_e is the radius of influence.

The time factor for one-dimensional consolidation is given by Eq. (4.8).

Solutions to Eq. (4.16) is already discussed and solution for Eq. (4.17) was given by Rendulic (1935) for free-strain condition and by Barron (1948) for equal-strain condition. It has been reported by Richart (1959) for values of n (= r_e/r_w) greater than five, both the solutions give closer values. Hence, for all practical purposes Fig. 4.13 (*c*) (Barron, 1948) may be used regardless of the imposed boundary strain condition.

The above discussed theory is based on instantaneous loading. But in practice the loading is one of gradual loading and the prediction of the amount of consolidation during the loading stage

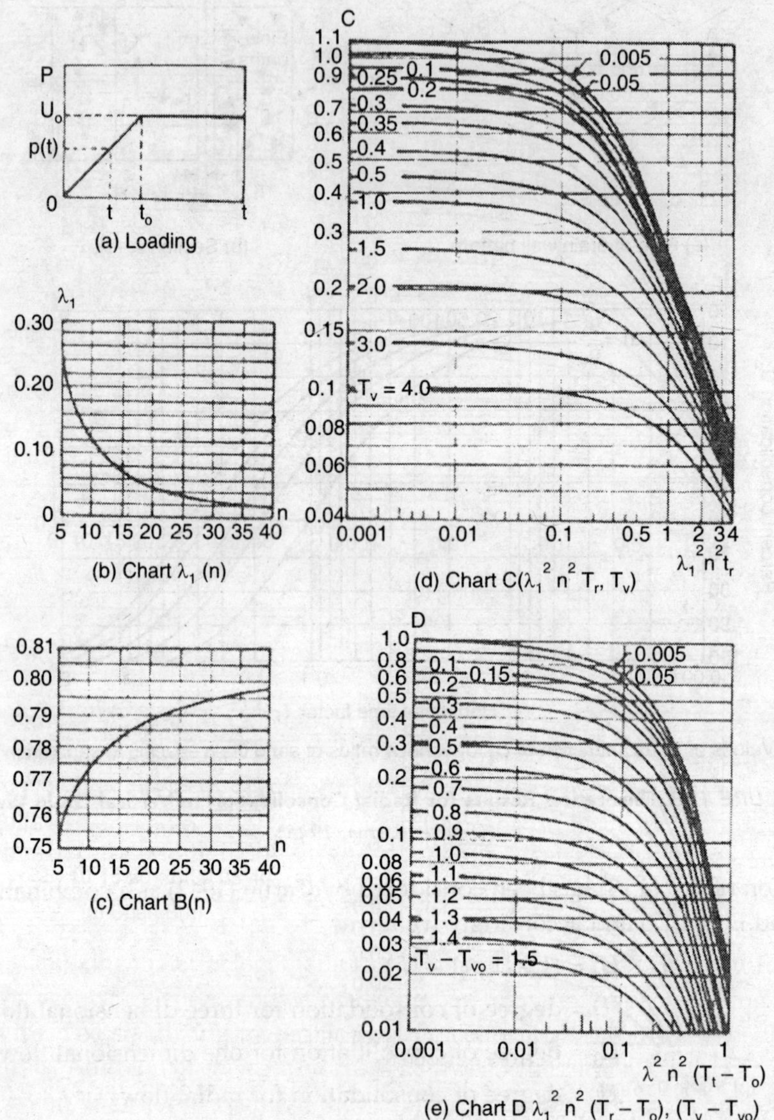

FIGURE 4.14. **Nomograms for the Functions Required for the Design of Vertical Drains Under Progressive Loading Conditions.**

(*Source: Chaput and Thomann, 1975*)

is of considerable importance. Chaput and Thomann (1975) have provided a solution which is presented in Fig. 4.14. If the average degree of consolidation U, is defined as the ratio of the settlement at time t when the load reaches $p(t)$ [Fig. 4.14 (*a*)] to the ultimate settlement that would result from $p(t)$ applied instantly, then during the load application ($0 \le t \le t_o$)

$$U = 1 - B(n)\, C(\lambda_1^2 n^2 T_r, T_v) \qquad \ldots(4.20)$$

where λ_1, $B(n)$ and C are presented in Fig. 4. 14(*b*), Fig. 4.14(*c*) and Fig. 4.14 respectively when the total load has been applied, $t > t_o$ then

$$U = 1 - B(n)\ C(\lambda_1^2 n^2\ T_r,\ T_{vo})\ D[\lambda_1^2\ n^2\ (T_r - T_o),\ (T_v - T_{vo})] \qquad \text{...(4.21)}$$

B and C are obtained for $t = t_o$ and D is given in Fig. 4.14 (*e*).

All design procedures for vertical drains require a proper estimate of the coefficient of radial consolidation c_h. This parameter links vertical compressibility and horizontal permeability and controls the radial flow of water into the drain. Hence, the accurate determination of this parameter is essential. Coefficient of radial consolidation varies two to ten times that of coefficient of vertical consolidation. Both laboratory and field methods have been proposed for determination of c_h. In the laboratory method a consolidometer is used in which the clay specimen is placed with a vertical drain at its centre. Conventional consolidation test procedure is followed with this set up and the

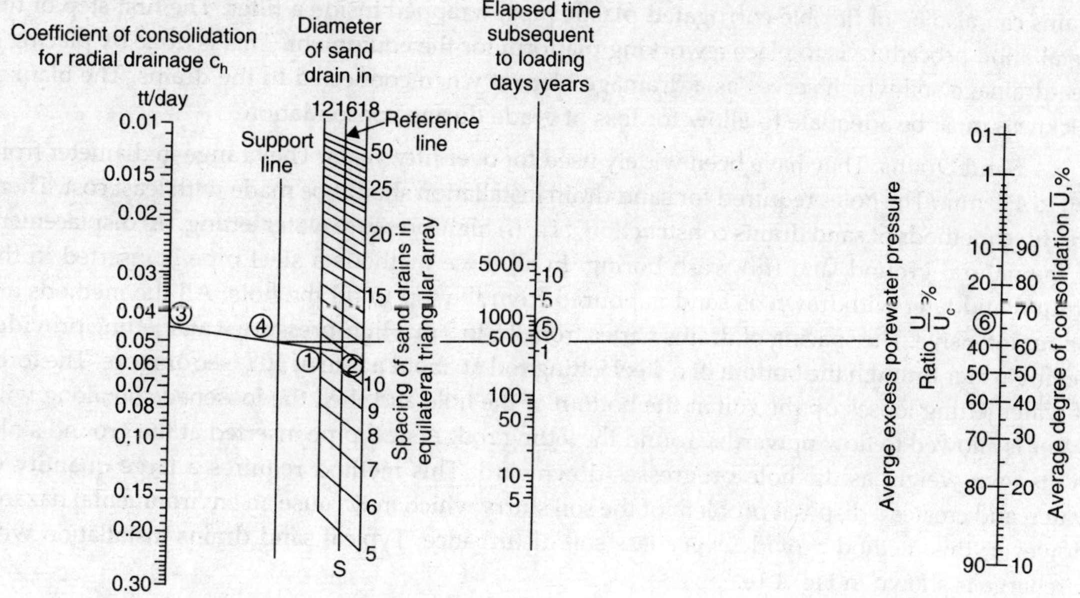

Directions

1. Locate point ① corresponding to assumed diameter and spacing of sand drains arranged in an equilateral triangular array (If sand drains are arranged in a square array multiply spacing by 1.072 to convert to an equivalent triangular array)
2. From point ① extend a line horizontally to reference line to locate point ②
3. Connect point ② and assumed value of the coefficient of consolidation for radial drainage (Point ③) with a straight line to locate point ④ on the support line
4. Pass a straight line from point ④ through point ⑤ the elapsed time after instantaneous loading to determine point ⑥ corresponding to the desired values of the average degree of consolidation or average excess pore water pressure ratio

Notes. Nomograph applies to radial flow to vertical sand drains assuming equal strain conditions throughout compressible stratum.

FIGURE 4.15. **Nomograph for Consolidation with Radial Drainage to Vertical sand Drains**

(*Source: NAVFAC, 1982*)

consolidation rate is measured. This laboratory method has been extensively used (Hansbo, 1960 ; Shield and Rowe, 1965; Rowe and Barden, 1966; Rowe, 1968; Berry and Wilkinson 1969; Paute, 1973). It has been reported (Pilot, 1981) that the laboratory measurement of c_h is quite reasonable and however large specimens should be used to account for stratification of soil stratum. In-situ measurements are based on permeameter tests (Torstonsson, 1975; Mieussens and Ducasse, 1977). A monograph for estimating the average degree of percent consolidation for various values of c_h drain diameter and spacing and time is given in Fig. 4.15.

4.5.3 Types and Construction of Vertical Drains

Vertical drains are mainly of two columnar types: (*i*) sand drains, and (*ii*) prefabricated drains. Sand drains are made by filling a cylindrical hole with sand. Prefabricated drains are also known as wickdrains or wicks and if sand is packed in filter stockings they are called sandwicks. These drains can also be of flexible corrugated plastic pipe, wrapped inside a filter. The first step of the installation procedure is to place a working platform for the equipment. This is done by placing a free-drainage soil which serves as a drainage blanket when connected to the drains. The blanket thickness must be adequate to allow for loss of grade during consolidation.

Sand Drains: They have been widely used for over fifty years. They range in diameter from 180 to 450 mm. The holes required for sand drain installation should be made with least cost. There are three methods of sand drains construction, *viz.*, (*i*) high-pressure water jetting, (*ii*) displacement of the natural ground, and (*iii*) wash boring. In all these methods a steel pipe is inserted in the ground and later withdrawn as sand is poured from the top to fill the hole. All the methods are labour intensive. The spacing of drains varies from 2.5 to 5 m. High-pressure water jetting provides forcing water through the bottom of a steel jetting rod at about a rate of 50 l/sec or more. The force of water jetting looses up the soil at the bottom of the hole and then the loosened soil along with water is allowed to flow upwards around the jetting rod. A steel pipe inserted at the ground sinks by its own weight as the hole progresses downward. This method requires a large quantity of water, and creates a disposal problem of the soil slurry which may cause an environmental hazard. However, this method provides very less soil disturbance. Typical sand drains installation with surcharge is shown in Fig. 4.16.

In the displacement method a close mandrel consisting of a steel tube closed at the lower end by a loose cap is used. The mandrel is driven by percussion or vibration or sometimes jetted into the place. When the pipe has reached the desired depth sand and water are introduced at the top and the pipe is withdrawn. Due to the forcing of the pipe the soil is displaced upwards and sideways by shear and compression creating a severe soil disturbance. The disturbance causes high initial pore pressure, low permeability around the hole and decrease in shear strength. However, this is a simple inexpensive method which is very popular.

In the wash boring method the hole is advanced by circulating water into the hole at a rate of 1 to 2 l/sec. The soil slurry from the hole are allowed to settle and from which clean water is siphoned back into the circuit. This method causes less soil disturbance and it can be performed with simple subsurface investigation equipment. This method is slower and more expensive than the previous two methods. Still it has provided to be quite useful in small projects. Table 4.5 shows the disturbance effects from sand drain installations (Johnson, 1968).

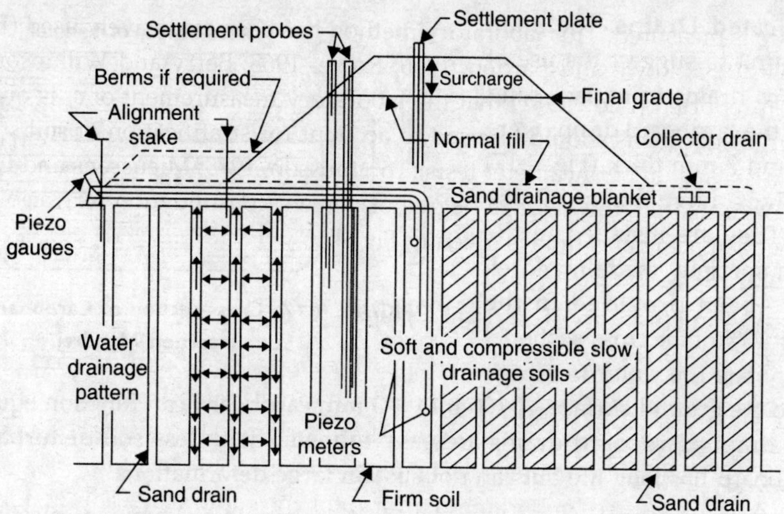

FIGURE 4.16. **Typical Sand Drain Installation with Surcharge.**

The sand used for filling the holes should not contain fines and be uniformly graded. The sand should not be coarse also as to over facilitate the migration of fine particles from the soil into the drain. For this reason particles larger than 4 mm should not be used. However, before using necessary tests the gravel-sand mixture is compatible with the surrounding soil. The voids in sand created by filling the hole should be sufficiently pervious to allow the unobstructed flow of water from the soil into the drain and from the lower part of the drain to the top.

Table 4.5. Disturbance Effects from Sand Drain Installations

(*Source: Johnson, 1968*)

Disturbance effect	*Remarks*
Smear	Can be caused by any method of installation, even by thin sharp knife drawn through soft clay.
Soil displacement and remoulding effects	Outward soil displacement caused by driven closed-end mandrels Inward soil displacement resulting from jetting methods or withdrawal of solid stem augers Either outward or inward soil displacement may result from hollow stem auger methods, depending on the rotation and advance rates of auger.
Grouting of thin sand layers	Caused by natural drilling mud formed by jetting methods.
Thin film of mud on sides of drain	Caused by natural drilling mud formed by jetting methods.
Contamination of sand backfill in drain	Possible in jetting methods if washing during jetting is insufficient; also possible when withdrawing driven mandrel if sand stick in mandrel.
Distortion of thin sand layers	Most likely with driven mandrel method, possible with solid or hollow stem augers ; may be severe in varved or thinly bedded deposits.

Prefabricated Drains. Kjellman (1948) was the first to suggest the use of driven cardboard drains to replace sand drains. Initially the cardboard drains were 100 mm wide and 3 mm thick (Fig. 4.17) and varied sizes have been in use subsequently. They are installed in the ground using a specially made mandrel. Channels have been provided in the cardbord (about 10% of the cross-sectional area) which facilitate the water to escape from the clay to the ground surface. A 100 mm × 3 mm cardboard can function equal to a 50 mm diameter sand drain. Cardboard drains are easy to install with less soil disturbance. Specially processed cardboard has long life but can not sustain large deformations.

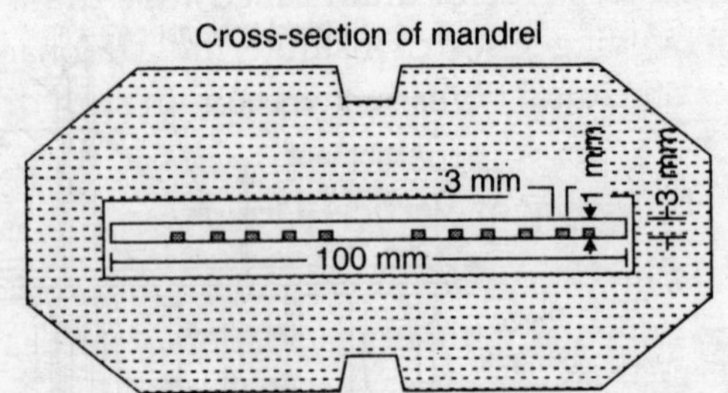

FIGURE 4.17. **Cross-section of Cardboard Drain and Insertion Mandrel.**

Cardboard drains have been replaced by plastic drains (also known as plastic band-shaped drains) and has been manufactured in a large scale. The well-known brand is Geodrain which consists of a 100 mm wide and 3 mm thick paper-covered polythene strip which contains channels along both sides (Boman, 1973 ; Terrafigo, 1976). The configuration is so chosen such that a drainage channel of more than 70% of the total drain area is available. In order to give a long life the exterior filter paper is, chemically impregnated. The cross-section of plastic Geodrain and insertion mandrel are shown in Fig. 4.18. The plastic drains have been widely used (Hansho and Torstensson, 1977) and showed relatively little disturbance and could sustain large settlement without break of drain continuity. Geodrains have been claimed to have the following advantages: (*i*) low cost, (*ii*) fast installation, (*iii*) ensured drain continuity, (*iv*) clean site, (*v*) light-weight installation equipment, (*vi*) high permeability, (*vii*) negligible subsoil disturbance, and (*viii*) positive drainage. Common types of plastic drains (90 to 100 mm wide by 3 to 5 mm thick) are usually installed by a lance about 140 mm in cross section. Open or closed mandrel are fitted to the rigs. A downward force of 200 kN can be executed by the rigs and they provide for the simultaneous driving up to four wicks at a time. Installation speed is of the order of 0.3 to 0.6 m/sec and driving depth could be up to 45 m (Nicholls and Barry, 1983). Plastic drains are wound around reels, and as the mandrel is lowered into the ground, the band is released by unwinding. The spacing is commonly in the order of one and half to two wick drains per one sand drain.

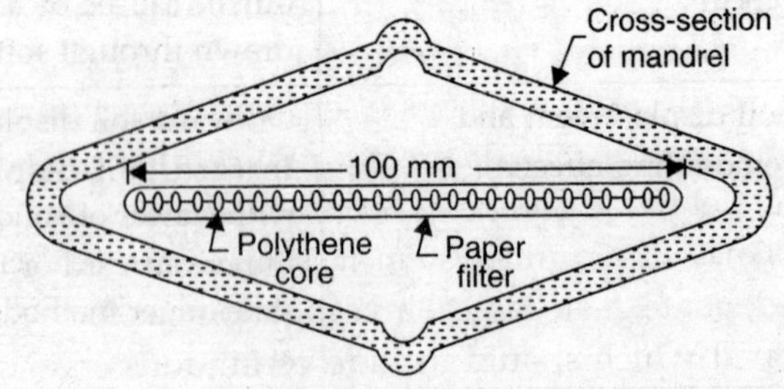

FIGURE 4.18. **Cross-section of Plastic Geodrain and Insertion Mandrel.**

For want of modern manufacturing and installation techniques these type of drains are not yet available in India. However, the sandwick drains were first used in India by Dastidar *et al* (1969) and has been used successfully in several installations (Subbaraju *et al.*, 1973; Som, 1975). Sandwicks are ready made small diameter (of about 100 mm) sand drains which are contained in long filter stockings. Sandwicks are installed by the closed-mandrel technique adopted for sand drains. This type of vertical drain is economical where labour costs are low.

Another type of drain called rope drain has been developed and used in several jobs by Central Building Research Institute, India (Mohan *et al.*, 1977; Sengupta *et al.*, 1980). The rope material consists of natural fibres such as coir. In such drains the drainage capacity is the major constraint.

4.5.4 Efficiency of Vertical Drains

The efficiency of system of vertical drains is assessed with reference to the primary consolidation attained with and without installation of vertical drains has been expressed by Bjerrum (1972) as

$$\eta = \frac{\rho_c}{\rho_c + \rho_{sc}}$$

where ρ_c and ρ_{sc} are the primary compression and secondary compression respectively. Values of ρ_{pc} and ρ_{sc} are evaluated as shown in Fig. 4.19. According to Bjerrum (1972), a value of the range of 0.6 to 0.8 may show a satisfactory performance of drains.

The efficiency of drains-system depends on drain effects such as smear disturbance and drain resistance. In the installation of sand drains the soil is disturbed in two ways: (*i*) being compressed and sheared especially when closed-mandrel is used, and (*ii*) by being smeared due to remoulding. Both action reduces the soil permeability around the drains. If the change in permeability and the thickness of the smear zone were known, smear could be accounted for (Barran, 1948; Richart, 1954). As an approximation, the effect of smear is sometimes taken as equivalent to halving the well radius. The drain should be capable to collect the water coming from the consolidating soil and also to conduct it to the surface. For water to flow upwards in the drain, there should be a head difference from bottom to the top with the maximum head at the bottom.

Buckling or folding of sand drains or prefabricated drains does not generally affect the performance of the drains even for a prolonged service period (Hansbo, 1983; Sinclair *et al.*, 1983). Clays exhibiting horizontal stratification with seams of sand or silt may affect/effect in the following manner (Rowe, 1968): (*i*) thin lenticular strata of high permeability greatly increase the efficiency of drains, since they act as horizontal drains attached to main arteries, (*ii*) continuous thick seams of high permeability material at sufficiently close spacings often render vertical drains unnecessary or greatly reduce their real effectiveness. In order to accurately assess the efficiency of a drain-system it is necessary that an adequate geotechnical investigation has to be planned which should include continuous core sampling of the strata, in-situ permeability measurements at low hydraulic heads and laboratory consolidation tests on large diameter specimens (Rowe, 1968).

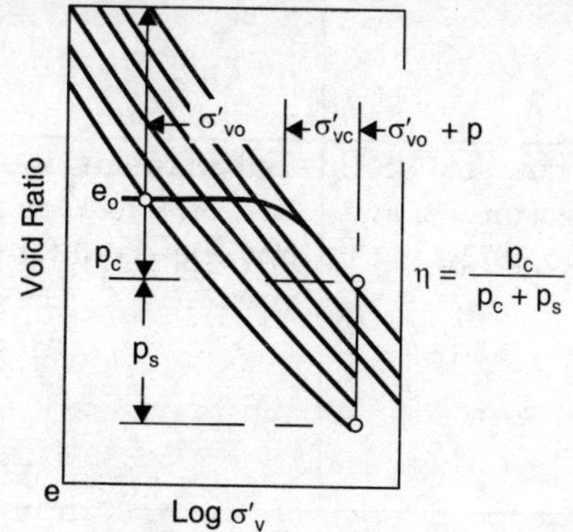

FIGURE 4.19. **Definition of Vertical Drain Efficiency.**
(*Source: Bjerrum, 1972*)

4.5.5 Applications

The concept of vertical drains was known more than a century ago. But it has been applied only after 1930s. Until about 1950 most installed vertical drains were sand drains. Since 1980 prefabricated drains have become popular because of their less cost and fast installation. It has been reported that the cost of prefabricated drains is about one-thirds of the cost of sand drains. However, in India only sand drains have been widely used (Datye and Nagarajan, 1975, 1976). Datye (1982) has reported that in India over 10,000 drains of 200 to 400 mm diameters have been successfully installed at spacings of 2 to 5 m adopting conventional equipment used for pile driving.

Johnson (1970), Bjerrum (1972), and Stamatopoulos and Kotzias (1985) have reported a number of case histories involving the use of vertical drains. A compilation of case histories reported by Pilot (1981) is presented in Table 4.6.

Table 4.6. Case Histories of Vertical Drain Applications
(*Source: Pilot, 1981*)

Location and reference	*Structure and soil description*	*Drain System*	*Comments on performance*
Fukuroi, Japan (Ueda, 1971).	Embankment 10 m high (of which 3 m were over load), on 5 to 8 m peat and 7 m soft clay (s_u = 20 to 60 kPa).	Sand drains, 0.4 m diam. at 1.2 m centres.	Failure occurred at an embankment height of 6 m.
Massachusetts, USA, (Aldrich and Johnson, 1972).	Test embankment 6 m high, on about 18 m silty to organic clays.	1. Three test areas withsand drains 0.45 diam.at 3 to 4.2 m centres: (*a*) auger-bored drains, (*b*) driven drains, (*c*) jetted drains 2. Test area with the same drains arranged differently.	With drains at 3 m, in-situ c_r values lower than for 4.2 m spacing: concluded that disturbance increased from jetted drains to auger-bored drains to driven drains. Settlements in driven-drain area were greater than in other areas.
San Francisco, USA, (Margason and Arango, 1972).	Embankment in the Bay, on 6 to 12 m soft clay. Observed for two years.	Jetted sand drains, 0.3 m diam. at 3 to 4.5 m centres.	Good results. Area with drains at 3 m settled 20 to 25% faster than that with drains at 4.5 m. Settlements agreed well with conventional theories, but not so with pore water pressures.
Ska-Edeby, Sweden, (Holtz and Broms, 1972).	Test embankments, on about 10 m soft clay (s_u 5 to 10 kPa). Observed for 14 years.	1. Driven sand drains, 0.18 m diam. at 0.9 m, 1.5 m and 2.2 m centres. 2. Drainless reference area.	Disturbance due to the drains were small: close spacing more efficient. Results suppose Hansbo's theory with n = 1.5.

(Table Contd...)

Table 4.6. *(Contd...)*

Location and reference	*Structure and soil description*	*Drain System*	*Comments on performance*
Massachusetts, USA, (Ladd *et al.*, 1972).	Approach ramp 10 m high, on 10 m chiefly soft and sensitive clay. Observed for two years.	1. Sand drains (jetted open-mandrel), 0.3 m diam. at centres from 2.7 to 4.8 m according to test area. 2. Drainless reference area.	Good results. Consolidation virtually completed within 200 to 400 days in various test areas, but $U < 70\%$ in drainless area after 700 days. c_r/c_v in-situ found to be equal to 2, values of c_r and c_v being much higher than laboratory.
Napa River, USA, (Su *et. al.*, 1973).	Access ramp to a structure 7 m in height, on 15 m of soft silty clay. Observed for four years.	Driven sand drain, 0.45 m diam.	Good results. Greatly increased strengths compatible with increased pore water pressure and settlements.
Hampton Roads, USA. (Kuesel *et al.*, 1973).	Precompression embankment 9 m high, on about 20 m soft clay. Observed for about two years.	Jetted sand drains with layout pattern from 1.8 × 2.1 m to 2.7 × 3 m.	Good results. c_r found to be equal to c_v and, on this basis, measured settlements agreed with design calculations.
Fiuminico, Italy (Croce *et al.*, 1973).	Test embankment, on about 30 m more or less organic clay.	Driven sand drains, 0.4 m diam. at about 3 m centres.	Driving the drains brought about excess pore pressure of approx. 20 kPa that virtually vanished within 70 days.
Oxford, England, (Murray, 1973).	Embankment 3.4 m high, on 18 m soft clay. Observed for one year.	1. Driven sand drains, 0.3 m diam. at 1.8 m centres. 2. Drainless reference area.	Settlements occurred as quickly with or without drains, laboratory testing having under-estimated c_r and c_v.
Vizag, India. (Subbaraju *et al.*, 1973).	Stockpiled ore 9 m high, on 12 m soft clay.	1. Test area with open-mandrel sand-drains, 0.36 m diam. at 2.14 m centres. 2. Test area with sand-wicks, 0.06 m diam. at 2.14 m centres. 3. Drainless reference area.	Grains in strength were largest with sand-drains, lower with sand-wicks, negligible in reference area.

(Table Contd...)

Table 4.6. *(Contd...)*

Location and reference	*Structure and soil description*	*Drain System*	*Comments on performance*
Cran, France. (Paute, 1973 a).	Access ramp to structure about 3 m high, over 17 m soft clay. Observed for 3.5 years.	Driven sand drains, 0.4 m diam. and 13 m deep at 4 m centres.	Good results. Settlements slightly less than calculated using method of finite differences. Decreased moisture content and increased cohesion noteworthy in upper 5 m.
Palavas, France. (Bourges *et al.*, 1973).	Test embankment 7 m high, on 22 m soft clay (s_u = 10 to 30 kPa). Observed for three years.	1. Sand drains (jetted open-mandrel), 0.3 m diam. at 4 m centres. 2. Drainless reference area.	Good results. Settlements much more rapid in drain area (2.8 m against 1.9 m). Appreciable uniform cohesion increase throughout layer.
Marbonne, France. (Mieussens and Ducasse, 1973).	Test embankment 5.5 m high, on 15 m various soft clays inter-stratified with sand. Observed for 1.5 years.	1. Test area with cardboard drains at 3 m centres. 2. Test area with sand; drains (jetted open-mandrel), 0.3 m diam. at 4 m centres. 3. Drain less reference area.	Unsatisfactory results. Pore water pressures dissipated slightly faster in drain areas, but settlements much the same in three areas.
Bordeaux, France. (Queyroi *et al.*, 1977).	Embankment 6 m high, on 12 m soft organic clay. Observed for two years.	Auger-bored sand drains, 0.3 m diam. at 2 m centres.	Good results. Measured settlements agreed with those estimated on basis of c_r measured in-situ.
Ska-Edeby, Sweden. (Hansbo and Tosensson, 1971).	Test embankment 1.5 m high, on 10 m soft clay. Observed for three years.	1. Geodrains at 0.9 m centres. 2. Sand drain reference area. 3. Drainless reference area.	Results for Geodrains equivalent to results from sand drains 0.18 m diam. at 0.9 m centres.

Based on the compiled data Pilot (1981) has drawn the following conclusions:

(*i*) Vertical drains are generally very effective except where they are installed in organic soils, in highly stratified soils, or where serious stability problem exist.

(*ii*) Installation of sand drains by the open-mandrel method results in more efficient drains than does the closed mandrel method.

(*iii*) Plastic drains are increasingly being used.

(*iv*) Predictions of rates of consolidation where drains are installed can not be made reliably, because of the difficulties of determining a representative value of c_h and of accounting for the effects (disturbance and smear) of the drain installation.

4.6 DYNAMIC CONSOLIDATION

The densification of soils by dynamic consolidation, (dynamic compaction, heavy tamping) has been described in an earlier section. It has also been discussed that this method could be used as a means for improving silts and clays. It has also been noted that the secondary compression is also reduced. The time required for treatment is less than, that for surcharge loading with sand drains.

The application of this method is same as that needed for cohesionless soils but more time is required. Several blows are applied at each location followed by one to four week rest period, then the process is repeated. Several repetitions may be needed and in each repetition immediate settlement occurs followed by drainage of pore water. Drainage is facilitated by radial fissures that form around impact points and by the use of horizontal and peripheral drains. As there is a need for time lapse between successive repetitions of heavy tamping when treating silts and clays, a minimum treatment area of 15,000 to 30,000 m^2 is necessary for economical use of the method (Mitchell, 1975). Menard and Broise (1975) suggested that this technique is effective in soft clays for the following reasons:

(*i*) Many soft clays are not fully saturated, and the small percentage of gas in the voids is dissolved in the pore water under the hammer impact, thus reducing the void volume.

(*ii*) Soft clays often liquefy under impact.

(*iii*) The permeability is increased during heavy tamping because of liquefaction, fissuring and shearing, and rapid dissipation of the pore pressure occurs.

(*iv*) Thixotropic strength gain follows the shear strength decrease caused by compaction.

The compression of a soft clay during dynamic compaction and its subsequent consolidation characteristics under static loads have been demonstrated in the laboratory by Magnam and Dang (1977). Case histories on the application of this method for soft clays and clayey silts of low shear strength has been reviewed and reported (Table 4.7) by Pilot (1981). Out of the six case histories reported three dealt on soft clays and three on clayey silts of low shear strength.

4.7 CONSOLIDATION BY ELECTRO-OSMOSIS

Electro-osmosis has been used as a means for consolidation of certain soils. The electro-osmosis process, by which the pore water moves under the influence of an electrical potential, effects in the decrease of water content in fine grained soil and thereby brings the consolidation of the soil. This is possible under certain soil conditions and limited volume of soil to be consolidated. The flow of water from anode to cathode ceases when a hydraulic gradient induced by water content variation tending to cause flow from cathode towards anode exactly balances the electrically induced hydraulic gradient causing flow from anode towards cathode. At this condition there will be an increase in effective stress ($\Delta\sigma'$) due to pore pressure dissipation. Esrig (1968) showed that the pore

Table 4.7. Case Histories of Dynamic Consolidation (*Source: Pilot, 1981*)

Location and reference	Foundation Soil							Surface fill thickness (m)	Dynamic consolidation details				Results of the dynamic consolidation					
	Soil type	Depth (m)	w_p (%)	I_p (%)	w_n (%)	k (m/sec)	c_c (m²/sec)		Weight (kN)	Drop (m)	Specific power (kN/m²)	Compaction time (day's)	Settlement mm		plor s_u (kPa)		E_m(kPa) (kPa)	
													total	found	before	after	before	after
Embourge, Belgium (Van Manbeke and De Beer, 1973)	Silt	3	22	13	30	1.3 × 10^{-9}		2	80	5.15	8.80	15	210 150 } 360	150 average for 1st stage	110	420 for 1st stage 480 for 2nd stage	700	5500 for 1st stage, 4800 for 2nd stage
										10	+410							
									80	10	880	15	240 130 } 370					
										10	+740							
									80	5	860	15	330 100 } 430	60 average for 2nd stage				
										10	+790							
									80	5	550	15	220 70 } 290					
										10	+410							
Pont de-Clichy, France (Gigan, 1977)	Organic silt (beneath 7 m fill)	3	70	20	63	1.3 × 10^{-5}	5 × 10^{-6}	1	120	22	700 +550 +400 +200	10 7 4			520	730 670 730 650 (after 10 months)	3700	5000 4700 5000 4400 (after 10 months)
Ambes, France, (Menard, 1976)	Soft clays	10 +8	21 22	25 39	48 64		3 × 10^{-7} to 2.5 × 10^{-8}	0.5	120 120 120 170 120	16 16 16 20 8	384 +307 +235 +204 +120		180 150 120 190 90 } 730		120 (−2m) to 450 (−16 m)	210 (−2 m) to 520 (−16 m)		

(Table Contd...)

Table 4.7. (*Contd...*)

Ogawara, Japan, (Menard, 1976)	Softclays (volcanic origin)	7	80	40				0.5	90 and 120 90 and 120	27 (max) 27 (max)	500 + 500 + 200 1000 + 1000 + 200	30 2 30 2	320 } 340 20 } } 20		s_u 50 60	s_u 100 970		
Libson, Portugal, (Menard, 1976)	Silt	27	35	35	66			1 +1	160	22	1750 + 1750	60	460 } 700 240 }		50 (–4 m) 100 (–7 m)	420 (–4 m) 300 (–7 m) 300 (– 15 m)	500 (–4 m) (–15 m) (after 8 months)	4500 (–4 m) after 8 months
Ilha dos Asores Santos, Brasil, (Project document)	Silt	20	40	95		10^{-9}		2	200	20	350 +750	25	800 } 1360 560 }		s_u 3 to 15 *pl* 3 to 4	s_u 12 to 25 *pl* 10 to 12	?	?

Notation: s_u = undrained shear strength: *pl* and E_u = pressuremeter parameter.

pressure varied linearly between the electrodes after a long period of time in a soil undergoing electro-osmotic consolidation. The magnitude of the pore pressure difference (u) between the two electrodes is given as

$$u = \frac{k_e}{k_h} \gamma_w V \qquad \text{...(4.22)}$$

where, k_e = coefficient of electro-osmotic permeability

k_h = coefficient of hydraulic permeability

γ_w = unit weight of water

V = voltage difference applied across the electrodes (a function of position).

Thus the increase in effective stress is equal to pore pressure difference between the electrodes.

Hence,

$$\Delta\sigma' = (k_e/k_h)\, \gamma_w V \qquad \text{...(4.23)}$$

The amount of consolidation associated with this effective stress increase is obtained from a void ratio versus pressure relationship for the soil determined in the usual manner.

The rate of consolidation is governed by the same relation as that for conventional loading. The time ***t*** for a given degree of consolidation is

$$t = \frac{TL^2}{c_v} \qquad \text{...(4.24)}$$

where T = a time factor, dependent on the degree of consolidation and boundary conditions

L = electrode spacing

c_v = coefficient of consolidation.

Values of T for different degrees of consolidation for the case of parallel plate electrodes are given in Table 4.8 (Mitchell, 1976).

Table 4.8. Time Factor for Various Degrees of Consolidation by Electro-Osmosis between Parallel Plate Electrodes

(*Source: Mitchell, 1976*)

Degree of Consolidation U%	*Time Factor T*	*Degree of Consolidation U%*	*Time Factor T*
0	0	50	0.294
10	0.050	60	0.384
20	0.102	70	0.501
30	0.157	80	0.665
40	0.221	90	0.946

Electrodes are generally placed in a particular pattern, as shown in Fig. 4.20 and not as parallel plates. Thus the time factor given in Table 4.6 has to be treated as approximate values. It has been reported (Mitchell, 1976) that a hexagonal arrangement of electrodes is efficient in terms of power consumption, average voltage and anode to cathode ratio. Well points are generally used for cathode and reinforcing bars or aluminium rods for anodes which cost less than well points. Further iron and aluminium anodes decompose during treatment and participate in electrochemical hardening. Thus for the above reason increasing the number of anodes relative to cathode is generally beneficial.

Electro-osmotic consolidation will be effective and economical under the following conditions: (*i*) a saturated silt or silty clay soil, (*ii*) a normally consolidated soil, and (*iii*) a low pore water electrolyte concentration. The efficiency of the method may be reduced if gas generation, drying and fissuring at the electrodes are taking place. Non-uniform changes in properties of soils between electrode is possible because the induced consolidation depends on the voltage and the voltage variation between anode and cathode. In order to attain a more uniform stress conditions, reversal of electrode polarity may be adopted (Fig. 4.21). Electro-osmosis may also be combined with preload or surcharge fill to accelerate consolidation.

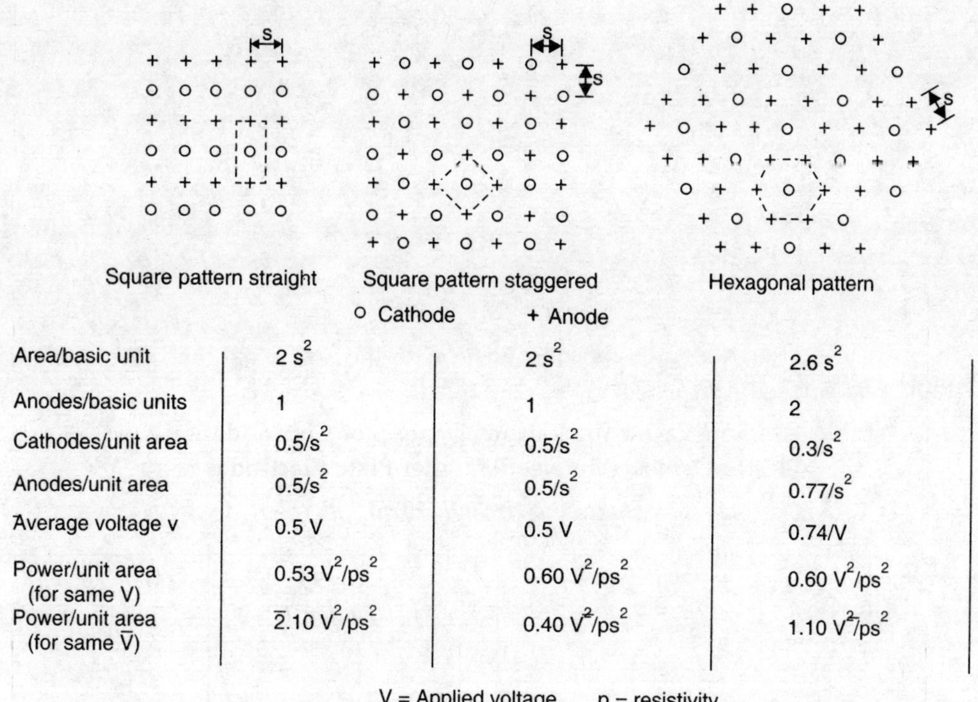

	Square pattern straight	Square pattern staggered	Hexagonal pattern
Area/basic unit	2 s^2	2 s^2	2.6 s^2
Anodes/basic units	1	1	2
Cathodes/unit area	$0.5/s^2$	$0.5/s^2$	$0.3/s^2$
Anodes/unit area	$0.5/s^2$	$0.5/s^2$	$0.77/s^2$
Average voltage v	0.5 V	0.5 V	0.74/V
Power/unit area (for same V)	$0.53\ V^2/ps^2$	$0.60\ V^2/ps^2$	$0.60\ V^2/ps^2$
Power/unit area (for same $\bar{V}$)	$2.10\ \bar{V}^2/ps^2$	$0.40\ \bar{V}^2/ps^2$	$1.10\ \bar{V}^2/ps^2$

FIGURE 4.20. **Characteristics of Different Electrode Patterns in Electro-osmosis.**
(*Source: Mitchell, 1976*)

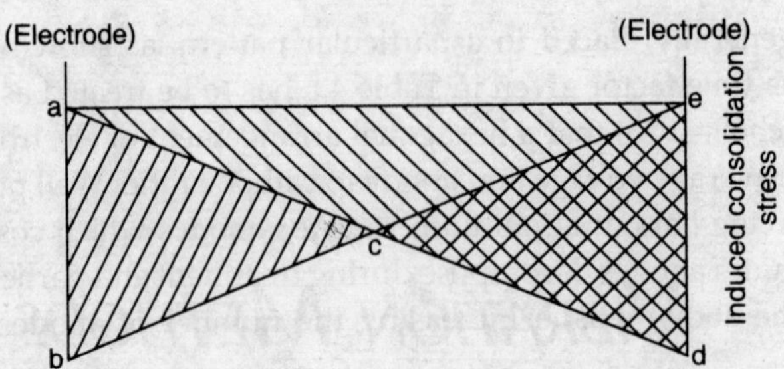

ade = Equilibrium consolidation pressure with *ed* as anode

abe = Equilibrium consolidation pressure with *ab* as anode

abc = Additional consolidation owing to electrode reversal

cde = Rebound after electrode reversal

FIGURE 4.21. **Consolidation Stress Induced by Electro-osmotic Consolidation, Before and After Electrode Reversal.**

(*Source: Mitchell, 1976*)

Chapter 5 VIBRATION METHODS

5.1 INTRODUCTION

Inertia forces become significant in comparison to static forces when the applied loads on the soil mass changes rapidly causing excessive deformation of the soil. This characteristic behaviour could be used in any ground improvement technique by way of adopting some form of vibration which could bring in deformation and displacement resulting in densification. These techniques for compacting cohesionless soils, arranged in the order of decreasing effectiveness are vibration, watering and rolling. In practice combinations of these techniques have been used in improving the properties of in-situ soils. Cohesionless soils get densified largely by fracture and reorientation of the grains. Static forces are not effective in this process. But both vibration and shock are helpful in reducing the wedging and aiding densification in cohesionless soils. Vibration and shock are less effective in cohesive soils as they are offset by the increased cohesive resistance due to dynamic loading. This Chapter deals with ground improvement techniques which utilize the principles of vibration.

5.2 VIBRO-COMPACTION

Vibro-compaction method is a rapid densification technique which could be used effectively in saturated cohesionless soils. In such loose deposits vibration or shock waves cause localized spontaneous liquefaction followed by densification and settlement. The load due to shock is temporarily transferred to the liquid and the soil particles take a much denser pattern aided by the soil particles. Even in dry soils due to shock and vibration, the particles move from the original position and take a more compact pattern. Once the particles are free or loosened due to the shock, even a small pressure is enough to put them to a more compact mass. The new density or the compactness attained is permanent and not reversible.

The effectiveness of these methods decrease with increase in the percentage of fines in the soil, because the permeability is too low to prevent rapid drainage following liquefaction. More than 20% of silt, or 5% of clay may reduce the effectiveness of the method (Mitchell, 1976). Further their application in partially saturated soils is restricted because compressive stresses from the presence of air-water menisci act to prevent the soil particle movements necessary for densification.

Three methods which are in use are : blasting, vibratory probe and vibratory compactors.

5.2.1 Blasting

In this technique a certain amount of explosive charge is buried at a certain depth of a cohesionless soil required to be compacted and is then detonated. As explained above the shock waves produced by the blasting cause densification.

A pipe of 7.5 to 10 cm is driven to the required depth in the soil strata. The sticks of dynamite and an electric detonator are wrapped in the water proof bundles and lowered through the casing (Fig. 5.1). The casing is withdrawn and a wad of paper or wood is placed against the charge of explosives to protect it from misfire. The hole is backfilled with sand in order to obtain the full force of the blast. The electrical circuit is closed to fire the charge. A series of holes are thus made ready. Each hole is detonated in succession and the resulting large diameter holes formed by lateral displacement are backfilled. The surface settlements are measured by taking levels or from screw plates embedded at certain depth below the ground surface.

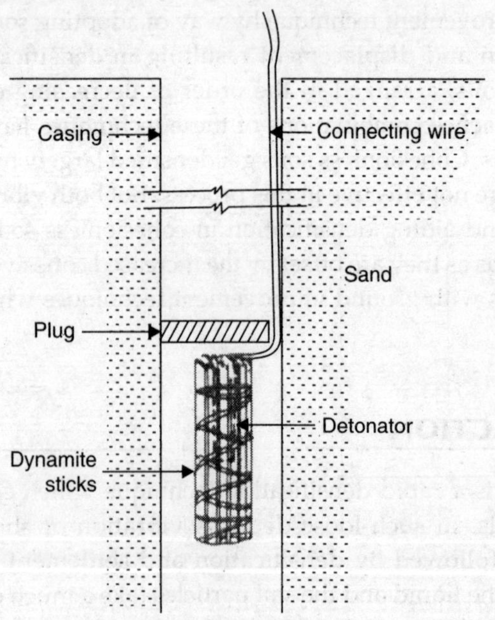

FIGURE 5.1. **Installation of Explosives.**
(*Source: Mitchell, 1970*)

Usually the explosives are arranged in the form of a horizontal grid. The spacing of the charges is decided by the depth of strata to be densified, the size of charge and the overlapping of the charges. A spacing of 3 to 8 m is typical and a spacing less than 3 m should be avoided.

Compaction is carried out in a single tier only if the depth of stratum to be densified is 10 m or less. In such a case the depth of explosive charge should be below half the depth of the mass or stratum to be densified (approximately at 2/3 point). More than one tier should be planned if the depth of stratum to be densified is more than 10 m. Generally the depth of charge should be greater than the radius of sphere of influence (R).

Successive blasts of small charges at appropriate spacings are likely to be more effective than a single large blast (Hall, 1962; Mitchell, 1976). Theoretically, one charge densifies the surrounding adjacent soil and the soil beneath the blast. Charges should be timed to explode such that the bottom of the layer being densified upwards in a uniform manner (Koerner, 1985). The uppermost portion of the stratum may be less densified which may be compacted by vibratory rollers. The amount of charge to be used should be optimal such that it is just enough to shatter the soil mass uniformly but not to create permanent surface craters. A carefully placed charge with required amount and depth shall not create a surface heave more than 0.15 m (Mitchell, 1976). As a rough guidance, the weight of charge required is computed from the following relationship.

$$W = 164\, C\, R^3$$

where W = weight of explosive (N)

C = coefficient (0.0025 for 60% detonator)

R = radius of influence (m).

Charge masses of less than 2 kg to more than 30 kg have been used (Mitchell, 1976). A typical firing pattern (Mitchell, 1976) is shown in Fig. 5.2.

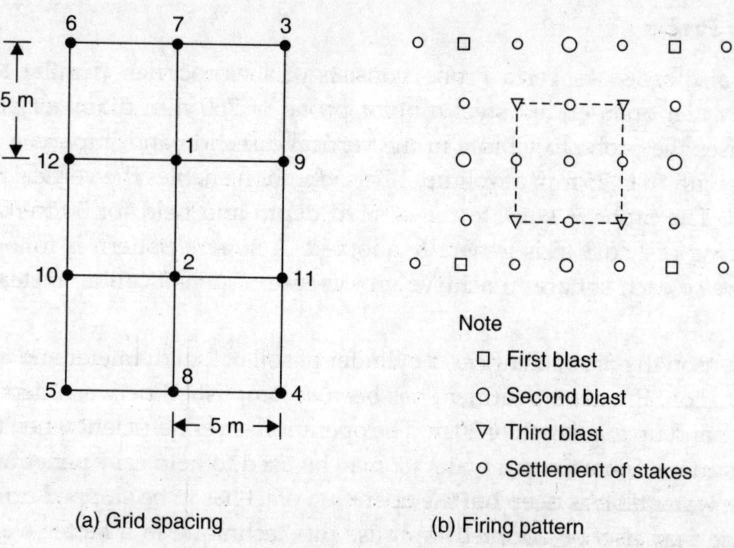

***FIGURE 5.2.* Typical Charge Spacing and Firing.**

(*Source: Mitchell, 1976*)

Blasting technique involves less time, labour and expense. This technique needs no special equipment and could be successfully used for densifying soil at a great depth. Further this could be effectively used to compact a large volume to a substantial depth up to 20 m and in small areas where the use of other methods would be impractical. Relative densities of the order of 70 to 80% can be achieved. In remote areas where vibrations are favourable, this technique may prove most cost effective. Invariably the blasting work should be executed only by an experienced contractor under special supervision.

Although blasting is one of the most economical ground improvement techniques, it suffers the disadvantages of non-uniformity, potential adverse effects on adjacent structures and the danger associated with the use of explosives in populated areas. Very fine grained soils which have high cohesive forces can not be compacted by this method. Maximum compaction is obtained only when the soil is dry or completely saturated. In case of dry loose sand, good results are obtained due to free fall of small-size particles into the voids between the soil grains thus making a dense soil. In case of saturated soils shock waves cause liquefaction leading to expulsion of water resulting in a better arrangement of particles. But in partially saturated soils due to capillary tension between the soil grains, less densification is achieved. Pre-flooding of the site is desirable, if blasting is to be resorted to for densification of partially saturated sands or loess. Theoretically there is no limit for the depth of densification by this technique. However, if the depth is more than 10 m, compaction should be done in more than one tier for which a careful planning is needed to achieve the result.

Thus it is emphasized that adequate data with regard to type of soil, degree of saturation, depth of deposit to be densified and degree of densification required should be collected. A preliminary test may be necessary to ascertain the spacing, depth, amount of charge and sequence of operation. In order to evaluate the economical feasibility of this technique, the above details are needed.

5.2.2 Vibratory Probe

Vibratory probe, also called as Terra Probe, consists of a vibrodriver (similar to vibratory pile driving) coupled to an open-ended steel tubular probe of 760 mm diameter and 15 m length. Vibrodriver activises the probe to vibrate in the vertical direction and imparts a vertical impulse normally at 15 Hz with 10 to 25 mm amplitude. This vibration enables the vertical pipe to penetrate the loose material. The probe is sunk to the desired depth and held for 30 to 60 seconds before extraction. A spacing of 1 to 3 m is generally adopted. A square pattern is followed with a fifth probe at the centre of each square to achieve an increase of densification, instead of adopting a reduced spacing.

In each insertion the probe densifies a cylinder of soil of 1 m diameter and about 1 m deeper than the probe location. Effective treatment has been accomplished between depths of 4 m below the ground surface and up to a depth of 20 m. The operation is very efficient when the groundwater is 2 to 3 m of the surface. Sometimes a water jet may be used to help easy penetration of the probe for sites where the water table is deep but the operation of jet has to be stopped during withdrawal. Ponding of the site may also be adopted to utilise this technique in a better way. Saturated soil conditions are necessary and underlying soft clay layers may damp vibrations thereby hampering the densification process.

In order to evaluate the effectiveness of this technique and to decide a proper spacing, test sections of the order of 10 to 20 m may be utilised. As this technique needs no backfilling, the cost of densification by this method is moderate. Thus this technique is preferred for densifying offshore locations. Compared to vibroflotation (discussed in the next section) this technique needs a more closely spaced penetration to improve a given area uniformly as only a lesser lateral soil zone is improved by this technique. The terra probe technique is considerably faster than the vibroflotation procedure. Although both the methods give a fairly good result, the vibroflotation method will achieve somewhat higher relative densities (McCarthy, 1982).

5.2.3 Vibratory Compactors

These compactors are available as vibrating drums, vibrating pneumatic-tire, and vibrating plate equipment. They are operated with a frequency range between 1500 and 2500 cpm which is within the natural frequency range of most of the soils. Compacting at natural frequency gives the maximum effect because of high amplitude resulting in a more dense arrangement. Good results are achieved when the compactor travels at a slow speed of 3 to 6 km/hr. Smooth drum and rubber tired vibratory compactors have proven very effective in compacting cohesionless deposits of limited thickness, *e.g.*, 2 to 3 m or where cohesionless fills are placed. The vibratory plate compactors generally have a limited depth of effectiveness. Details of vibratory rollers are dealt in Chapter 2.

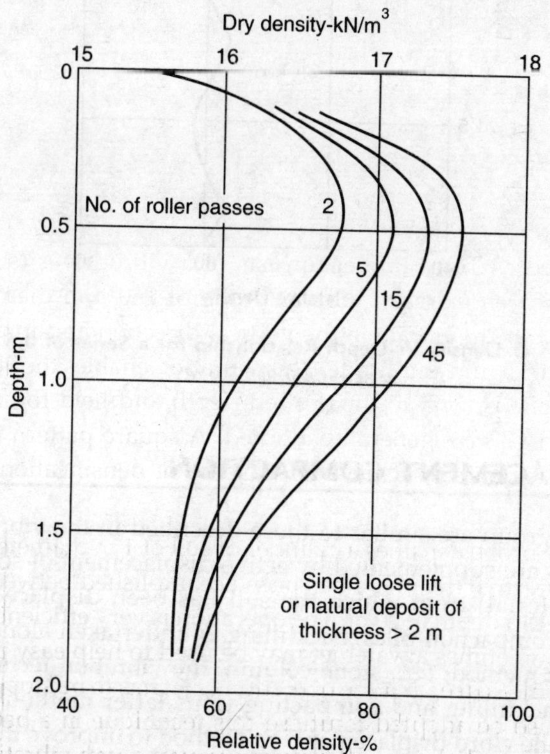

***FIGURE 5.3.* Density Vs Depth for Different Number of Roller Passes.**
(*Source: Mitchell, 1976*)

Heaviest vibratory rollers can densify above 2 m depth. Figure 5.3 shows field results from surface compaction tests (D' Appolonia *et al.*, 1968). Soil nearer to the ground surface will have low density because of over-vibration and lack of confinements. Figure 5.4 shows a density-depth densification for a fill placed in successive lifts (Mitchell, 1976).

In order to achieve best results it is essential that lift thickness, roller type and soil type should be matched. High lift thickness yields loose density layers with high density layer alternatively and low lift thicknesses cause waste of energy through repeated over-compaction of near surface layers. Relative densities as high as 85 to 90% and a lateral earth pressure coefficient of one or more can be achieved if the system is matched properly.

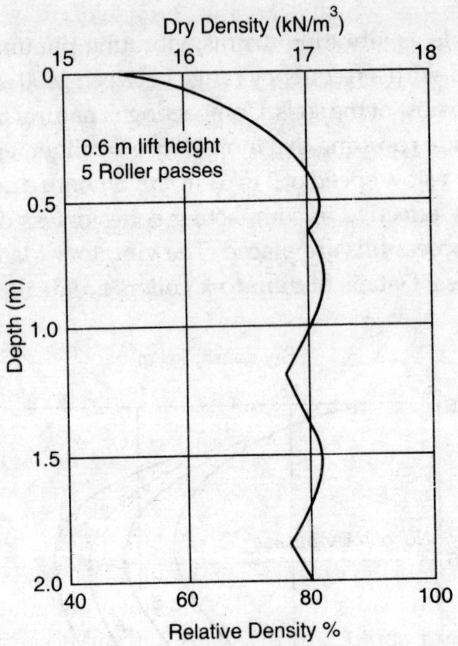

FIGURE 5.4. **Density Vs Depth Relationship for a Series of 0.6 m Lift.**
(*Source: Mitchell, 1976*)

5.3 VIBRO-DISPLACEMENT COMPACTION

These techniques in this group are similar to those described in the vibro-compaction technique except that the vibrations are supplemented by active displacement of soil and by backfilling the zones (by a suitable material) from which the soil has been displaced. In the methods, *viz.*, vibroflotation and sand compaction piles, backfilling is undertaken along with the displacement of soil grains. In another method, *viz.*, stone column, the vibration technique is used to make a borehole followed by backfilling and compacting. This latter method is referred to as vibro-replacement method. While vibro-displacement is a method to improve the density of cohesionless granular soils, vibro-replacement is used to improve cohesive soils with deep vibratory methods. Heavy tamping method (Dynamic Compaction) discussed in Chapter 2 also utilises basically the principle of vibration.

5.3.1 Displacement Piles

Natural ground and existing fill can not be compacted in layers. The simplest method among the vibro-displacement techniques is by driving piles which is very effective in densifying cohesionless sands up to reasonable depths. Such piles are referred to as displacement or compaction piles. When the piles are driven into loose sand, the ground surface between the piles commonly subsides in spite of the displacement of the sand by the piles. Hammer vibration coupled with the displacement force of the pile is ideal for densifying cohesionless soils, but the method is slow and expensive.

Compaction piles can also be used in sandy soils with some cohesion and in existing cohesive fills. Compaction in such soils due to pile driving is not caused by the vibrations associated with the driving, but by static pressure which decreases the size of the void spaces. If the soil is located above the water table and the voids are largely filled with air, the compacting effect of pile driving is commonly very satisfactory (Terzaghi and Peck, 1967). The effect of compaction of such soils below water table decreases with decreasing permeability of the soil. In clays displacement piles may cause heaving of the ground surface. The ratio of heave volume to pile volume is about 50% for clays and 30% for silty clays.

Displacement piles cause densification of sand for distances as large as eight diameters away from the piles. Because of lateral displacement of soil, the horizontal stress acting on the pile increases resulting in large coefficient of earth pressure at rest value of the stratum being compacted. Timber piles are extensively used for compaction of soils.

5.3.2 Vibroflotation

Vibroflotation is an efficient technique for densifying cohesionless soils, with simultaneous vibration and saturation. This technique was originated in Russia in mid thirties and was applied in 1939 in Germany for improvement of building foundation soils.

The vibroflotation equipment comprises of a vibroflot probe, accompanying power supply, water pump, crane and front-end loader (Fig. 5.5). Vibroflot probe consists of a cylindrical penetrator, about 400 mm in diameter and about 2 m in length with an eccentric weight inside the cylinder developing a horizontal centrifugal force of about 100 kN at 1800 rpm. A typical vibroflot is formed of two parts. The lower part is the horizontal vibrating unit which connects to the upper part, a follow-up pipe, the length of which can be varied depending on the compaction depth (Fig. 5.6). The devise provided for water flow from jets at top and bottom at the rate of 225 to 300 litres/min at a pressure of about 400 to 600 kPa. The water jets the vibroflot into the ground as it is lowered with the crane. The front-end loader is used to supply the backfill material as the in-situ soils are densified.

The probe is freely suspended from a crane. Each compaction sequence has four basic steps (as suggested by Brown (1976) and Vibroflotation Foundation Co. USA). They are (Fig. 5.7):

(*i*) Vibroflot is positioned over the spot to be compacted and its lower jet is then fully opened.

(*ii*) Water is pumped in faster than it can drain away into the subsoil. This creates a momentary "quick" condition beneath the jet which permits the vibroflot to settle of its own weight and vibration.

(*iii*) Water is switched from the lower to the top jets and the pressure is reduced enough to allow water to be returned to the surface, eliminating any arching of backfill material and facilitating the continuous feed of backfill.

(*iv*) Compaction takes place during the 0.3 m per minute lifts, which return the vibroflot to the surface. First, the vibrator is allowed to operate at the bottom of the crater. As the particles densify, they assume their most compact form. By raising the vibrator step by step and simultaneously backfilling with sand, the entire depth of the soil is compacted into a hard core.

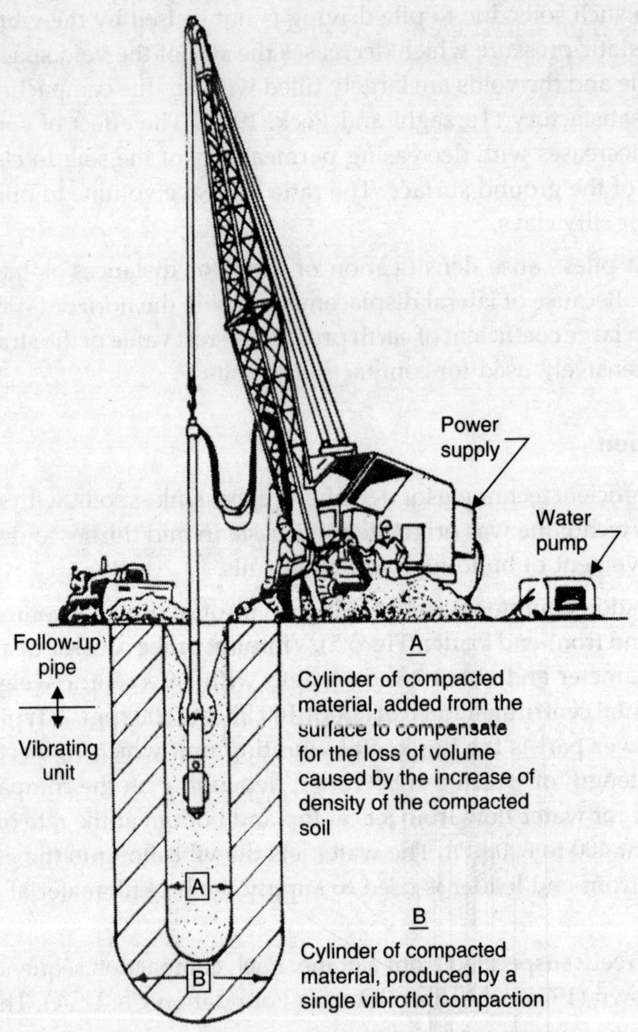

FIGURE 5.5. **Vibroflotation Equipment.**

(*Source: Brown, 1976*)

In partially saturated sands water jets at the top of the vibroflot can be opened to facilitate liquefaction and densification of the surrounding ground. Liquefaction occurs to a radial distance of 300 to 500 mm from the surface of the vibroflot. In clean and free draining soils, backfill consumption is at a rate of 0.5 to 1.5 m^3 per meter of compacted depth. On an average backfill sand added will be equal to 10% of the total volume compacted.

In one penetration of vibroflot a cylindrical column of 2.5 to 3 m in diameter can be compacted.

Densification results should be verified by continuous monitoring (such as cone penetration resistance or standard penetration resistance) or discrete monitoring (such as nuclear methods). A typical continuous monitoring results from a project (Davis, *et al.*, 1981) is shown in Fig. 5.8.

The successful application of this technique depends on the following factors : (*i*) equipment capacity, (*ii*) probe spacing and pattern, (*iii*) type of soil to be compacted, (*iv*) backfill material, (*v*) vibroflot withdrawal procedure, and (*vi*) workmanship. For a given equipment and an excellent workmanship the end result depends on the other factors (*ii*) to (*v*).

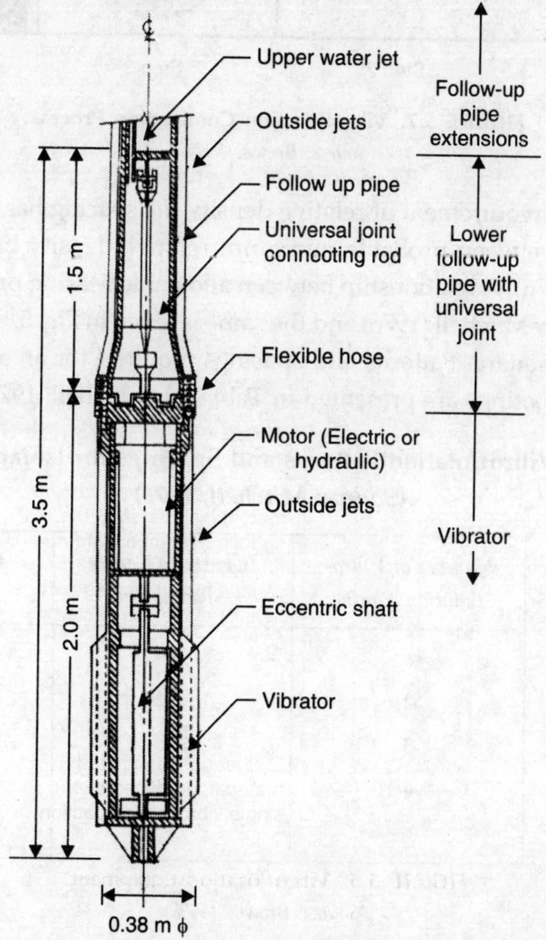

FIGURE 5.6. **100 HP Vibroflot.**

(*Source: Brown, 1976*)

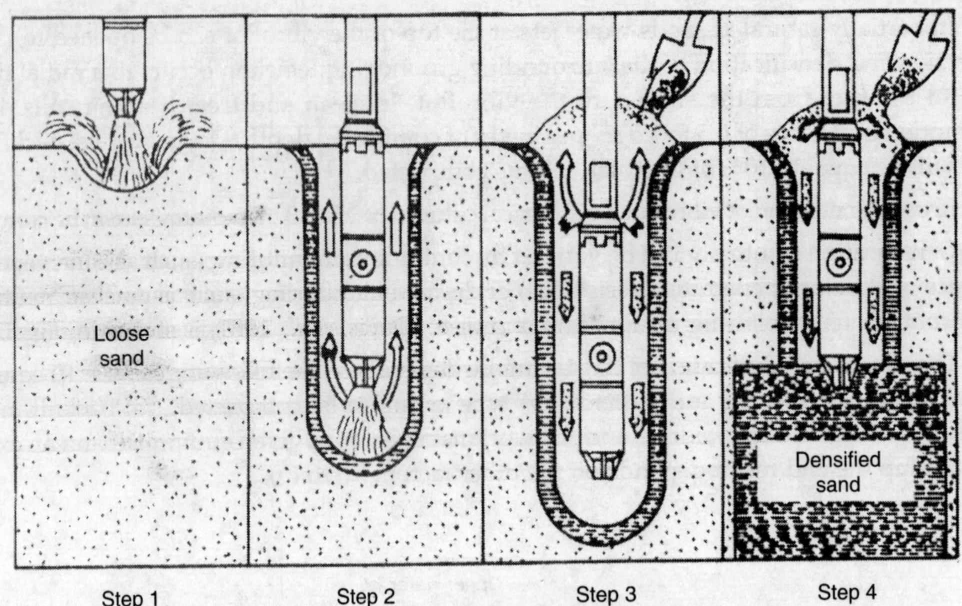

***FIGURE 5.7.* Vibroflotation Compaction Process.**
(Source: Brown, 1976)

Depending on the requirement of relative density the spacing has to be adopted. Figure 5.9 shows the relationship between probable minimum relative density between vibration centres and spacing (Mitchell, 1976). A relationship between allowable bearing pressure to limit settlement and spacing is reported by Mitchell (1976) and the same is given in Fig. 5.10. The pattern of spacings may be line, triangle or square. Patterns and spacings required for an allowable soil pressure of 300 kN/m^2 and square footings are presented in Table 5.1 (Mitchell, 1976).

Table 5.1. Vibroflotation Patterns and Spacings for Isolated Footings (*Source: Mitchell, 1976*)

Square Footing Size (m)	*Number of Vibro-flotation Points*	*Centre to Centre Spacing (m)*	*Pattern*
< 1.2	1	—	—
1.4 to 1.7	2	1.8	Line
1.8 to 2.1	3	2.3	Triangle
2.3 to 2.9	4	1.8	Square
3.0 to 3.5	5	2.3	Square + one at centre

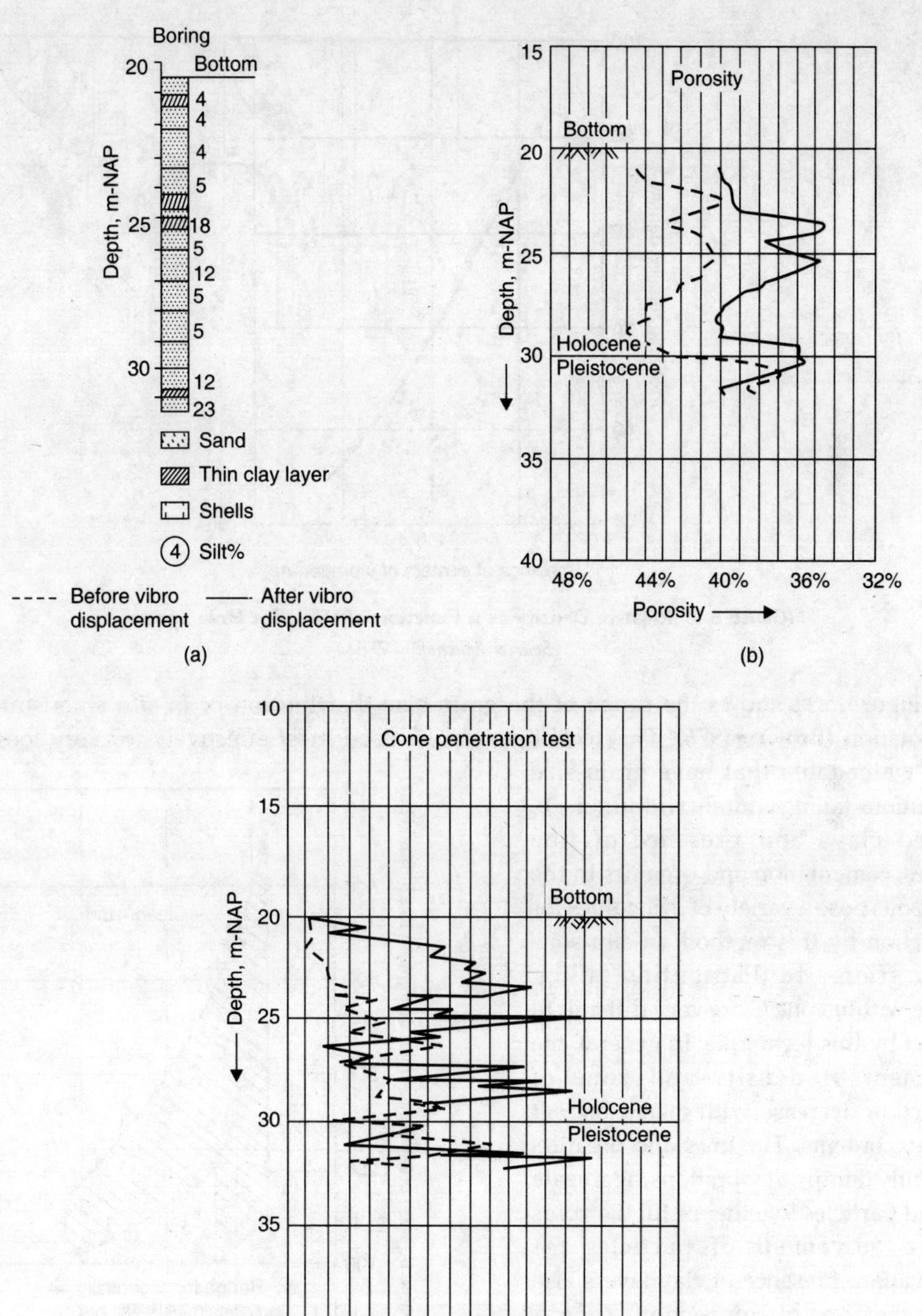

***FIGURE 5.8.* Porosity Decrease and Cone Penetration Resistance Increases Due to Vibro-displacement of Ocean Sand.**

(*Source: Davis et al., 1981*)

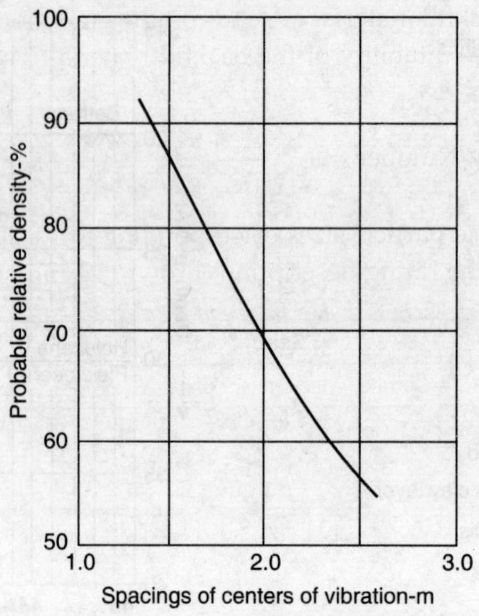

FIGURE 5.9. **Relative Density as a Function of Vibroflot Hole Spacings.**
(*Source: Mitchell, 1976*)

Figure 5.11 shows the range of the grain-size distribution of in-situ soils suitable for vibroflotation (Brown, 1976). This technique can be used most effectively for very loose sands below water table that have grain-size distributions falling entirely within zone B. Layered clays and presence of fine particles, cementation and organics in the in-situ soils pose a variety of difficulties for compaction by this method. In-situ soils having grain-size distribution falling entirely within zone C are very difficult to compact by this technique. In general, the attainment of density and zone of compaction decrease with increasing silt and clay contents. The fines and organics apparently damp out vibrations, aggregate the sand particles together or fill the voids relative movement of particles for densification. Presence of clay layers also reduce the zone of compaction. Zone A includes gravel, dense sand and cemented sands. The rate of probe penetration is less in these soils and the effect becomes still less when the water table is at a greater depth. Hence, under these conditions this technique may prove to be uneconomical.

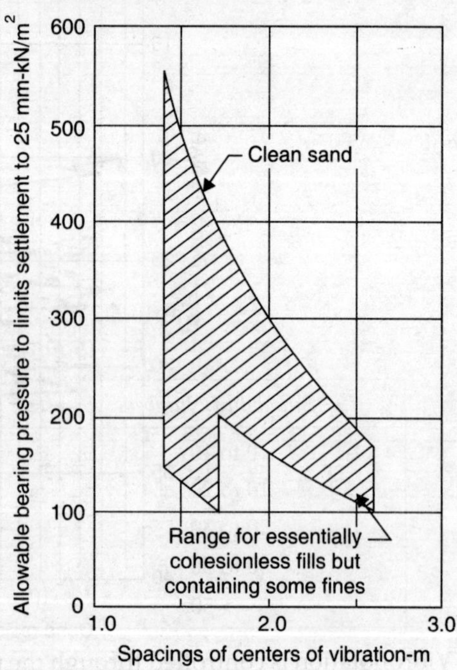

FIGURE 5.10. **Allowable Bearing Pressure on Cohesionless Soil Layers Stabilised by Vibroflotation.**
(*Source: Mitchell, 1976*)

The suitability of a backfill material depends on the gradation. Brown (1976) has developed a rating system to assess the suitability of the backfill material. The rating system is based on a suitability number defined as

$$\text{Suitability Number} = 1.7\sqrt{\frac{3}{(D_{50})^2} + \frac{1}{(D_{20})^2} + \frac{1}{(D_{10})^2}}$$

where D_{10}, D_{20} and D_{50} are the particle sizes corresponding to 10, 20 and 50% finer of the backfill material, Table 5.2 presents the rating description (Brown, 1976).

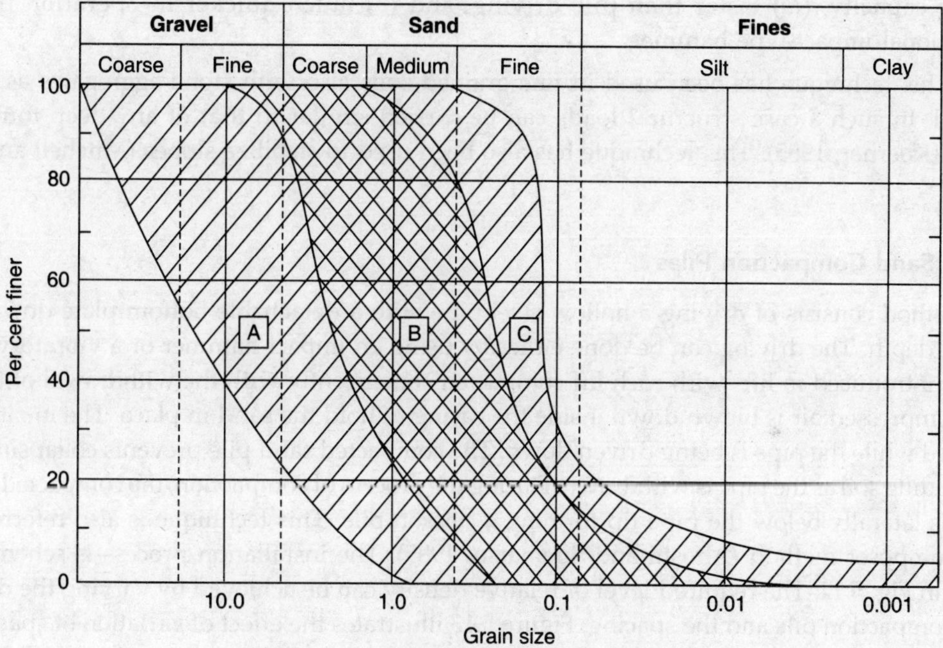

***FIGURE 5.11.* Soils Suitable for Vibroflotation.**
(*Soucre: Brown, 1976*)

Table 5.2. Backfill Evaluation Criteria
(*Source: Brown, 1976*)

Suitability Number	*Description of Rating*
0 to 10	Excellent
10 to 20	Good
20 to 30	Fair
30 to 50	Poor
> 50	Unsuitable

Vibroflotation is controlled through the power consumption of the vibrator within the probe and by the care with which withdrawal of vibroflot and backfilling the central portion of the hole. Withdrawal rates of 0.3 to 0.6 m/min are customary. This is of great concern since the density of the backfill zone must be assured.

Most vibroflotation applications have been adopted to a depth less than 20 m, although in a few cases depths up to 20 m have been densified successfully. The maximum is limited mainly due to the capacity of the crane to pull the vibroflot out of the ground. It has been reported in literature that in soils treated by vibroflotation, relative densities of at least 70% can usually be obtained at points midway between compactions. By this treatment bearing capacities of 250 to 400 kPa can be obtained (Bowels, 1988).

This technique claims its merit based on the following : (*i*) no material cost except the backfill material, (*ii*) complete uniformity in density and hence better control on settlement, (*iii*) gives high bearing capacity, (*iv*) faster than pile driving, and (*v*) much quicker in operation than the conventional impact-type hammer.

This technique has been used in fine grained soils also with stone aggregates as backfill material. In such a case structural loads can be carried similar to that of any deep foundation system (Koerner, 1985). This technique has also been used to stabilize slopes (Mitchell and Katti, 1981).

5.3.3 Sand Compaction Piles

This method consists of driving a hollow steel pipe with a detachable bottom plate down to the desired depth. The driving can be done either by using an impact hammer or a vibratory driver. Sand is introduced in lifts with each lift compacted concurrently with the withdrawal of the pipe pile. Compressed air is blown down inside the casing to hold the sand in place. The in-situ soil is densified while the pipe is being driven down. The compacted sand pile prevents collapsing of the surrounding soil as the pipe is withdrawn. During the process of compaction, the compacted column expands laterally below the pipe tip forming a caisson pile. This technique is also referred to as vibro-composer method (Aboshi and Suematsu, 1985). The installation process is schematically shown in Fig. 5.12. The required level of relative density can be achieved by varying the diameter of the compaction pile and the spacing. Figure 5.13 illustrates the effect of variation of spacing and the size of the compaction piles on relative density (Gupta, 1969).

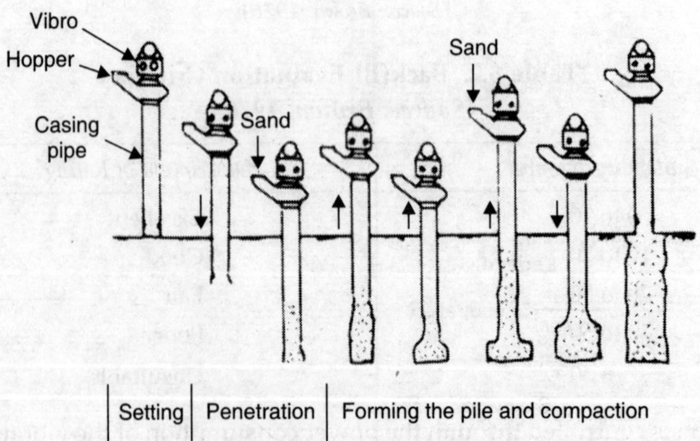

***FIGURE 5.12.* Vibro-compozer Method.**
(*Source: Aboshi and Suematsu, 1985*)

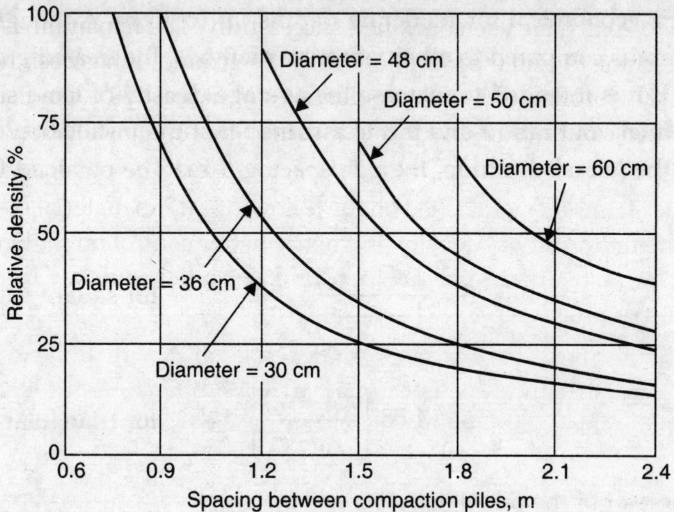

***FIGURE 5.13.* Effects of Size and Spacing of Sand Compaction Piles on Relative Density of In-situ Soil.**

(Source: Gupta, 1969)

The soil at shallow depth (less than 1 to 2 m) may have less density than that below due to lesser confinement. Further the density will decrease radially outward from the compaction piles. The increase in density may be computed based on the average surface settlement and the amount of backfill used. Typical densification of a loose sand surrounding a pile after driving and backfilling is shown in Fig. 5.14 (Meyerhof, 1960). The density increase is characterised in Fig. 5.14 in terms of the ratio of final standard penetration resistance to initial penetration resistance.

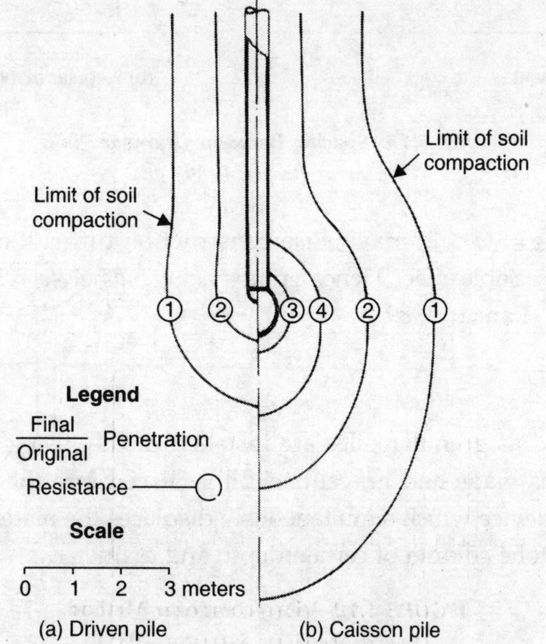

***FIGURE 5.14.* Compaction of Loose Sand Around Compaction Piles.**

(Source: Meyerhof, 1960)

This method is economical for moderate depths up to 15 m. Although this technique is costlier for deeper depths compared to other vibration methods, the treated ground generally has uniform properties. If it is intended to increase the average density of loose sand from an initial void ratio of e_0 to a final void ratio e and if it is assumed that the installation of sand pile causes compaction only in the lateral direction, the pile spacing S may be obtained from (Mitchell and Katti, 1981).

$$S = \left\{\frac{\pi(1+e_0)}{e_0 - e}\right\}^{1/2} d \quad \text{for square pattern}$$

and

$$S = 1.08\left\{\frac{\pi(1+e_0)}{e_0 - e}\right\}^{1/2} d \quad \text{for triangular pattern}$$

in which d is the diameter of the pile (Fig. 5.15 a and b).

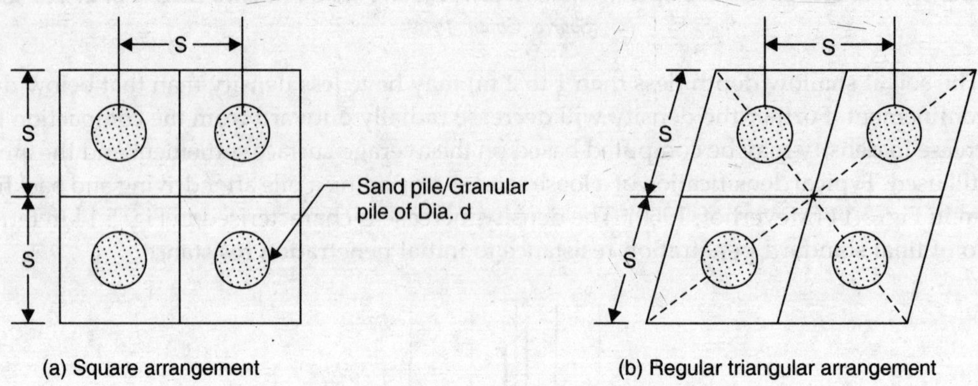

***FIGURE 5.15.* Spacing Between Granular Piles.**
(*Source: Ranjan, 1989*)

Closer pile spacing ($S/d \leq 2$) may cause construction difficulties where large spacings ($S/d > 4$) may have no appreciable effect. Hence a pile spacing S/d between 2.5 and 4 may be adopted with reasonable accuracy (Ranjan, 1989).

5.3.4 Stone Columns

Stone columns also called as granular piles are installed mostly using vibration techniques. A cylindrical vertical hole is made and gravel backfill is placed into the hole in increments and compacted by a suitable device which simultaneously displaces the material radially. This results in a densely compacted stone column of certain depth and diameter.

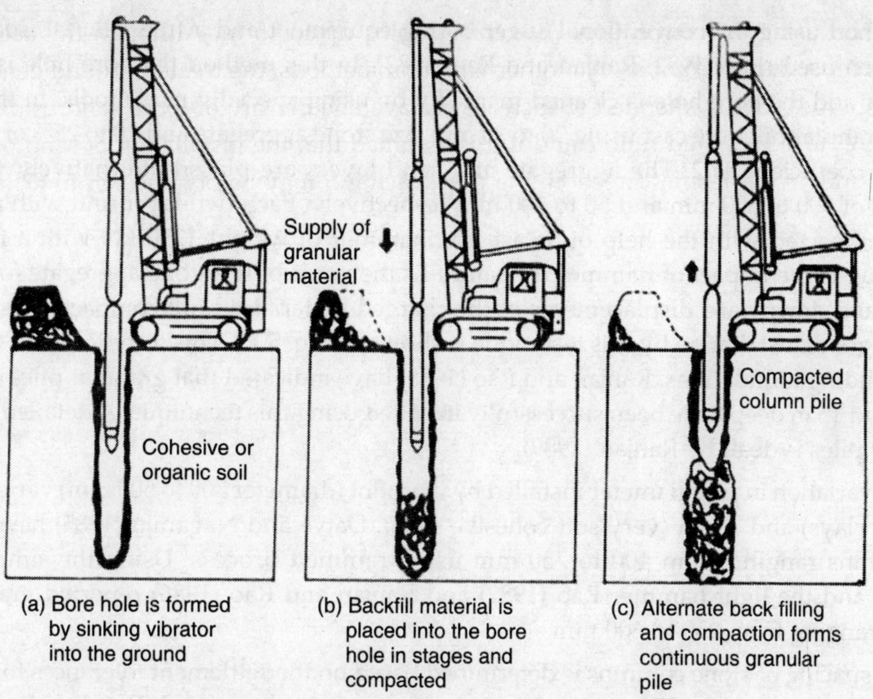

FIGURE 5.16. **The Vibro Replacement Process.**
(*Source: Baumann and Bauer, 1974*)

Formation of stone columns using a vibroflot is quite suitable techniques for improvement of cohesive soils (Engelhardt and Kirsch, 1977 ; Thornburn and McVicar, 1968). The vibroflot is allowed to sink into the ground due to its own weight, assisted by water or air as a flushing medium, up to the required depth. The soil surrounding the vibroflot is disturbed or remoulded and the softened material can be removed by jetting fluid (water or compressed air). Water is used as a jetting fluid for fully saturated soils while compressed air is used for partially saturated soils (Engelhardt and Kirsch, 1977). By this process a borehole of larger diameter than the vibroflot is formed once the vibroflot is withdrawn (Fig. 5.16). So formed borehole is backfilled with gravel of 12 to 75 mm size or furnace slag (Thornburn and McVicar, 1968). With the repenetration of the vibroflot the backfill material is displaced into the sides of the borehole and compacting underneath its tip. While backfilling the vibroflot is raised and lowered. This procedure is repeated till the hole is completely filled and compacted which forms a cylindrical granular pile.

Unlike in vibroflotation, wherein water jets are used, some dry processes are also in use to make the hole. A closed end pipe mandel is driven to a desired depth and the gravel is allowed to fill after opening trip valve. A rammer is used to pack the soil through the pipe as it is withdrawn and gravel added.

Apart from the use of vibroflot and hollow pipe, different simple techniques are being used for constructing stone columns. Datye and Nagaraju (1975, 1985) proposed a method which primarily uses a hammer weighing 15 to 20 kN falling through a height of 1 to 1.5 m for compacting stone aggregates placed in pre-bored holes. The resulting stone column is referred to as rammed stone column and the technique has been claimed to be economical than vibrator compaction. A simple method using the conventional auger boring equipment and a free fall cast iron hammer has also been used (Rao, 1982; Ranjan and Rao, 1983). In this method the bore hole is made by spiral auger and the bore hole is cleaned manually by using specially made tools. In the cleaned bore hole granular piles are cast using 20 to 30 mm size stone aggregate and 20 to 25% of sand with uniformity coefficient of 2. The aggregate and sand layers are placed alternatively with layer thicknesses of 300 to 500 mm and 50 to 100 mm respectively. Each two-layer unit with sand layer at top is compacted with the help of a cast iron hammer of weight 1250 kN with a free fall of 750 mm. Due to the impact of hammer the sand fills the voids of the stone aggregate followed by the lateral and downward displacements of the charged material till full compaction is achieved. Various stages of installation by this technique is shown in Fig. 5.17. This technique can be applied to small building foundations. Ranjan and Rao (1988) have indicated that granular piles of 600 mm diameter and 15 m deep have been successfully installed using this technique. A detailed treatment of granular piles is dealt by Ranjan (1989).

The variation in pile diameter installed by vibroflot (diameter 300 to 500 mm) varies between 0.6 m (stiff clays) and 1.1 m (very soft cohesive soils). Datye and Nagaraju (1985) have reported stone columns ranging from 400 to 750 mm using rammed process. Using the simple boring equipment and the light hammer Rao (1982) and Ranjan and Rao (1986) reported granular pile diameters ranging from 250 to 600 mm.

The spacing of stone columns is determined based on the settlement tolerances for the loads to be applied (Greenwood, 1970) and the degree of improvement required (Engelhardt and Kirsch, 1977). Stone columns are spaced from 1.2 to about 3 m on centre over the site (Bowles, 1988). Spacing ranges recommended for sand piles can be adopted for granular piles also.

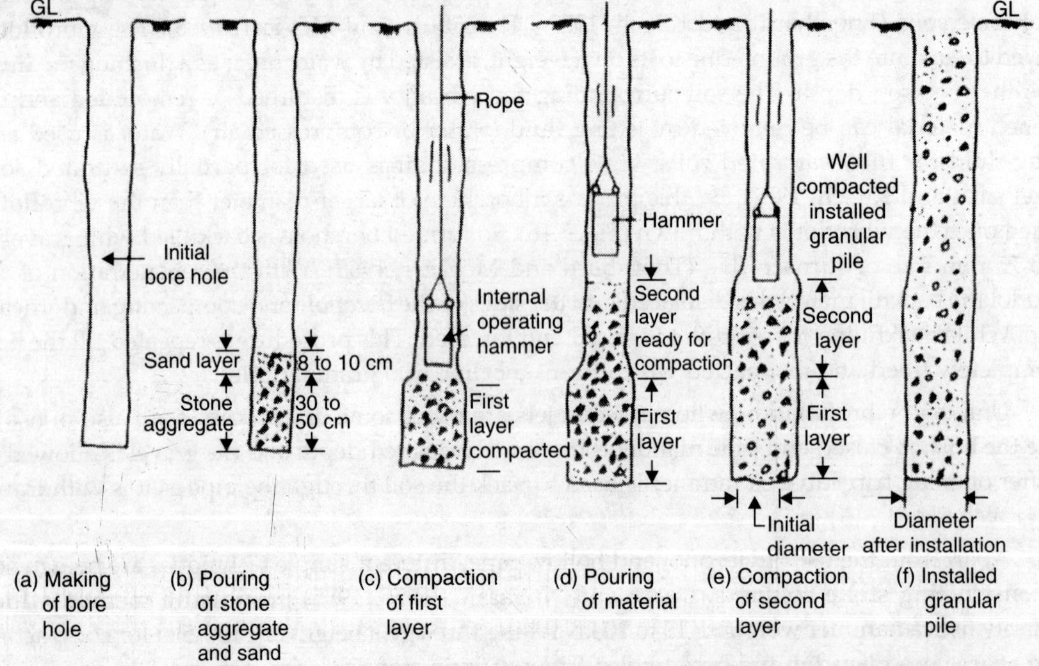

***FIGURE 5.17*. Granular Pile Installation Method Using Indigenous Know How.**

(*Source: Rao, 1982*)

The length of the stone columns is sufficient either to extend below the depth of significant stress increase caused by the foundation or should extend through the soft clay to firm strata to control settlements. The resistance is derived by stone columns only by the perimeter shear and not by end-bearing. On this basis the stone column length (L_c) should be greater than (Bowles, 1988)

$$L_c \geq \frac{P - A_c(9\, c_{pt})}{\pi\, dc}$$

where P = total load on the stone column

A_c = cross-sectional area of the stone column

d = average diameter of the stone column

c, c_{pt} = side and point cohesion.

There is no theoretical procedure available to accurately predict the combined improvement of the ground. Thus it is usual to assume the foundation loads are carried only by the total number of stone columns with no contribution from the intermediate ground (Bowles, 1988).

Stone columns may be arranged to support isolated footings, strip footings or mat foundations. The entire foundation area should be covered with a blanket of sand or gravel at least 0.3 m thick to help distribute loads and to facilitate drainage of water conducted out of the soft soil through the columns which act as vertical drains as well as reinforcing elements.

The load capacity of a stone column is controlled by the passive resistance of the soft soil that can be mobilised to withstand radial bulging and on the friction angle of the gravel. An approximate formula for the allowable bearing capacity of stone columns is given by

$$q_a = \frac{k_p}{SF}(4c + \sigma'_r)$$

where $k_p = \tan^2(45 + \phi'/2)$

ϕ' = drained angle of internal friction of stone

c = either drained cohesion (suggested for large areas) or the undrained shear strength, s_u.

σ'_r = effective radial stress as measured by a pressure meter (but may use $2c$ if pressure meter data is not available)

SF = safety factor-used about 1.5 to 2.

As bearing capacity of stone columns is generally high, settlement is the important criterion. For want of theoretical equations to predict settlement of stone columns, empirical methods are used. The settlement of stone column foundation depends on column spacing. The settlement of a single column of a group in a load test may be in the range of 5 to 10 m at the design load. The settlement of large group is usually about 5 to 10 times, the single column settlement (Mitchell, 1976). Figure 5.18 shows estimated settlement of the treated ground as a function of soil strength and column spacing (Greenwood, 1970).

Stone columns are very much suitable for soft, inorganic cohesive soils. They also can be used in loose, sand deposits to increase the density. Stone columns are capable of dissipating excess pore water pressure in the in-situ soil and thereby reducing the void ratio in the zone of influence.

As they are constructed on volume displacement basis, there is an intimate contact between soil and column. Stone columns can not be used effectively in thick deposits of peat or highly organic silts or clays.

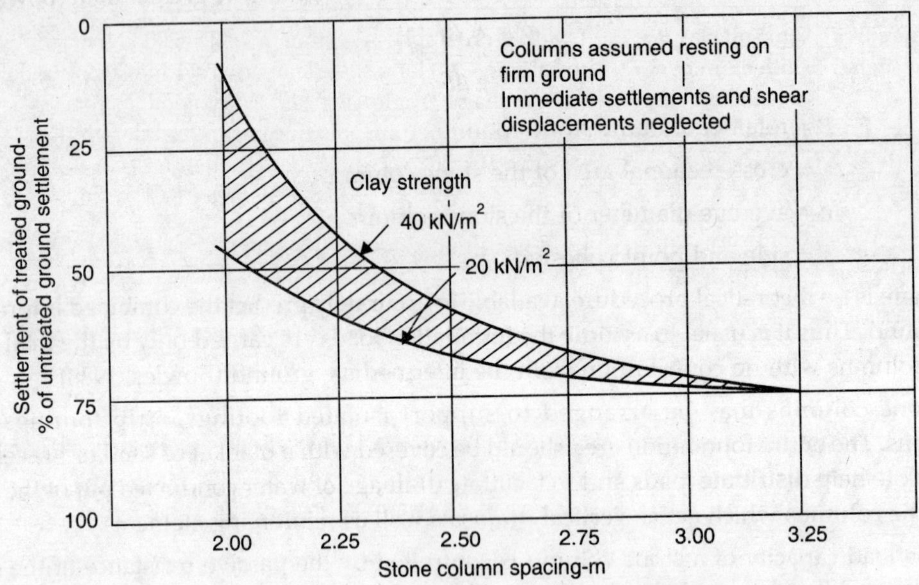

FIGURE 5.18. **Effect of Stone Columns on Anticipated Foundation Settlement.**
(*Source: Greenwood, 1970*)

Stone columns using vibroflot has been installed in India in a big way for strengthening soft soil for foundation of steel cylindrical storage tanks (more than 100) all around the coast of India for refineries and chemical plants from seventies (Varadarajulu and Gupta, 1993). The tanks sizes varied from 2.5 to 79 m in diameter and up to 15 m height. It has been reported that the subsoil conditions around coast starting from Haldia to Kandla through Madras on the east coast and Cochin on the west coast consist of deposits of thick soft clay/clayey soils of very high compressibility. Table 5.3 (Pilot, 1981) gives a few case histories of stone column applications. The results reveal the following (Pilot, 1981):

(*i*) Stone columns are generally employed in thick deposits of soil of very low shear strength.

(*ii*) The undrained shear strength has increased which is not taken into account at the design stage.

(*iii*) Settlements are reduced and stability of embankments and slopes are increased.

(*iv*) A good installation technique is important to ensure efficient functioning of the columns.

5.3.5 Heavy Tamping

This technique also called as Dynamic compaction or Dynamic consolidation has been dealt in Section 2.4.6. This technique basically utilizes the vibration and shock caused by the dropping of a heavy weight. Evidently the densification takes place by displacement of soil grains. As explained earlier this technique can be adopted for all soils.

Table 5.3. Case Histories of Stone Column Applications
(*Source: from Pilot, 1981*)

Building site	*Foundation soil*	*Column data*	*Results*
Canvey Island, England (Hughes *et al.*, 1975)	7 m soft clay under 2 m over consulated clay (s_u = 15 to 30 kPa)	Loading test applied to column 10 m long, 0.66 m diam. Angle of internal friction (assumed) = 30°	Ultimate load = 200 kN; maximum settlement = 0.17 m (0.01 m under 100 kN). Very good agreement between measurements and design data of Hughes *et. al.* on basis of column's actual diameter (0.73 m).
Freyming A 34, France (Bachy, 1976)	8 m peaty clay. (Menard pressure meter: P_1 = 270 kPa, E_m = 2000 kPa).	Columns under foundations embankment. 1 column per m^2	Settlement under loads less than 100 mm. After improvement process. P_1 = 20,000 kPa, E_m = 54,000 kPa.
Argenteuil RB 311, France. (Jardin, 1974)	3.5 m soft organic clay (s_u = 10 to 33 kPa).	Columns under embankment slope, 5.5 m long × 0.8 m diam. 1 column per 3.5 m^2	No settlement or stability problems encountered; after improvement process, s_u = 40 kPa.
Cagnes AB, France. (Cassan, 1974)	36 m soft silty clay (s_u = 10 to 70 kPa).	Columns under viaduct abutment and approach embankment 18 m long × 0.9 m diam. 1 column per 2.8 m^2	Average soil strength increase of 70% Trouble-free behaviour of abutment.
Konstanz, Germany. (Engelhardt *et al.*, 1974)	15 m soft and low consistency silty clay.	Test area with 12 columns, 12 m deep × 0.8 diam. 1 column per 1.1 m^2	Greatly decreased settlements: Under stress of 200 kPa, 55 mm prior to and 10 mm after improvement process.
East Brent, England. (MacKenna *et al.*, 1975)	12 m silty soft clay (s_u = 15 to 50 kPa) over 16 m silty sand.	Columns under embankment abutment, 11.3 m long × 0.8 m diam. 1 column per 5.7 m^2	No reduction in settlements, nor in pore water pressures (might be due to columns being clogged by mud).
Rouen. France. (Vautrain, 1977)	About 10 m soft soils: clay, peat and chalk (s_u = 20 to 50 kPa).	Columns under reinforced-earth embankment 10 m long × 0.9 m diam. 1 column per 3 m^2	Greatest settlement measured = 0.4 m (0.9 m without improvement treatment), greater than anticipated. Test apparently carried out to columns ultimate strength.

Chapter 6

GROUTING AND INJECTION

6.1 INTRODUCTION

Grouting is a process of ground improvement attained by injecting fluid like material into subsurface soil or rock. The technology of grouting is not new, yet it is constantly developing with innovation of new materials and construction techniques. The modern grouting was first started in mining works for arresting seepage and strengthening in civil engineering works. Grouting, in particular, is valuable in foundation works before construction (*e.g.*, to control water problem, to infill voids to control settlement, to increase soil bearing capacity, *etc.*), during construction (*e.g.*, to control groundwater flow to stabilise loose sand against liquefaction, to provide adequate lateral support, *etc.*), and after construction (*e.g.*, to reduce machine foundation vibrations, to eliminate new seepage, to apply in underpinning work, *etc.*). Grouting is usually limited to zones of relatively small volume and special problems. Grouting is adopted both for temporary and permanent works and the following applications have been in use (Harris, 1983):

(*i*) Sealing pockets and leaks of permeable or unstable soil or rock prior to excavation of a tunnel heading or alternatively grouting a stratum from ground level.

(*ii*) Sealing the base of structures (such as cofferdams or caissons) founded on pervious ground.

(*iii*) Fixing ground anchors for sheet pile walls, concrete pile walls, retaining walls, stabilising rock cuttings, tunnels, *etc*.

(*iv*) Repairing a ground underneath a formation or cracks and structural defects on building masonry or pavement and sunken slabs or damp proof course.

(*v*) Filling the voids between the lining and rock face in tunnel works.

(*vi*) Forming grout curtains in layers of permeable strata below a dam.

(*vii*) Fixing the tendons in prestressed post tensioned concrete.

(*viii*) Sealing the gap between the surface of a concrete foundation and the base plate of a stanchion.

(*ix*) Producing mass concrete structures and piles.

All types of grouts are used, including cement, cement and sand, clay-cement, slag-cement, resin gypsum-cement, clays, asphalt, pulverised fuelash (PFA) and a large number of colloidal and low viscosity chemicals. However, in most of the cases cement and water is the most widely used group because of its relatively low cost. Although cement has been and is widely used in grouting, it has the disadvantage that it can not penetrate soils of low permeability. In such situations injection of chemicals have been in use. Recent development is compaction grouting which is relying not on infilling, but on densification of suitable soils by displacement.

In grouting different types of grouts are used, different ground conditions have to be satisfied and different functions have to be served, no generalization about grouting equipment and methods can be made.

In sum, it may be stated that grouting today is an invaluable tool in civil engineering practice and whose importance is constantly increasing.

6.2 ASPECTS OF GROUTING

The principle of grouting is to introduce a substance into rock fissures or into a soil by pumping fluid (called a grout) down a small diameter tube in the required location. It is essential that a particular grout should penetrate satisfactorily into the permeable materials or rock fissures and seal all voids. Before implementing any grouting technique in the field, the basic function it is intended to do should be realised. The three basic functions involved in soil and rock grouting (Fig. 6.1) are the following (Koerer, 1985):

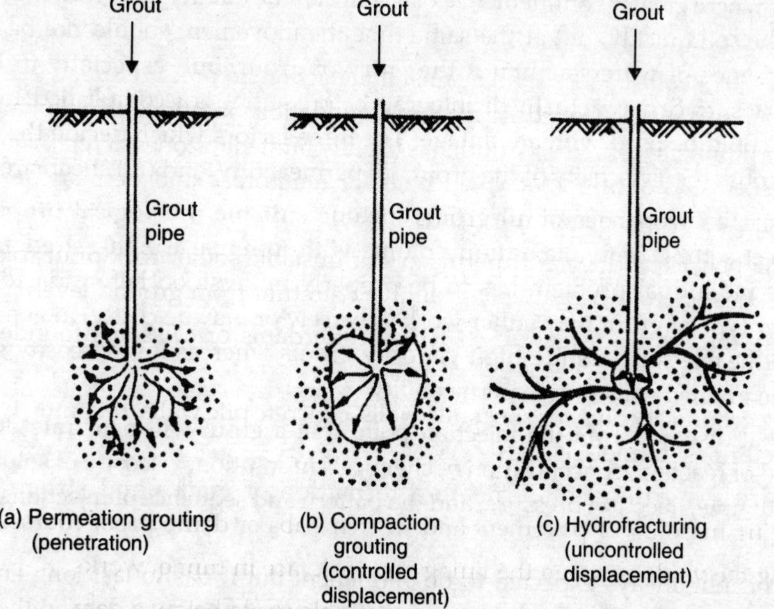

***FIGURE 6.1.* Various Functions Involved in Soil and Rock Grouting.**

(*Source: Koerner, 1985*)

(*i*) **Permeation or penetration.** In this situation the grout flows freely with minimal effect into the soil voids or rock seams.

(*ii*) **Compaction or controlled displacement.** In this case the grout remains more or less intact as a mass and exerts pressure on the soil or rock.

(*iii*) **Hydraulic fracturing or uncontrolled displacement.** In this condition the grout rapidly penetrates into a fractured zone which is created when the grouting pressure is greater than the tensile strength of the soil or rock being grouted.

6.2.1 Groutability

In order to obtain a satisfactory performance of a grout, its grain-size distribution should be known because it shall show the relationship between the suspended particles and the void dimensions. Should the latter be larger than the particles, a plug may form in air opening and ultimately block it. Grouts with coarser particles in them, on the other hand, may also tend to build up a block leading to a low rate of flow. In such cases, treatment of drainage filters may be adopted. In order to prevent seggregation of agglomeration and to increase stability fine grained particles like bentonite may be added.

Some preliminary work is needed on the location before the actual injection of grout is taken up. Initial washing of the surface openings is needed to make them clean and clay-free such that grout may adhere. In fractured rock, prior injection of dilute silicate can make subsequent grout penetration easier.

Best results are obtained when the pumping pressure is not allowed to grow large enough for the particles of soil to be disturbed and moved. To prevent blowout during grouting, the grouting pressure is generally limited to about 20 kN/m^2 unless grouting under heavy structures or in other situations where greater confinement exists. Further the quality of a grout must be sufficiently fluid to enter the soil quickly, but at the same time the movement should not be too fast. In dry ground, the absence of water facilitates the entry of grout, but, especially in loose soils, the penetrating grout surface may actually disintegrate before setting is accomplished. In such situations viscous grouts could be used with advantage. The three factors which decide the effect of rate of injection of a grout are (*i*) viscosity of the grout, (*ii*) permeability, and (*iii*) shear strength of the soil.

The desirable properties of all grouts include suitable rheological properties with low viscosity, correct setting time, maximum volume with minimal weight, strength, stability and durability. The individual problem has to be properly understood before deciding the type of grout. Improper type of grout may either work adversely or may be costly. In general coarser and moderately permeable soils require high viscosity grouts whereas low viscosity grouts are quite suitable for fine grained soils of low permeability.

The radial distance from the injection point that a grout will penetrate depends on such factors as speed of reaction, concentration of chemicals in solution, viscosity of solution and change in viscosity with time, injection pressure, and the pattern and sequence of injection points (Mitchell, 1976).

A grout is not always expected to be permanent but it has to last long enough to fulfil a necessary operational function. The relevant life of a grout is based entirely on the requirement of the individual job. For example, a grout curtain under a dam is expected to be long last for so many decades whereas grout has to last only for a few weeks, if it is used to temporarily arrest a quicksand condition during execusion.

A life of a grout depends on the geological and hydrogeological situations. For example, a water-soluble chemical grout may not be suitable in a location where there is movement of water. In order to ascertain the durability level attained by a grout a short or long time check is essential. For long-term jobs new boreholes may be drilled in the grouted zone for observational purposes. In alluvial deposits grouting efficiency is checked using boreholes and water tests. Fissures in rocks which have been grouted can be checked by driving a gallery into it but care has to be taken that no new fissures are formed due to stress release because of the creation of the gallery.

6.2.2 Grouting Materials

Grouting materials may be grouped under two basic types (Bowen, 1981):

(*i*) **Suspension grouts.** These are multi-phase systems capable of forming subsystems after being subjected to natural sieving processes, with chemical properties which must be carefully scrutinised so as to ensure that they do not militate against controlled properties of setting and strength. Water in association with cement, lime, soil, *etc.*, constitute suspensions. Emulsion (asphalt or bitumen) with water is a two-phase system which is also included under suspension. Suspension grouts are also referred to as particulate grouts (Mitchell, 1976).

(*ii*) **Solution grouts.** These are intimate one-phase system retaining an originally designed chemical balance until completion of the relevant reactions. Solutions in which the solute is present in the colloidal state are known as colloidal solutions. Chemical grouts fall into this category.

6.2.3 Suspension Grouts

When grouts (particularly suspensions) are injected into the soil formation the relationship between the grout particle size and the soil void size should be considered. A groutability ratio (*GR*) has been defined as a rough guide (Kravetz, 1958)

$$GR = \frac{D_{15}\text{ (formation)}}{D_{85}\text{ (grout)}} > 20$$

D_{15} = particle size at which 15% of the soil is finer (of the formation being grouted).

D_{85} = particle size at which 85% of the soil is finer (of the grout being injected).

This criterion decides the limit for suspension grouting.

Grouting with Soil. Soil itself can be used to fill up some of the voids in coarse-grained soils. Even fine sands and silts may be used for this purpose which would settle out quite quickly after injection. The soil to be used as a grout should be a very fine-grained soil. Clay is a complex compound with particles < 0.002 mm and is thus suitable for injection into medium coarse sands and other soils with permeability of the order of 1 to 10^{-1} mm/s. Clay grout behaves like a Bingham fluid and gels when undisturbed. This type of grout will exhibit low shear strength and hence can be used to reduce permeability. Kaolinite or illite based clays produce low viscosities and are preferred as filler grouts. Bentonite clay is a commonly used material whose structure is such that water is readily adsorbed on its surface and enables reasonably to control the viscosity, strength and flow properties.

No flow of soil-grout occurs when the water-to-soil ratio is kept very low. Pressure is then exerted by the grout against the soil mass from the grouting pipe causing densification and movement of adjacent areas. This technique has been known earlier as mudjacking and quite often used to raise pavement slabs or to underpin shallow building foundations. Nowadays, it is also used to strengthen in-situ soil by forming compaction piles.

Grouting with Cement Mixes. A satisfactory cement based grout depends on the water-cement ratio, the rate of bleeding and subsequent ultimate strength of the grout. Although water may be giving mobility for a grout, if in excess may cause separation of ingredients of the grout leading to high bleeding. Low strength water-cement ratio with a proper proportion of admixtures or fillers allows a maximum physico-chemical reaction, giving rise to good ultimate strength.

Cement grouts are usually formed from Ordinary Portland Cement (OPC) and water. The water-cement ratio may be varied from about 0.5 : 1 to 5 : 1 depending upon the ground conditions and required strength. The OPC grouts are suitable only in fissured rocks, gravels and coarse sands. Rapid hardening cement which is finer than OPC has a quick setting time and high early strength and therefore may be preferred to OPC in ground with high flowing water. High alumina cement also has rapid strength gain and resists attack by sulphates and dilute acids. Super-sulphated cement is very finer and therefore suitable for penetrating finely fissured rocks. When selecting a particular cement grout for use, one would obviously like to know its final strength, flow rate, set time, shrinkage, permeability, and durability. Cement grouting has been widely used, more often in seepage cut-off beneath dam, but also in ground water control in certain cases.

Other ingredients sometimes used in cement mixes are clay, fine sand, fly ash, fluidizers, accelerators, and retarders or expansion additives. Clay is considered as a filler in cement grouts when used in amounts less than 3% by weight. It can also be used to control bleeding by holding the cement in suspension while it is setting. Sand is a true filler in cement grouts and is used to reduce the overall cost of the grout material. Pozzolans such as fly ash or ground slag are also sometimes used as fillers. Additives or admixtures are meant to impart controlled features to the final composition of the grout.

Soil in combination with a stabilising material, *e.g.*, cement would do better than soil alone. Grouts may have different properties depending on the amount and type of soil (say clay, silt or sand), cement, and water they contain. Further the viscosity of grout depends on solid-to-water ratios and different cement-to-soil ratios. In soil-cement systems, volumes of soil between four and six times the loose volume of cement are common. The volume of mixing water varies from about three-fourth to twice the volume of clay per bag of cement in cement-clay grouts, and from about one-third to one time the loose volume of sand per bag of cement in cement-sand grouts. Water-cement ratios in the range of 0.5 : 1 to 5 : 1 have been in use. The lower this ratio, the less likely will be cement segregation and filtering, but the more difficult will be injection and the greater will be friction losses in the pumping system. Bentonite clay-cement mixes have an interesting synergistic properties and has been widely used as a permeation grout. Low-water-content soil-cement mixtures, like soil alone, can serve as displacement grouting. The advantage of these soil cement mixes over soil alone is permanence of the grout and their disadvantage is increased cost. The other factors, such as equipment, pressure, and pumping rates, however, are roughly the same for that of soil alone.

Water-insoluble and chemically active materials, natural possolans comprise volcanic glass, opal, pumicite, clay minerals, zeolites and hydrated oxides of aluminium. The most useful from the grouting point of view is pulverised fuel ash or fly ash which is often employed in conjuction with cement. PFA can retard alkali aggregate reactions and reduce the heat generation. The chemical interaction produced by PFA with cement is much more effective than that between sand and cement. Addition of fly ash markedly reduces the consistency at a constant water-cement-fly ash ratio. Fly ash with OPC provides lower viscosity than special cements. It has been reported (Somanathan, 1968) that about 20 to 30% of fly ash reduces gelation time, afflux time, bleeding and strength. Certain additives are used with the cement in order to vary its characteristics depending on the individual job. Some are water-soluble, a few are colloidal and some produce gas. For example, calcium chloride is used as an accelerator, carbohydrates can be employed to reduce the water-cement ratio, expansion may be achieved using embeco, *etc.* Common additives to cement grout used to impart specific properties to the final products are given in Table 6.1. (Little John, 1982).

Table 6.1. Common Additives to Cement Grout Used to Impart Specific Properties to the Final Product
(*Source: Little John, 1982*)

Admixture	*Chemical*	*Optimum dosage, % cement weight*	*Remark*
Accelerator	Calcium chloride	1 to 2	Accelerates set and hardening
	Sodium silicate	0.5 to 3	Accelerates set
	Sodium aluminate	0.5 to 3	Accelerates set
Retarder	Calcium lignosulphonate	0.2 to 0.5	Also increases fluidity
	Tartaric acid	0.1 to 0.5	
	Sugar	0.1 to 0.5	
Fluidiser	Calcium lignosulphonate	0.2 to 0.3	
	Detergent	0.05	Entrains air
Air entrainer	Vinsol resin	0.1 to 0.2	Up to 10% of air entrained
Expander	Aluminium power	0.005 to 0.02	Up to 15% preset expansion
	Saturated brine	30 to 60	Up to 1% postset expansion
Antibleed	Cellulose ether	0.2 to 03 (for $w < 0.7$)	Equivalent to 0.5% of mixing water
	Aluminium sulphate	Up to 20% (for $w < 5$)	Entrains air

Grouting with Asphalt Emulsions. Residues of oil refining industry in an emulsifying plant is processed to produce nearly circular droplets of asphalt suspended in a water matrix. By choosing the proper emulsifying agent one can obtain negatively charged anionic asphalt globules (for maximum flow and penetration) or positively charged cationic asphalt globules (to attach to negatively charged clay soils). Further, both anionic and cationic types of rapid, medium or slow setting globules can be manufactured. These globular spheres of 1 to 2 μm in diameter, along with water can be used as a grout to fill soil voids or rock fissures. Rock fissures of 10 μm size and middle silts have been grouted using these globules. The choice of type of emulsion setting needed is based on the flow of groundwater in the voids. Slow-setting emulsions are generally recommended since they can travel the longest distance into the material being grouted.

6.2.4 Solution Grouts

There are numerous solution grouts are available. Solution grouts can generally permeate finer soils than can suspension grouts. Many of the solution grouts are termed as chemical grouts. Such chemical grouting is done using "one-shot" system or "two-shot" system. In the "one-shot" system where all chemicals are injected together after pre-mixing. Setting times are controlled by varying the catalyst concentration according to the grout concentration, water composition, and temperature (Fig. 6.2). In "two-shot" system (called Joasten process) wherein one chemical is injected followed by injection of a second chemical which reacts with the first to produce a gel which subsequently hardens. Two-shot systems are slower and require higher injection pressure and more closely spaced grout holes. Although chemical grouting costs high, it has several advantages, *viz.*, (*i*) absence of particulate material, (*ii*) low viscosity, and (*iii*) control over setting time.

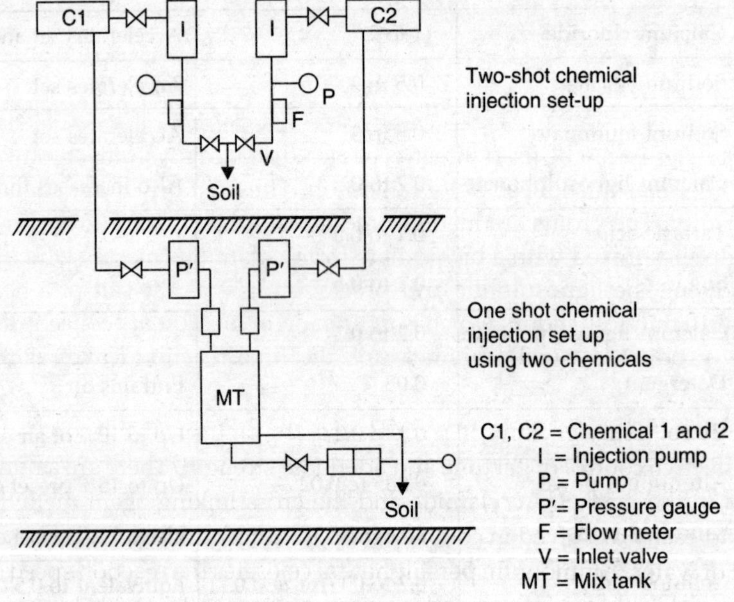

FIGURE 6.2. **Two-shot and One-shot Methods of Chemical Grouting.**

The following detailed classification on solution grouts have been given by Tallard and Caron (1977):

(*a*) Aqueous solutions—Silicate derivatives, Other mineral gels, Lignosulphite derivatives, Other plant derivatives, Arcylamides, Phenoplast Resins, Aminoplasts, combination of above.

(*b*) Colloidal solutions—Organic solutions, Mineral solutions.

(*c*) Nonaqueous solutions—Synthetic resins, Vulcanizable oils, Bitumen and other materials, Solvent systems.

(*d*) Emulsions—Bituminous, Others.

(*e*) Products reacting with the ground—Reaction with ground or groundwater salts, Reaction with groundwater.

(*f*) Combined systems.

Some of the more commonly used solution grouts are discussed below:

Silicate Derivatives. Sodium silicate, also called as water glass, is a commonly available and relatively an inexpensive aqueous solution. Here the sodium silicate is allowed to react with carbonic acid, by the two-shot system, to form salt (sodium carbonate, a solid) and acid (silicic acid, a liquid) which further breaks down to water and silicon dioxide (a solid precipitate). Thus the two original liquids react to form two solids which remain in the soil voids or rock cracks as the grout. The above reaction is slow which can be accelerated by adding calcium chloride. Instead of the two-shot system, a delayed and controlled reaction has also been developed using an organic reagent, usually ethyl acetate or formamide, which slowly breaks down the acid and produces a colloidal silica gel. The gelation period can be varied allowing greater coverage of the area by the grout. The common silicate based grout adopted nowadays contains sodium silicate, a reactant, an acceleration and a water carrier in the proportions roughly of 40, 10, 10 and 40 respectively (Koerner, 1985).

Lignosulphite Derivatives. This grout can be used similar to silicates and possessing a viscosity increasing with time. It is basically made from lignin, a residual product from wood industry. The raw lignin can be used directly, or as a dried, catalyst-mixed powder to which other additives are sometimes introduced before injection. These additives are added for control of gel time and uniformity. The grouts are made from lignosulphites and a hexa chromium compound. Soluble lignosulphites mixed with a bichromate form gelatinous masses which are quite firm, the bichromate oxidising the lignosulphite and precipitating it as the salt of a heavy metal. Setting time varies from 10 min to 10 hours which is a function of bichromate concentration. Although use of the waste by product is a welcome measure, the final reaction leaves as residual amount of toxicity.

Acrylamides. The acrylamide grouts are characterised by constant viscosity, good penetrability, effective control of gel time and adequate strength. There are a number of acrylamide grout pattented in the market. Acrylamide and the cross linking agent methylene-bisacrylamide can be mixed and easily dissolved in cold water up to concentration of 20% viscosities of the mixes are close to that of water. Ammonium persulphate accelerates the reaction speed so that, if required, a gel can be formed within a few seconds. Concentration of 0.5 to 1% by weight of ammonium persulphate provide gel times from 6 sec to 20 min according to water temperatures and ground temperatures. The gel times are affected by other factors such as variations in the p_{11} of the soil which could be controlled by using appropriate buffers or regulators. Acrylamide are mostly used in controlling the flow of water and rarely used for stabilisation of rock masses or soils. Acrylamide grouts should not be used near potable water sources.

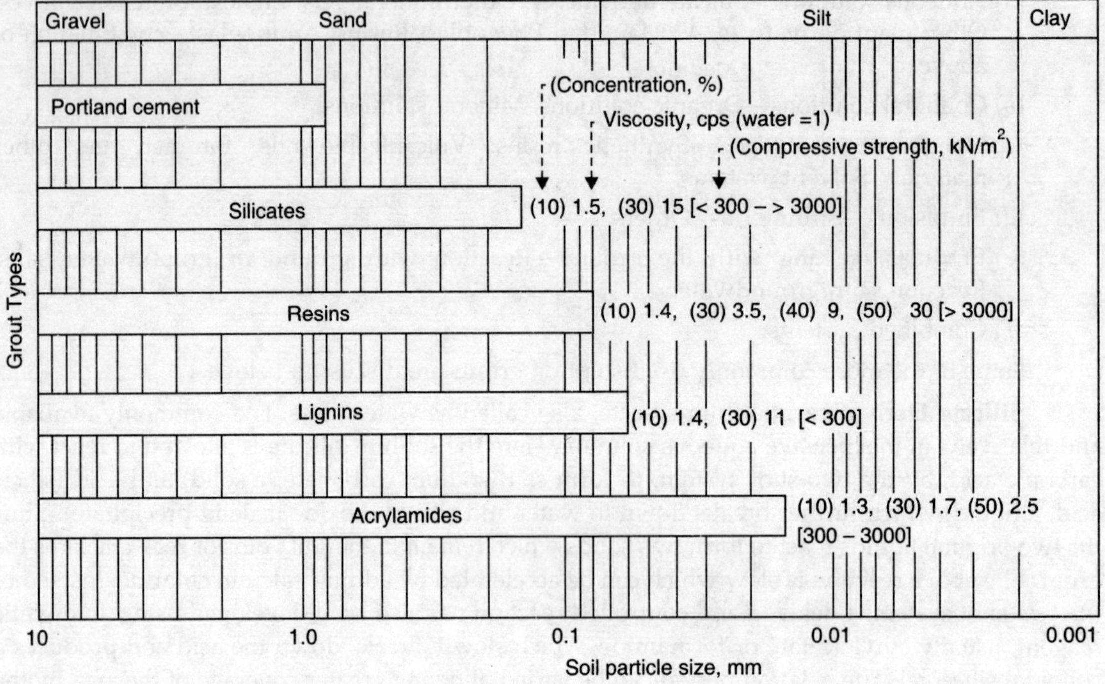

***FIGURE 6.3.* Soil Particle Sizes Suitable for Different Grout Types.**

(Source: Mitchell, 1976)

Phenoplast Resins. The resorcinol-formaldehyde combination is a resin results from the polymerisation of resorcinol (meta-dihydroxybenzene) with formaldehyde inaqueous solution when the *pH* is changed. This gives higher strength than acrylamide. Resorcinol is a phenol derivative so that this grout might be considered as a phenoplast.

Table 6.2. Relative Ranking of Solution Grouts as to their Toxicity, Viscosity, and Strength

(*Source: Karol, 1982*)

Grouts	*Corrosivity or toxicity*	*Viscosity*	*Strength*
Silicates			
Joosten process	Low	High	High
Siroc	Medium	Medium	Medium high
Silicate-bicarbonate	Low	Medium	Low
Lignosulfates			
Terra Firma	High	Medium	Low
Blox-all	High	Medium	Low

(Table Contd...)

Table 6.2. *(Contd...)*

Phenoplasts			
Terranier	Medium	Medium	Low
Geoscal	Medium	Medium	Low
Aminoplasts			
Herculox	Medium	Medium	High
Cyanaloc	Medium	Medium	High
Acrylamides			
AV-100	High	Low	Low
Rocagel BT	High	Low	Low
Nitto-SS	High	Low	Low
Polyacrylamide			
Injectite 80	Low	High	Low
Acrylate			
AC-400	Low	Low	Low
Polyurethane			
CR—250	High	High	High
CR—260			

A comparison between the relative degrees of toxicity, viscosity and strength of various solution grouts has been made by Karol (1982) which is presented in Table 6.2. Particle size ranges over which each of these grout types are effective is presented in Fig. 6.3 (Mitchell, 1976).

6.2.5 Compaction Grouting

The basic concept of compaction grouting technique is that of injecting an expanding bulb of high viscous grout with high internal friction into a compressible soil or into a soil mass containing large voids. The injected grout acts as a radial hydraulic jack which compresses the surrounding soil and thus achieving controlled densification. This is also called a displacement grouting. It is essential to distinguish between conventional penetration grouting and the compaction grouting. Conventional penetration grouting basically involves in filling openings in soils and rocks by a fluid grout, to reduce permeability or to increase the strength. Compaction grouting does not depend upon grout entering openings but involves displacement and compaction of soils as a result of the intrusion of a mass of thick grout. The difference between these two techniques is shown in Fig. 6.4. Compaction grouting is applicable to partially saturated cohesive or organic soil masses, silts, sands and soils containing void pockets. It is also used to stabilise soil under residences and light commercial buildings and sometimes foundations of large structures. The following advantages and disadvantages have been focussed by Shroff and Shah (1992) based on the studies by Graf (1969) and Warner and Brown (1974):

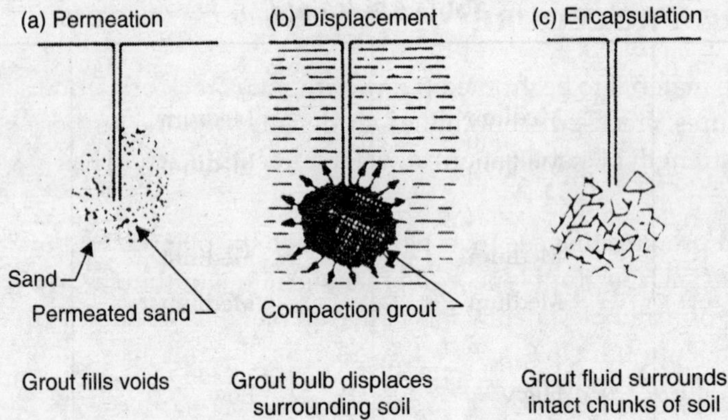

FIGURE 6.4. **Types of Grouting.**
(*Source: Graf, 1969; Mitchell, 1970*)

Advantages:

(*i*) Minimum disturbance to the structure and surrounding ground during repair.
(*ii*) Minimum risk during construction.
(*iii*) Greater economy.
(*iv*) Supports all portions of structures.
(*v*) Reduced need for extensive exploration.
(*vi*) Greater flexibility.
(*vii*) Groundwater not affected.

Disadvantages:

(*i*) Relative ineffectiveness in stabilising near-surface soils.
(*ii*) Prohibitive cost for some structures if the soil to be treated is excessively deep.
(*iii*) Grouting adjacent to unsupported slopes may be ineffective.
(*iv*) Difficulty of analysing results.
(*v*) Not suitable in decomposable materials.
(*vi*) Effectiveness questionable in saturated clays.
(*vii*) Danger of filling underground pipes with grout.

Boulanger and Hayden (1995) have given seven well-defined case histories on the compaction grouting treatment of liquefiable soils. From their analysis the following important lessons are derived:

(*i*) Construction procedures significantly affect the improvement achieved.
(*ii*) Significant increases in penetration resistance could be achieved in soils ranging from slightly silty sand to silt and the effect in silt is particularly promising.
(*iii*) Grout takes required to produce a specified increase in penetration resistance (near treatment and grid centre) will be greater than expected based on simple design calculations. The apparent discrepancy is attributed to the concentration of volumetric strain near the injected grout columns.
(*iv*) Significant economy may be achieved by properly quantifying the distribution of volumetric strain around a grout mass.

6.3 GROUTING PROCEDURE

Depending upon the material to be grouted (*i.e.*, whether it is rock or natural soil or fill) its quality (*i.e.*, in terms of fissures, cracks, discontinuities, or density) and the purpose of grouting (*i.e.*, for seepage control, or strength of stabilization), a well planned procedure has to be adopted to attain the desired result.

A full fledged pre-grouting site investigation has to be planned and undertaken in order to decide several factors such as grout type, equipment, grouting technique, *etc.* Based on the nature of work the number of drill holes, depth and pattern have to be decided. Further, an appropriate injection or compaction grouting has to be followed. Proper grout injection measurements should also be incorporated in the grouting procedure.

6.3.1 Pre-grouting Site Investigation

It is essential to conduct a full site investigation before commencing grouting operations so as to decide the method of grouting and the extent of grouting required in a particular location. The investigation should include a geological survey and investigation drilling.

Geological survey consists of studying the general geology of the area with the help of mapping methods and exploratory drilling, to establish fissures, faults, folds, *etc.* The final study reveals the extent of soil and rock formations, zones of weakness, the dip and strike, *etc.*, of the ground.

Detailed explorations are carried out by drilling boreholes and collecting samples for laboratory tests, to determine shear strength, grain-size and permeability, *etc.*, in the precise locations where grouting has to be effected. In addition to these a field permeability test is a must to facilitate more precise estimation of the required pumping pressures, grouting pattern, and type of grout.

6.3.2 Grout Holes Pattern

Grouting may be of shallow depth with low pressure or deeper depth with high pressure. The shallow grouting is done first to seal all major crevices in and consolidate rock and deep grouting is undertaken later. The field permeability log helps in deciding the zone to be grouted, the type of grout to use, and the possible extent of grout penetration under a given pressure. Ideally, the spacing of grout holes should follow a grid pattern such that the radius of penetration is sufficient to cause slight overlapping between adjacent holes. A second and subsequent half-size grid is then injected to fill the spaces between adjacent columns (Fig. 6.5). Sometimes a third and quarter-size grid is required to achieve the target.

As a preliminary guidance the spacing of grout holes may be as given in Table 6.3 (Harris, 1983). However, this will very much depend on the viscosity of the grout. After fixing the length, breadth and depth of the zone the quantity of grout is calculated. The total volume of grout required depends on the void ratio of the zone to be grouted, and the conditions prevail during execution such as ground water, fissures, *etc.* Thus an exact quantity of grout requirement can not be planned at the beginning of the project.

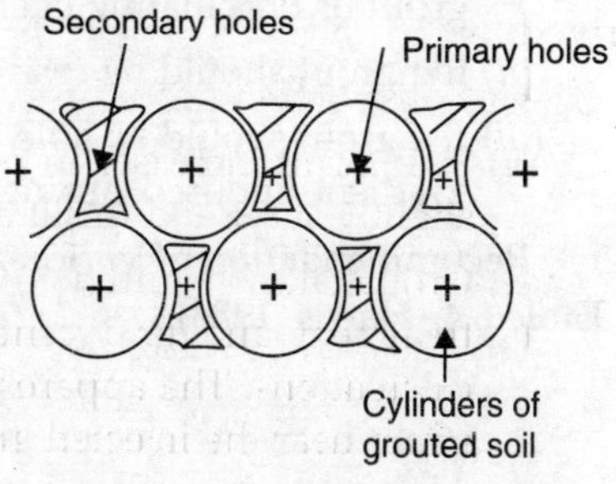

FIGURE 6.5. **Grouting Pattern.**

Table 6.3. Grout Hole Spacing
(*Source: Harris, 1983*)

Coefficient of Permeability of the Zone (mm/s)	*Grid Spacing (m)*	*Soil Type*
> 1	6	Fissured rocks
1 to 1 x 10^{-1}	3	Medium/coarse sands and gravels
< 10^{-1}	0.5 to 1	Fine sands

Grout holes are generally vertical but sometimes inclined holes are made to meet some objectives. Inclined holes, are made to ensure that maximum number of joints or bedding planes are intersected by a specific length of grout hole and beneath building foundations. Oblique holes are usually made with an angle not exceeding 15 degree or so to the vertical. Depths of grout hole varies from 20% of the hydrostatic head in hard rock to 70% of this in poor rock (Krynine *et. al.*, 1957).

Grout holes are drilled with rotary drilling or percussion drilling. In these hole drilling no samples are taken. However it is advisable that a small percentage of the grout holes be cored or cased so that sample can be obtained. In rock, the grout holes are always washed out and pressure tested.

6.3.3 Grout Characteristics

In order to obtain the desired effects a grout should not set quickly because this may affect the pumping but should set and harden after the completion of pumping. The viscosity and subsequent rate of hardening are regulated frequently during pumping by addition of additives. The fluid viscosity gradually changes with time as chemical reactions take place to transform the grout from a liquid stage into a gel (initial set) stage. Some grouts react in minutes while others may take even 24 hours to develop the maximum viscosity. Clay suspensions with bentonite have thixotropic properties, *i.e.*, they behave like a fluid when in movement and form as a jelly when stationary and this transformation is reversible. Suspensions of solid particles in water (such as clays, cements, bentonite, plaster, PFA, lime, *etc.*), emulsions (such as bitumen in water) and solutions (which react after injection form insoluble precipitates) exhibit the above characteristics.

The principles to follow in choosing the grout are (Harris, 1983):

(*i*) the grout must be able to penetrate the voids of the mass to be injected (*e.g.*, a cement grout or flocculating chemical grout may get filtered in a fine sand).

(*ii*) the grout should be resistant to chemical attack when in place.

(*iii*) the grout should be able to develop sufficient shear strength to withstand the hydraulic gradient imposed during injection and on flowing groundwater.

Recommendation of various grout types for application in different situation is presented in Table 6.4 (Harris, 1983).

Table 6.4. Grout Types and Applications
(*Source: Harris, 1983*)

Grout Type			*Application*
Suspensions	PFA	:	Mass filling in very coarse soils and rock fissures.
	Cement	:	Mass filling in very coarse soils and rock fissures plus ground strengthening.
	Clays	:	Mass filling in medium coarse soils and impermeability improvement.
	Clay/Cement	:	Similar to clays plus added strength.
Emulsions		:	Impermeability improvement.
Solutions, one shot		:	Impermeability and/or strength improvement in medium coarse soils.
Solutions, two shot		:	As for one-shot with additional control of gel time. Also suitable in fine soils.

6.3.4 Grouting Plant and Equipment

Both suspension and solution grouts use the same mixing plants and delivery system and they differ mainly in their storage and mixing configuration. A grouting plant includes a mixer, an agitator, a pump, and piping connected to the grout holes. A typical cement suspension grouting plant is shown in Fig. 6.6 (Houlsby, 1983). Two systems, *viz.*, single-line type and circulating type, for piping are shown in the Fig. 6.6. In the circulation type, the unused grout is returned to the agitator and in the single-line type the grout refused is wasted.

For solution grouts, separate ingredients are stored in stationary tanks or tank trucks and metered out on a flow volume basis. They are mixed at junction points and delivered to the intended grout pipe. The grout pipes are connected in a manifold system, but each is separately valved so that complete control is obtained over their flow.

The basic items required for a grouting plant and their functions are:

(*i*) Measuring tank—to control the volume of grout injected.

(*ii*) Mixer—to mix the grout ingredients.

(*iii*) Agitator—to keep the solid particles in suspension until pumped (not required for chemical grouts).

(*iv*) Pump—to draw the grout from the agitator to deliver to the pumping line.

(*v*) Control fittings—to control the injection rate and pressure so that the hole can be regularly blend with water and thin grout.

It is always necessary that a thoroughly-mixed mixing is done in three stages (Houlsby, 1982) : (*i*) formation of a vortex which acts as a centrifugal separator-thicken grouts and unmixed cement are pushed to the periphery of the vortex and passed to a mixing rotor, (*ii*) treatment of thicker fraction and unmixed cement from the vortex - these are subjected to a violent shearing action in a mixer rotor, which breakes up thicker fraction and lumps of cement and wets and produces a grout resembling like a colloidal solution rather than a mechanical suspension, (*iii*) circulation of the treated fraction back into the vortex and the vortex continues to spin till all the thicken fraction of cement lumps are broken and the entire grout reaches an uniform consistency. Houlsby (1983) advocates the following precautions while mixing a grout:

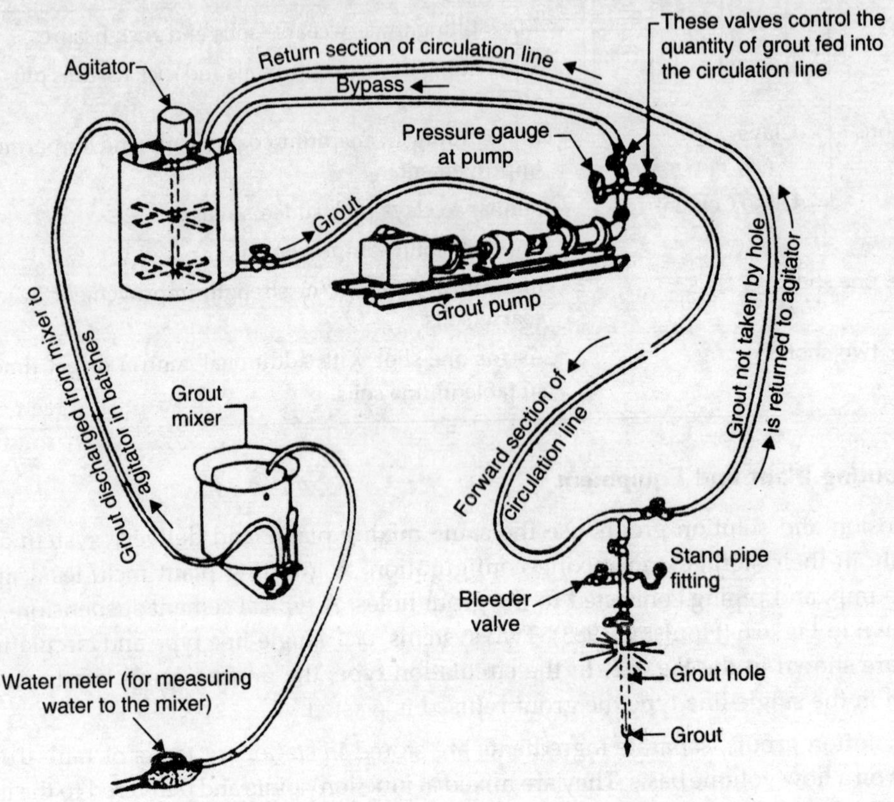

***FIGURE 6.6.* Typical Layout of a Grouting Plant.**
(Source: Houlsby, 1983)

(*i*) Water is placed first in the mixer.
(*ii*) Mixer is run at the maximum speed before adding the cement.
(*iii*) Grout is mixed in batches.
(*iv*) Ingredients have to be measured by volume.
(*v*) Enough water should be maintained to cover the rotor while it is functioning.
(*vi*) Mixer should not be allowed to run for more than a few minutes between batches.
(*vii*) Mixers should be cleaned thoroughly after the day's work.

A barrel type grout mixer (Fig. 6.7) consists of a cylindrical drum placed either horizontally or vertically with an axial shaft fitted with paddles or blades. The axial shaft is rotated normally or by power. Cement and water are thoroughly mixed with the help of paddles. Vertical type of mixer is used to handle small quantity of grout.

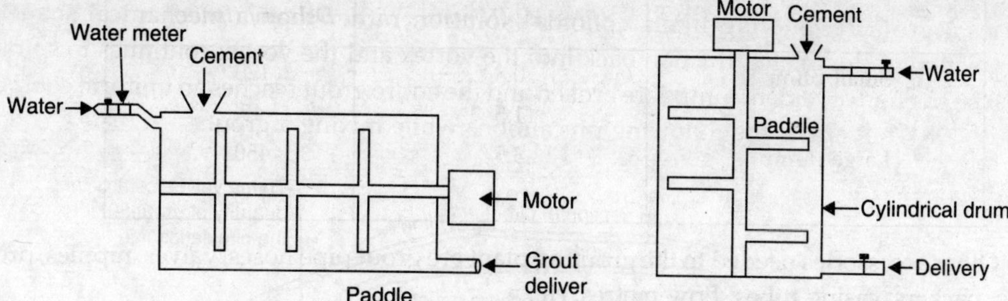

FIGURE 6.7 **Vertical and Horizontal Barrel Type Mixer.**
(*Source: Houlsby, 1982*)

A grout mix should be continuously agitated to prevent setting. This is achieved by an agitator sump between the mixer and the grout pump. An agitator sump is a tank which has an agitating mechanism consisting of a vertical shaft to which horizontal blades are connected which is revolved at 30 to 100 rpm (Fig. 6.8). The grout mix from the mixer is passed through a wire screen to remove pieces of sack, strings, and other foreign matters. Also another screen is fixed to the agitator near the delivery pipe, to prevent entering of lumps to the pump. A graduated (in litres) dipstick is used to measure the quantity of grout in the agitator.

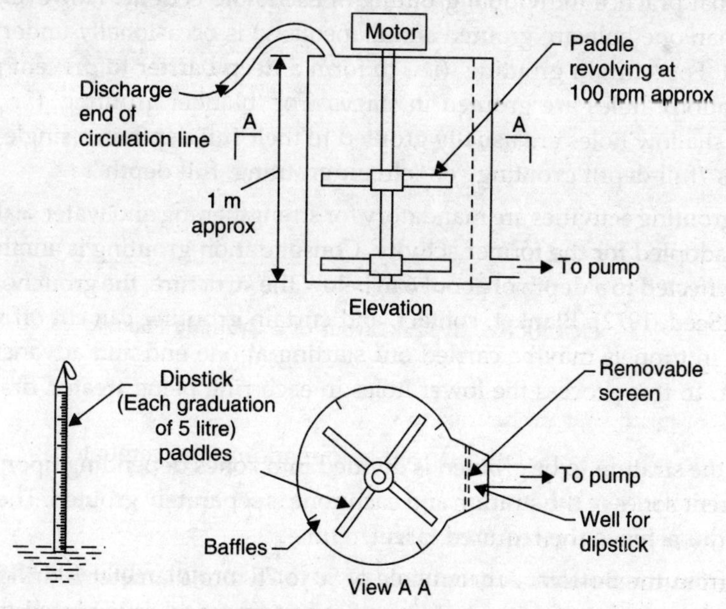

FIGURE 6.8. **Grout Agitators.**
(*Source: Houlsby, 1982*)

For most situations, pumps should be able to provide pressures of 2800 kN/m^2 or under, and rates of displacement around 0.007 m^3. Pumps may be of piston or diaphragm type. The best all-purpose pump should be able to displace the wide variety of grout consistencies actually employed in practice including very low slump mixtures. The following capacities of pumps may be preferred depending on the type of grout (Harris, 1983).

	Pressure (N/mm^2)	*Delivery (l/min)*
Small pump	0.8	120
	1.5	45
Large pump	3.5	450
	10.0	130

Other accessories needed to the grouting plant are grout pipe hoses, valves, nipples, pressure gauges, packers, casing tubes, flow metres, *etc.*

6.3.5 Injection Methods

The successful implementation of a grouting project very much depends on the skill and precision with which the grouting process is executed adopting an appropriate injecting technique. Generally the grouting methods to be adopted in a project are usually specified on the grouting project design. In selecting a grouting method the following aspects have to be satisfied: (*i*) soils or rocks of different characteristics should be treated individually, (*ii*) it should be possible to treat short sections of boreholes in any desired sequence and repeat the injection, if necessary, and (*iii*) leakage around the borehole should be prevented.

In the normal practice individual grouting of each hole is done. However, group grouting (in which more than one hole are grouted simultaneously) is occasionally undertaken when rock conditions permit. For 'curtain grouting' (*i.e.*, to form a deep barrier to prevent passage of water through a foundation), holes are grouted in stages. For 'blanket grouting' (*i.e.*, grouting of the surface area), the shallow holes are usually grouted to their full depth in a single operation by the method known as "full depth grouting" or "circuit grouting, full depth".

Basically, grouting activities are mandatory for strengthening and water sealing. Compaction grouting may be adopted for the former activity. Consolidation grouting is another approach and this is frequently effected to a depth of about 6 m below the structure, the grout holes being injected in a single stage (Seed, 1972). Blanket, contact, and curtain grouting can cut off water effectively. Contact grouting in tunnels may be carried out starting at one end and advancing continuously towards the other. In this process the lower holes in each ring being treated first and uppermost ones last.

In general the stratum to be grouted is divided into zones depending upon the permeability condition of different zones of the stratum and each zone is separately grouted. The several methods described below can achieve the required target.

Grouting from the Bottom. A grout hole of 50 to 75 mm diameter is drilled to full planned depth. In rigid soils or in intact rock strata a self expanding packer is placed directly above the lowest zone and grout is pumped in (Fig. 6.9). The procedure is repeated after the packer is raised and fixed to the next zone. Thus the drill hole is grouted successively upwards. In soft or unstable soils the drill hole must be supported by a casing and it also provides a good seal between the packer and borehole walls. The casing is raised progressively with packer. In fissured rock, a particularly permeable stratum can be isolated using a double packer (Fig. 6.10).

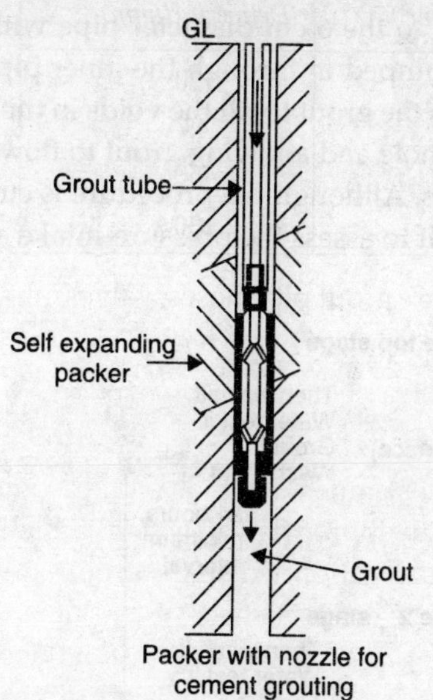

FIGURE 6.9. Self-expanding Packer.

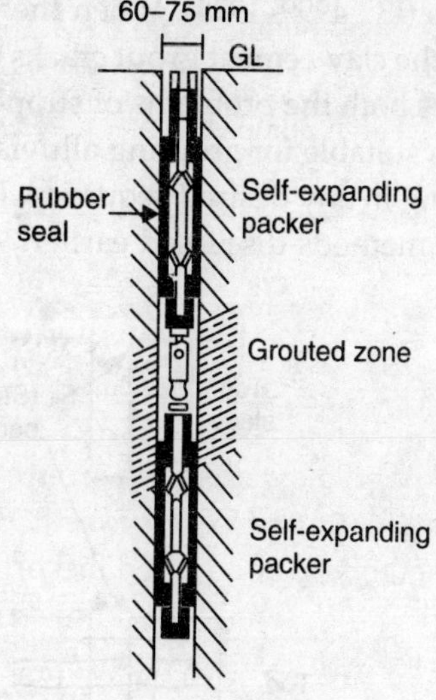

FIGURE 6.10. Grouting an Isolated Stratum Using Self-expanding Packer.

Grouting from the Top. In this method, holes are drilled down to the seam closest to the surface and grouting is carried out. Holes are then cleaned by washing and drilling continued to the next seam. The grouting process is then effected. Subsequent washing followed by further drilling and repeated grouting are done until the entire operation is completed [Fig. 6.11 (*a*)]. This method gives a low output. However, by this method, short passes can be treated individually and there is no risk of leakage along the hole into the underground zone. This method is quite useful for heterogeneous strata and provides improvement of the upper zones and should any weakness exist, deals with it automatically. During grouting if any sensitive surfaces are encountered, they should be thoroughly tightened before grouting of the lower stages is undertaken. In this method, rubber packers may also be used which could expand on pressure to about five to six times the diameter of the hole. Hence, grouting is facilitated in a particular zone by fixing packers on the top and bottom of it [Fig. 6.11 (*b*)].

Circuit Grouting. This method is based on the principle of grouting from the top downwards. A drill hole is bored to the depth of the bottom zone and grout is pumped down the grout pipe and returned up the drill hole (Fig. 6.12). By this process clogging is almost eliminated. The grout hole is then deepened and the procedure repeated.

Tube-a-Manchette Grouting. The double packer method adopted for fissured rocks was found unsuitable in alluvial soils as difficulties faced to effect a seal on the lower packer. This difficulty is overcome by this method. In this method a 12.5 to 15 cm diameter hole is drilled in the stratum and a 6 cm diameter pipe, called the tube-a-manchette, with a circumference row of holes, is lowered inside the drill hole. The drill hole is covered with 10 cm wide tightly fitting rubber sleeves. The 6 cm diameter pipe is sealed into the outer hole by a clay-cement grout. Grouting is effected from botton upwards through a smaller interior grout pipe lowered into the 6 cm diameter pipe (Fig. 6.13). The inner grout pipe is tightly fitted to the 6 cm diameter pipe with the help of rubber packers (IS : 4999–1978). When the grout is pumped in through the inner pipe, due to the high pressure the clay-cement grout cracks and allows the grout to fill the voids in the surrounding soil. This solves both the problems of supporting the hole and allowing grout to flow into the soil. This method is suitable for grouting alluvial type soils. Although the procedure is cumbersome, it enables grouting in any desired sequence. It is difficult to assess the pressure-intake and is costlier than the other methods discussed earlier.

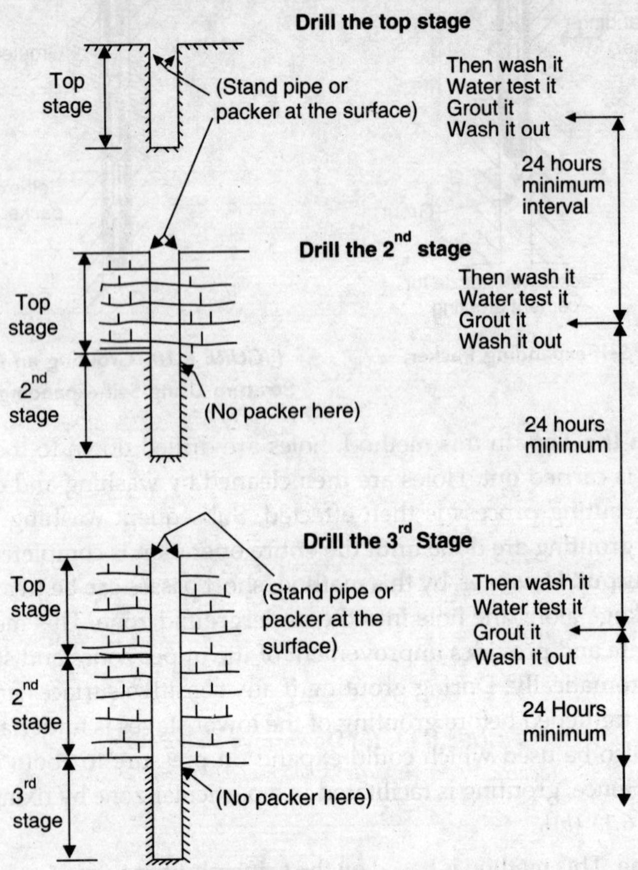

FIGURE 6.11. **(*a*) Downstage without Packer.**

(*Source: Water Resources Commission, Australia, 1981*)

Point Grouting. In shallow work of 10 to 12 m deep the grout is injected from the points of a driven or jetted lance. Injections are delivered at pre-determined positions along the line of drive and also on the return in systems where a second reacting grout ingredient is to be placed independently of the initial injection. This is widely used and has the limitation as regards to depth and penetrability of fine strata.

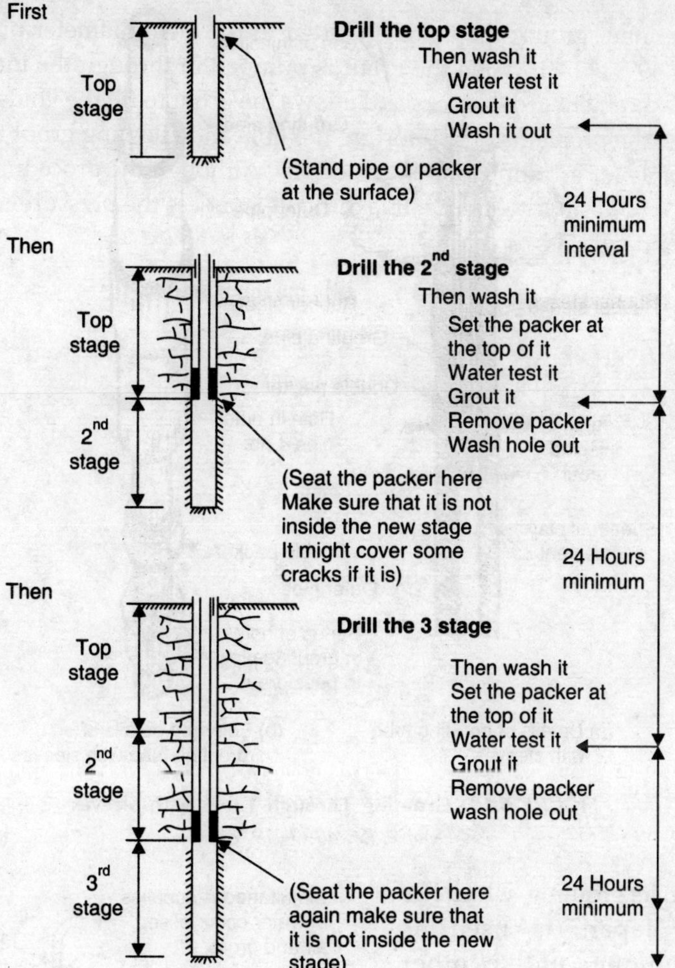

***FIGURE 6.11. (b)* Downstage with Packer.**

(*Source: Water Resources Commission, Australia, 1981*)

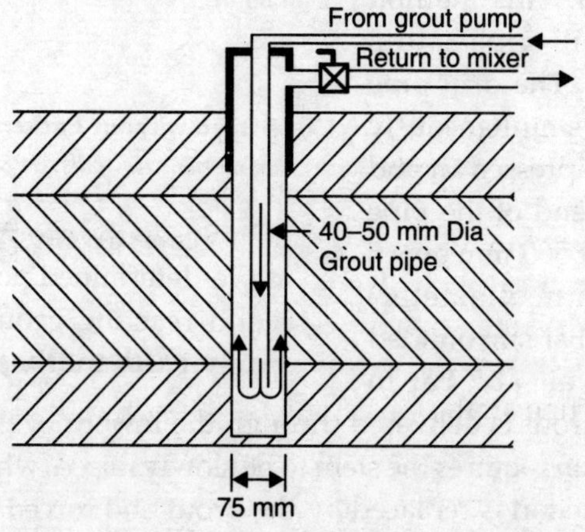

***FIGURE 6.12.* Circulation Grouting Method.**

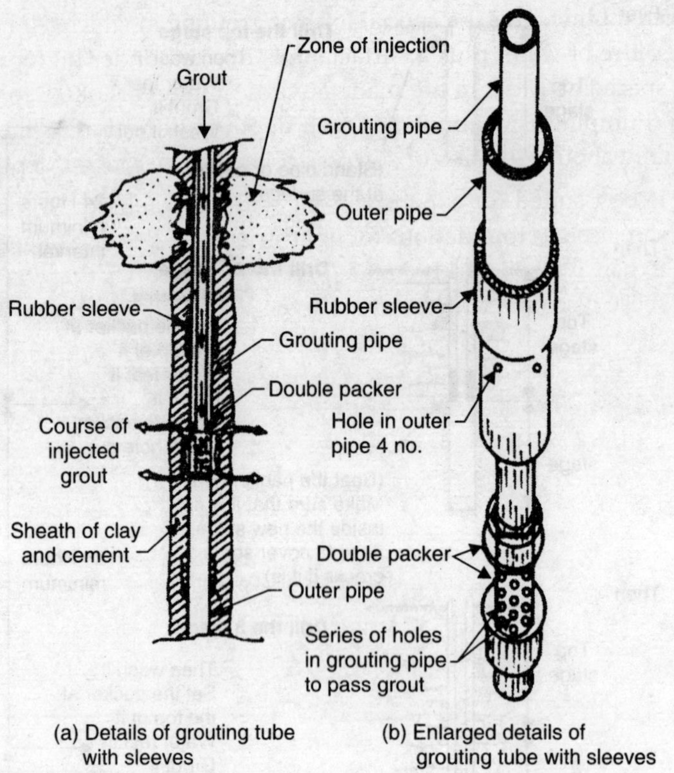

FIGURE 6.13. **Grouting Through Tubes with Sleeves**
(*Source: IS : 4999–1978*)

Jet Grouting. Jet grouting, which has originated from Japan, is used as replacement technique, unlike other conventional injection methods. By this method, soils ranging from silt to clay and weak rocks can be treated. This method consists of lowering a drill pipe into a 150 mm diameter borehole. The drill pipe is specially designed which simultaneously conveys pumped water, compressed air and grout fluid. At the bottom end of the pipe two nozzles are provided at 500 mm apart. The upper nozzle (1.8 mm diameter) delivers water at about 400 bar surrounded by a collar of compressed air at 7 bar to produce a cutting jet. The grout is delivered through the lower nozzle (7 mm diameter) at 40 bar (Fig. 6.14). The grouting action requires the stem to be slowly raised, whereby the excavated material produced from the jetting action is replaced by the grout and forced to the surface. The jet water could reach about 1.5 m and by rotating the stem a column of replaced earth may be formed.

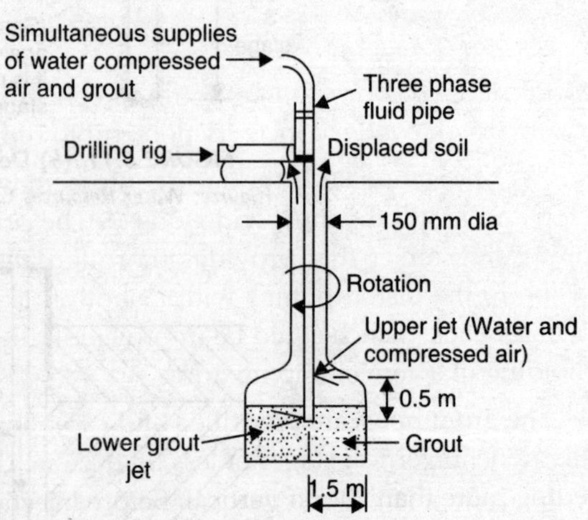

FIGURE 6.14. **Jet Grouting Method.**

Pressure Injected Lime. This is a special form of grouting in which a lime slurry, containing 0.3 to 0.4 kg lime per litre of water plus a surfactant, is injected under high pressure (350 to 1400 kN/m^2). Injections, spaced at 1 to 2 m are made at vertical intervals of 0.3 to 0.5 m to maximum depth of about 3 m. Pumping is continued at each depth until refusal or the slurry runs out at ground surface. Usually about 120 litres of slurry is consumed per meter depth.

This method is best suited for expansive soils with cracks, fissures, slickensides fractures, and root holes which are used as foundations for light structures. Because of high pressure adopted in this method, the expansive soil can be treated to a depth greater than the depth of seasonal moisture variation. These discontinuities existing in the soil, in conjunction with channels formed by hydraulic fracturing under the high injection pressure, provide channels for the slurry throughout the soil. Stabilization is achieved through reaction between the lime and the soil adjacent to the cracks which forms moisture barriers to protect the untreated soil blocks against volume change. As the lime particles are bigger than the pore size, significant penetration of the grout into soil pores does not occur. Further, migration of lime into the soil from the cracks by diffusion also is slow (Mitchell, 1976).

Truck-mounted injection rigs containing mixing tanks, pumps, hoses and injection pipes can treat areas of virtually any size or configuration. Cost of stabilization is generally low.

Electrokinetic Injection. Stabilization of silty soils may not be possible by chemical or admixture perhaps because of lack of confinement or the necessity to avoid disturbance of the ground. In such situations electrokinetic injection technique may be adopted. In this technique chemical stabilizers are introduced at the anode and carried towards the cathode by electro-osmosis. Direct current electrical gradients of the order of 50 to 100 volts per metre are required. Although the method is likely to be expensive, it can be effectively used in special cases (Mitchell, 1976). Two case histories of electrokinetic injection is presented in Table 3.2.

Compaction Grouting Techniques. Compaction grout work utilises portland cement clayey silty sand mixture. Normally, two to four bags of cement per cubic meter of clayey silty sand are used. Silts and clays are added to provide plasticity and cohesion to the mixture. The function of sand is to provide interparticle friction and to reduce the silts and clays to a semiplastic material.

Koehring mudjack is the basic equipment which enabled the compaction grout technique to develop and is still the most widely used. The principle behind the design of the equipment is a chopping pug mixer that provides thorough and uniform mixing of the materials and a feed provision to the piston pumps with a shut off arrangement (Graf, 1969). However, the mudjack has two set backs which could be improved in the refined models, *viz.*, limited pressure capacity, and holding of a zero slump mixture.

The grout holes are predrilled and cased up to the top of the zone to be treated. Injection is usually done in vertical stages of less than 2.4 m. Grout holes may be inclined with inclination not exceeding more than 20° off vertical. Before injection the holes are primed by injection of water or lime slurry.

Injection points are placed by drilling inside and ahead of a 3.8 or 5 cm pipe while simultaneously driving the pipe. The points are placed at different depths such that a staggered depth pattern is achieved (Graf, 1969). Generally shallow points are placed closer. Figure 6.15 shows a possible section showing grout point positioning varying with depth of grouting. Another pattern (Fig. 6.16) frequently used is angulation of the grout points so as to grout beneath a structure without going through it (Graf, 1969).

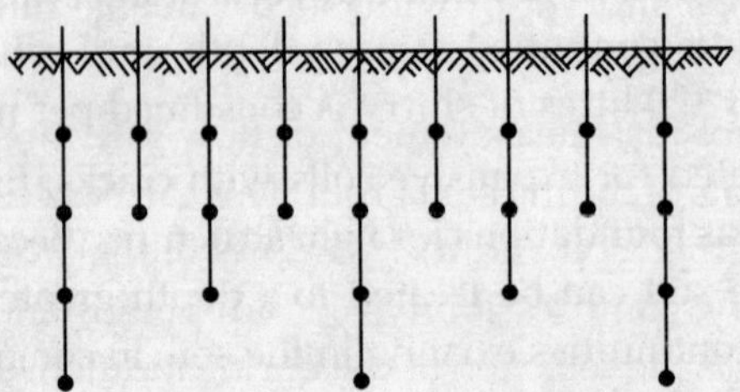

FIGURE 6.15. **Possible Section Showing Grout Point Positioning Varying with Depth of Grouting**
(Source: Graf, 1969)

In most of the compaction grouting work is effected by placing the points to full depth and pumping until (*i*) resistance to pressure is achieved, (*ii*) pressure loss or surface lift is observed, indicating maximum compaction for the bulb, or (*iii*) in the judgement of the grout foreman or engineer, pumping should be stopped (Shroff and Shah, 1992). The grout pipe is then withdrawn to some distance, varying with job conditions, and operation repeated.

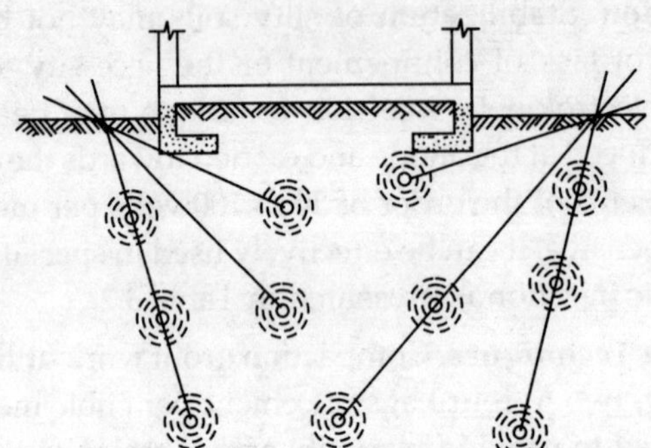

FIGURE 6.16. **Possible Section Showing Angulation of Grout Points and Bulbs**
(Source: Graf, 1969)

When maximum compaction of the soil surrounding each bulb is achieved, the pressure will cause a lift of the overlying soil. Lifting of light structures may be achieved at depths of 1.8 to 3.0 m below the foundation whereas for heavy structures the depth required will be about 3.6 to 6 m below the foundation.

6.3.6 Grout Injection Measurements and Monitoring

During the grouting process the basic information to be obtained is the weight for suspension grouts and volume for solution grouts and grout flow rate along with pressure for both. The measurements of weight or volume should be made accurately so as to confirm mix proportions. In solution grouting positive displacement meters may be used for their purpose (Baker, 1982).

The flow rate of the grout during injection should be continuously monitored and plotted against the grout pressure to ascertain the condition below the ground. Different types of flow meters are available in the market.

Pressures should also be monitored at the grout stations, if possible, continuously. Bourdon tube type gauges are also used to measure pressure periodically which usually take a dreadful battering on most grouting jobs and yet provide such valuable information that they are worth taking care of (Koerner, 1985).

Grout monitoring, is not just measurement of flow rate, pressure, *etc.*, but it is making a positive assessment of the results of the injected grout. This can be accomplished by the conventional methods by obtaining undisturbed soil and rock samples of the grouted material and then testing them for strength, permeability, compressibility, *etc.*, adopting standard laboratory methods. The constraint about this approach is the selection of test boring location, depth cf sampling and finance. A better approach is to use indirect methods such as geophysical or non-destructive testing methods which will provide a continuous trace of continuous before and after grouting either along the ground's surface or within adjacent grout pipes. Huck and Waller (1982) have recommended grout monitoring procedure which is presented in Table 6.5.

Table 6.5. Grout Monitoring Procedures at Various Times During the Grouting Process
(Source: Huck and Waller, 1982)

Grouting activity	*Minor monitoring effect only*	*Intense monitoring effort*
Prior to grouting	Inspect equipment Set elevation survey points Establish monitoring plan and procedures	Inspect equipment Set elevation survey points Establish monitoring plan and procedures
During drilling		Conduct pregrout radar and cross hole acoustic surveys
Grout Materials	Certificates of compliance Trial grout mixes	Certificates of compliance Trial grout mixes Independent laboratory tests
During grout	Monitor injection pressure and flow rate Grout samples for gel time and storage Plot grout-take log Heave measurements on survey points	Monitor and record injection pressure and flow rate (strip chart) Grout samples for gel time and storage Plot grout-take log Heave measurements on survey points In-situ deformation measurements as appropriate (inclinometer and/or insitu strain) Pore-pressure data In-situ resistivity Acoustic emission monitoring for hydrofracturing
After grouting	Final heave survey Final review and signoff	Final heave survey Postgrout radar and acoustic surveys Final review and signoff

6.4 APPLICATIONS

The major application of grouting is seepage control and strengthening. However, grouting has been in use for other allied engineering works such as underpinning, vibration control of machine foundation, slab jacking, anchor systems, *etc*. Some of the major applications of grouting are briefly explained below.

6.4.1 Seepage Control

Seepage Control in rock under dams is the oldest area of grouting application which has had considerable exposure. Based on the degree of fracturing in the rock near its surface a shallow blanket grout is placed, through which a deep curtain grout injected [Fig. 6.17(*a*)]. This curtain wall is a cut off wall which is on extension of the clay core of the claws and the junction is made more carefully to avoid a seepage path through the interface. Spacing and depth of the curtain grouting is shown in Fig. 6.17(*b*) and Fig. 6.17(*c*) shows the typical pattern which varies with location along the axis of the dam. Grout curtain configurations for seepage control of different rock foundations are shown in Fig. 6.18 (Wahlstrom, 1974).

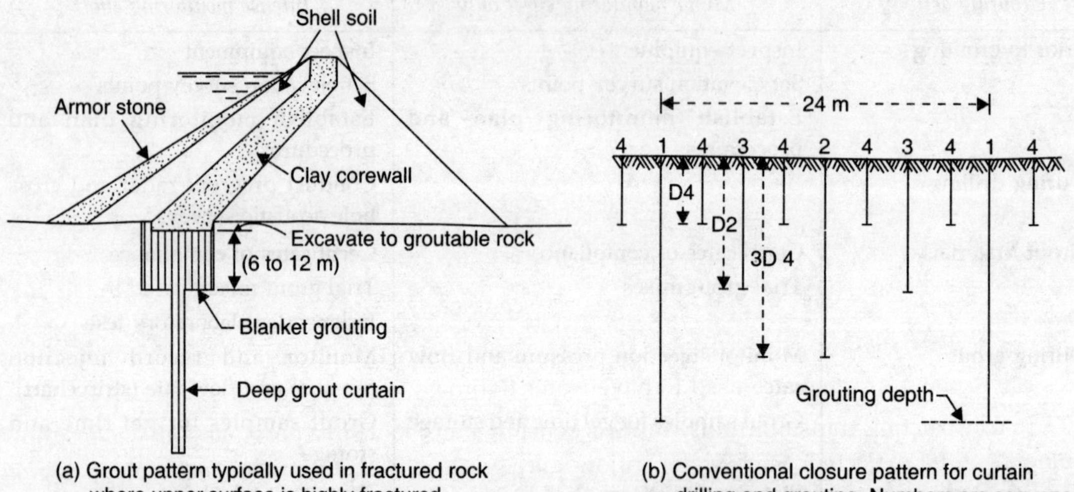

(a) Grout pattern typically used in fractured rock where upper surface is highly fractured.

(b) Conventional closure pattern for curtain drilling and grouting. Numbers are sequence of drilling and grouting.

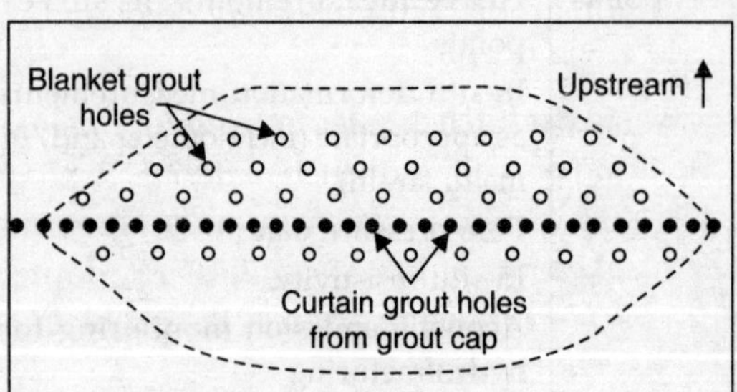

Plan view

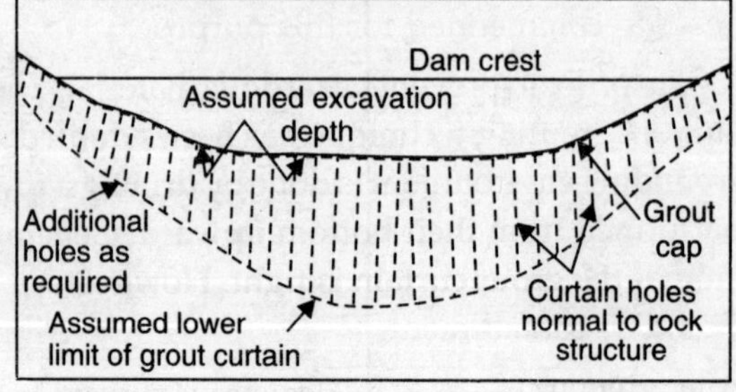

Cross section view

(c) Schematic locations of pattern blanket and curtain grout holes in bedrock of an earth dam of moderate size.

***FIGURE 6.17.* Blanket and Curtain Grouting for Seepage Control Beneath Dams on Rock Foundations.**

(*Source: Wahlstrom, 1974*)

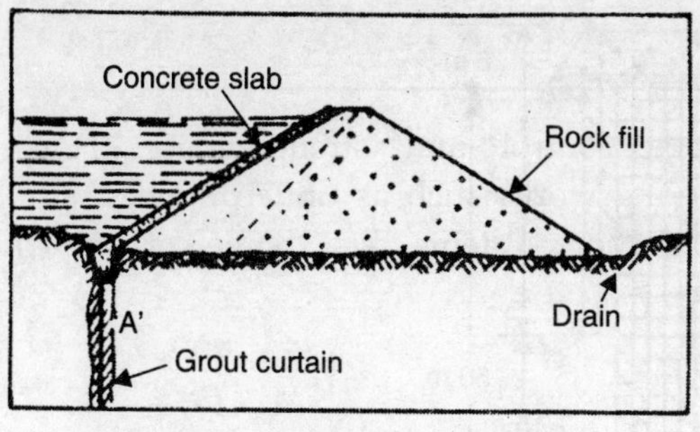

(a) Rock fill dam with concrete-lined face

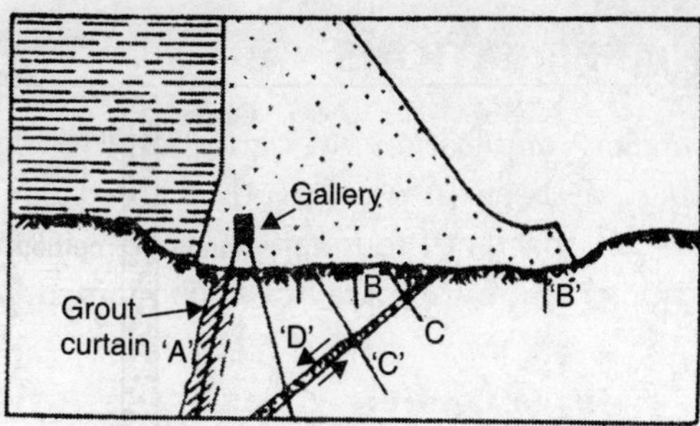

(b) Concrete gravity dam with intersecting fault zone

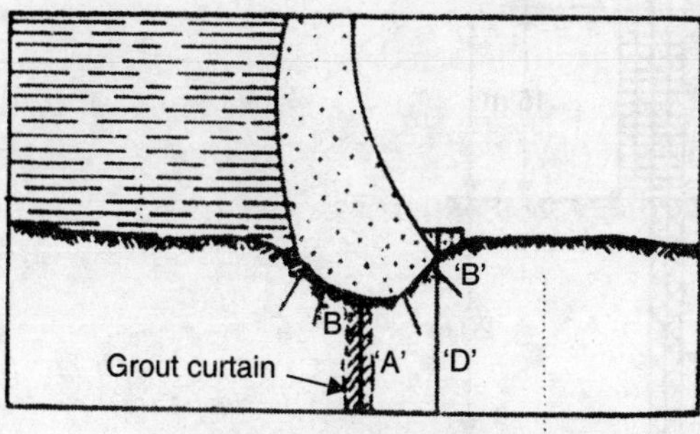

(c) Concrete arch dam

FIGURE 6.18. **Grout Curtain Configurations for Seepage Control of Rock Foundations.**

(*Source: Wahlstrom, 1974*)

Control of seepage of pervious dam foundations are accomplished by a variety of grouts and such grouts are usually placed deeply. It is customary to provide parallel rows of curtain grout lines in soils. A deep grout curtain used to limit the seepage loss in soil is shown in Fig. 6.19 (Bonazzi, 1965). In this situation, a flexible grout curtain was found suitable in certain places. Accordingly silicate solution grouting or clay suspensions were used in some places while in other places clay-cement grouts were used to have water tightness.

In constructing transit or utility tunnels beneath water table in a pervious soil, water inflow problems can be expected. Such water inflow may erode fine soil particles resulting in piping collapses and large surface subsidence. In such situations, one method of constructing the tunnel is to grout the soil in advance of tunneling. Chemical solution grouts, with relatively quick setting time are recommended for this purpose.

In order to prevent hazardous materials and toxic waste landfill leachates from environmental pollution, grouting technique has been adopted. The isolation of a pit with such materials from its surrounding environment is done by drilling a row of closely spaced grouts and continuous grouting is performed from their bottom upward. Generally chemical grouts of low viscosities are adopted such that the grout curtain is tight. However, jet grouting is well suited to address the problem of toxic waste containment.

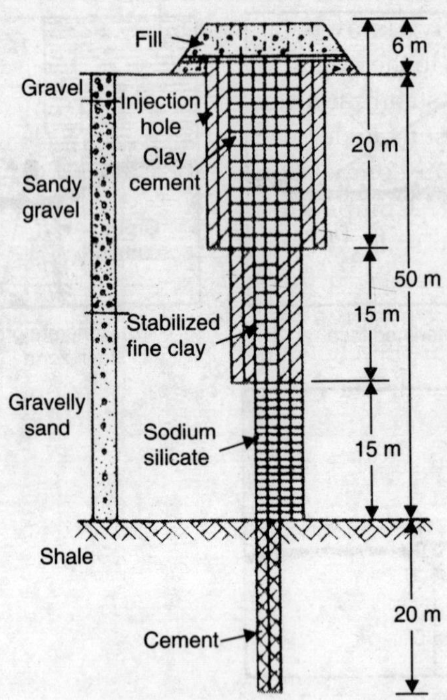

FIGURE 6.19. **Deep Grout Curtain (Various Types of Grout) Beneath Earth in France.**
(*Source: Bonazzi, 1965*)

6.4.2 Soil Solidification and Stabilization

Construction of transit and utility tunnels can cause settlement and stability problems for adjacent buildings and structures located above them. Such a situation for Prague subway was handled by Verfel (1979) by adopting suitable grout (Fig. 6.20). Different grouts were used depending on the nature and type of strata. By this grouting it was reported that adjacent buildings were raised by 12 to 17 mm. The volume of grout used was 35% of the grouted volume of the original soil. Stabilization grouting is also used as a method of under pinning of building foundations.

It may be necessary to treat ground destined to carry bridges with appropriate grouting materials in order to obtain stabilisation and ensure its ability to carry the projected load. Ground stabilization proved to be necessary especially under the piers. This is effected by appropriate chemicals or cement.

A mine may, from the hydraulic stand point, be regarded as a large scale drainage system entailing the necessity of pumping quantities of seepage water to the surface throughout its life which may be half a century or more. This is an expensive work which could be partially or completely obviated by grouting. The earliest grout used in mining operation is sodium silicate aluminium sulphate chemical grout. Some mining operations cause surface subsidence due to removal of subsurface support and the risk of occurrence of such a damage may be reduced or eliminated by appropriate grouting. In mining operation grouting is done for these basic reasons, support and water sealing, in every case, sometimes former, sometimes for the latter and, if necessary, for both.

Slabjacking, formerly known as mudjacking, technique relates to the raising of concrete pavements and the stabilization of slabs by means of infilling existing under slab voids. This is different from compaction grouting discussed earlier. Compaction grouting applies primarily to soils whereas slabjacking pertains to man-made structures and is particularly effective in highway maintenance. The use of pressure injection has been in use since Second World War and a variety of materials have been used. Some of the materials are: hot asphalt, soil-cement slurries and cement-sand-mixes, hydrated lime, fly ash, *etc*. Sufficient research and development have been directed in this technique and adequate refinement has been attained to achieve any desired result. Also, the terminology has been divided into the categories of slabjacking, where actual lifting or levelling is involved and pressure grouting, where the objectives is simply infill voids. The term mudjacking is no longer in use. The mudjack machine more or less in its original form is still employed for this type of work.

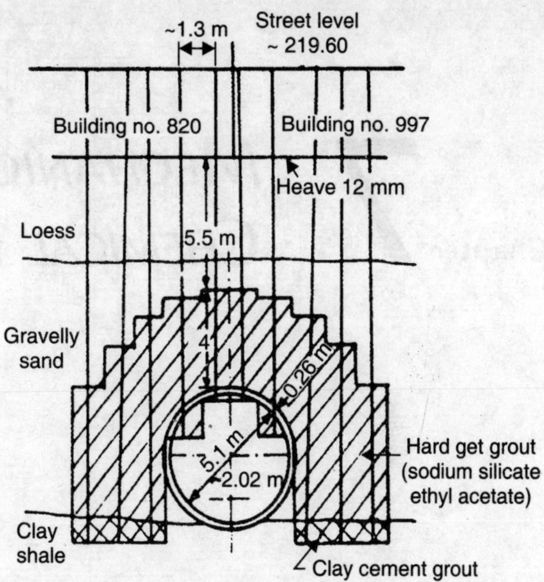

FIGURE 6.20. **Different Grout Types and How They Were Used to Stabilize Buildings Located Adjacent to and Above Site of Construction of Prague Subway.**

(*Source: Verfel, 1979*)

6.4.3 Vibration Control

Excessive vibration (amplitude or settlement) is caused in those cases where vibrating machine foundations have an operational frequency approaching the natural frequency of the foundation-soil material and induce resonance. If it were possible in some way to forecast the natural frequency accurately, the problem of resonance can be completely avoided. As the correct assessment of natural frequency is much more difficult, the problems of resonance may occur from time to time. Simple techniques are sometimes quite effective in eliminating the occurrence of resonance, *e.g.*, the foundation may be separated from adjacent floors or the water table may be lowered or mass may be added by underpinning the foundation. Grouting is an efficient corrective technique.

Chemical injection through machine foundation soil is a recent finding in decreasing vibration. Chemical grout alters the elastic properties of the soil and results in increased rigidity of the base. Increase in rigidity increases the difference between the frequency of natural vibration and the operating frequency of the machine resulting in decrease of amplitudes. The injection method has an added advantage of non-interruption of work of the machine during the treatment. It is sufficient to stabilize the soil near the foundation edges perpendicular to the plane of vibration.

Chapter 7 MECHANICAL, CEMENTING AND CHEMICAL STABILIZATION

7.1 INTRODUCTION

Soil stabilization is referred to as a procedure in which a special soil is proportioned/added/or removed, or a cementing material, or other chemical material is added to a natural soil material to improve one or more of its properties. One of the more common methods of stabilization includes the mixing of natural coarse-grained soil and fine-grained soil to obtain a mixture that develops adequate internal friction and cohesion and thereby provides a material that is workable during placement but will remain stable further. Improvement of soil property by proportioning of coarse and fine grained soils is commonly referred to as mechanical stabilization. Rearrangement of soil particles by some mechanical means (say by compaction) is also referred to as mechanical stabilization. On the other hand stabilization can also be achieved by mechanically mixing the natural soil and stabilizing material together so as to obtain a homogeneous mixture. After the soil and the stabilizing agent are blended and worked together, they are compacted using an appropriate compaction device. The stabilizing materials include cement, lime, bitumen/asphalt, polymers and other chemicals. Addition of chemicals causes a physico-chemical alteration and referred to as chemical stabilization. In order to augment the stabilization, additives which have the property of water-holding or water-resisting property are sometimes added.

7.2 REQUIREMENTS OF SOIL STABILIZATION

The mode of alteration and the degree of alteration necessarily depend on the character of the soil and on its deficiencies. In general the requirement is adequate strength. In the case of a cohesionless soils the strength could be improved by providing confinement or by adding cohesion with a cementing or binding agent. In the case of a cohesive soil, the strength could be increased by drying, making the soil moisture-resistant, altering the clay-electrolyte concentration, increasing cohesion with a cementing agent, and adding frictional properties. Compressibility can be reduced by consolidation, by filling the voids with an appropriate material, cementing the grains with a rigid material or by altering the (inter-particle electrical) forces. Swelling and shrinking can be controlled by adding cementing agents, by altering the double-layer thickness and property, and by preventing moisture changes. Permeability can be reduced by filling the voids with an impervious material or by creating a dispersed structure of the soil. On the other hand, permeability can be increased by removing fines or creating an aggregate grain structure. Different methods of stabilization are in use. Based on their function or effect on soil they may be classified as follows:

(*i*) Mechanical stabilization : improving soil gradation or arrangement.

(*ii*) Cementing: binding the particles together without their alteration.

(*iii*) Physico-chemical alteration: changing the clay minerals or the clay-water system.

(*iv*) Aggregants and Dispersants : alteration of electrical forces between soil particles in a modest way.

(*v*) Void filling: plugging in voids.

(*vi*) Consolidation.

Methods which are pertinent to the above function (*i*) to (*iv*) are discussed in this Chapter. Methods pertaining to function (*v*) and (*vi*) are discussed, in particular, in Chapters 6 and 4 respectively, although other methods also deal indirectly these aspects.

Every stabilization process will be satisfactory when it provides the required qualities and fulfils the following criteria:

(*i*) be compatible with the soil material,

(*ii*) be permanent,

(*iii*) be easily handled and processed, and

(*iv*) cheap and safe.

7.3 MECHANICAL STABILIZATION

As discussed earlier mechanical stabilization covers two methods of changing soil properties, *viz.*, (*i*) the rearrangement of soil particles, and (*ii*) the improvement of soil gradation. There are several examples of particle rearrangement, *i.e.*, the blending of the layers of a stratified soil, the remoulding of an undisturbed soil, and of most importance, the densification of soil. Methods of densification of soils are dealt in Chapters 2, 4 and 5. The remaining portion of this section will cover the aspects pertaining to improvement of soil gradation.

The use of correctly proportioned material is of particular importance in the construction of low cost roads. The principle of grading soil may be applied to the improvement of subgrade soils. The main application of the control of the grading of soils and low grade aggregates is to the construction of bases and sub-bases.

7.3.1 Mechanical Stability of Materials

In general weak aggregates are preferred for mechanical stabilization because they will break down under compaction to give a grain-size distribution more closely approaching that required for maximum dry density. The aggregates should be correctly proportioned before laying and should have sufficient mechanical strength so as to maintain the same grain-size distribution during compaction and subsequent use by traffic. Any material that is resistant to weathering shall be suitable for use as aggregate in a mechanically stabilized road. All kinds of natural rock, gravel, sand, artificial materials (such as slag, burnt shale, *etc.*) have been used in road construction with success.

In order to attain adequate mechanical stability it is necessary to have a well proportioned coarse material containing some clay binder. In the case of coarse materials it is assumed that the particle-size distribution giving the greatest density has the greatest internal friction. Fuller and Thompson (1907) showed that a granular mass has a relatively high dry density when its particle size distribution follows a certain law, which for practical purposes may be written as

$$\text{Percentage passing any sieve} = 100\sqrt{\frac{\text{The aperture size of that sieve}}{\text{The size of the largest particle}}}$$

It is observed that to obtain adequate cohesion, greater proportion of material less than 0.075 mm is necessary. Fuller curves have been widely used.

It has been found in practice that the plasticity index of material should be limited to a maximum of 6% for bases and should be between 4 and 9% for surfacings. The liquid limit should not exceed 25% for base material and 35% for surfacings. Higher liquid limits and plasticity indices are desirable for surfacing and quite undesirable for base courses.

Adequate compaction is necessary when construction with soil or low-grade aggregate mixtures. Experience has shown that good compaction of a base can be obtained if it is used by traffic for some months before a surfacing is applied. The type of compaction equipment and control methods are discussed in Chapter 2.

7.3.2 Proportioning the Materials

Natural materials are deficient in one or more of the particle-size fractions required. Thus a mechanically stable material can be produced only mixing two or more of the materials in appropriate proportions. Among the methods available, the method proposed by Rothfuchs (1935) has been in use, as it is reasonably quick and simple and can be applied to mixtures of any number of components. It consists essentially of the following stages (HMSO, 1974):

(*i*) The cumulative curve of the required aggregate particle-size distribution is plotted, using the usual linear ordinates for the percentage passing but choosing a scale of sieve size such that the particle-size distribution plots as a straight line. This is readily done by drawing an inclined straight line and marking on it the sizes corresponding to the various percentages passing.

(*ii*) The particle-size distribution curves of the aggregates to be mixed are plotted on this scale. It will generally be found that they are not straight lines.

(*iii*) With the aid of a transparent straight edge, the straight lines that most nearly approximate to the particle-size distribution curves of the single aggregates are drawn. This is done by selecting for each curve a straight line such that the areas enclosed between it and the curve are a minimum and are balanced about the straight line.

(*iv*) The opposite ends of these straight lines are joined together, and the proportions for mixing can be read off from the points where these joining lines cross the straight line representing the required mixture.

Laboratory tests, such as particle-size analysis, plasticity and compaction tests, are performed on the material or mixture of materials.

7.3.3 Addition or Removal of Soil Particles

The engineering behaviour of a soil also depends on the particle-size distribution and the composition of the particles. It is possible to significantly change the property of a given soil by adding some selected soil or by removing some selected fraction of the soil. Generally, the cost of addition-removal technique of stabilization can be very low. The construction procedure, cost, and results obtained from the addition-removal stabilization depend mostly on the type of problem and nature of soil at the field condition. Thus, it is not practically feasible to set certain principles and procedures for this method. In order to enlighten this method of approach three types of addition-removal technique are discussed (Lambe, 1962).

Addition of Binder to Gravel for Road Construction. The gradation for soil-aggregate materials have to be selected to give the densest mixture by supplying just the assortment of particles to minimise the amount of voids. The binder, a fine material, is intended to give cohesion to the mixture. The addition of fines, many a times, improves even an unacceptable material as a useful material. Addition of fines to road bases and subgrades should be done in a cautious way. Because by adding fines one should not change a free-draining, non-frost-susceptible material into a poor-draining, frost-susceptible soil. Soil stabilization by the addition of fines has proved a very cheap and powerful technique.

Addition of Material to Reduce Permeability. The properties of clay size materials may vary widely with the composition of the material and with the nature of the exchangeable ion on the material. In general, as discussed earlier, the properties of a soil can be altered by adding fines.

In order to reduce the permeability of a given soil, it is the common practice to add sodium montmorillonite (bentonite). For example, it is stated (Lambe, 1962) that the permeability of a silty sand was reduced from a value of 10^{-4} cm per sec to less than 10^{-9} cm per sec by addition of 10% of bentonite.

It is also possible to reduce the permeability of a given soil by addition of a suitable locally available fine grained soil. In general, being less sensitive and better graded, natural clays can be blended with pervious soils to result in a more nearly permanent blanket than can bentonite.

Removal of Fines from Gravel. Two of the most important uses of gravel are for pavement base courses and for filter courses. In order to use gravel for these purposes, the presence of fines should be less. An approximate upper limit of particle sizes finer than 0.02 mm for a non-frost-susceptible gravel is 3%. For a filter material, the maximum permissible amount of fines depends on the gradation of the neighbouring soil. The easiest way of removal of fines from gravel is by washing. Although the method appears to be simple it needs a large quantity of water. This technique is adopted more often for base-course materials.

7.4 PORTLAND CEMENT (CEMENTING) STABILIZATION

Binding of soil particles together without their alteration is referred to as soil stabilization by cementing. Portland cement and bitumen addition cause stabilization of soil by cementing.

Portland cement is one of the most successfully used soil stabilization. Cement and soil blended material is referred to as soil-cement. The mechanism involved in the process of stabilization of soil by cement is not fully known. It is generally accepted that cement reacts with the siliceous soil to cement the particles together. In a soil-cement more of coarse grained particles are cemented and the proportion of fine-grained soil cementation is small. The physical properties of soil-cement depend on the nature of soil treated, the type and amount of cement utilised, the placement and cure conditions adopted. Soil-cement has been employed for many applications and in particular for the bases of roads and airfields.

7.4.1 Nature of Soil

All inorganic soils which can be pulverised can be stabilized using cement. Soils should be low in organic matter for successful stabilization since this constituent tends to reduce the strength of soil-cement. About 2% of organic matter is considered to be the safe upper limit.

Soils with higher specific surface require more cement for stabilization. Presence of clay in soil causes problems in pulverisation, mixing and compacting the mixture. Further it is difficult to stabilize soils with clays of expanding type.

Exchangeable ions in a soil influence the response of soil treatment. Calcium is the most desirable ion for ease of cement stabilization. Lime or calcium chloride is sometimes added to clays being stabilized with cement.

Apart from organic matter, the chemical composition of the soil is of importance only if appreciable quantities of deleterious salts, such as a sulphates, are present. The harmful effect of these compounds is thought to be due not to a reaction affecting the setting of the cement, but to a subsequent disruption of the soil-cement structure caused by crystallization of highly hydrated salts in the pores.

Soils with the following limits can be economically stabilized (HRB, 1943):

Particle size distribution limits		Plasticity limits
Maximum size 75 mm		Liquid limits < 40%
Passing 4.5 mm I.S. sieve	> 50%	
Passing No. 40 I.S. sieve	> 15%	
Passing 75 μ I.S. sieve	< 50%	Plasticity index < 18%.

In general the best results are obtained with well-graded soils having less than 50% of its particle finer than 0.074 mm and a plasticity index less than 20%.

7.4.2 Amount of Cement

Soil-cement has been made with cement content varying from 5 to 20% for satisfactory stabilization, the following amounts of cement are usually required (Lambe, 1962):

For gravels, a cement level of 5 to 10% by weight

For sands, a cement level of 7 to 12% by weight

For silts, a cement level of 12 to 15% by weight

For clays, a cement level of 12 to 20% by weight

When the cement is hydrating satisfactorily in a mixture, an increase in strength is obtained with increasing cement content. Figure 7.1 illustrates the more the cement added to a soil stronger the resulting soil-cement. The cement content required is usually determined by measuring the

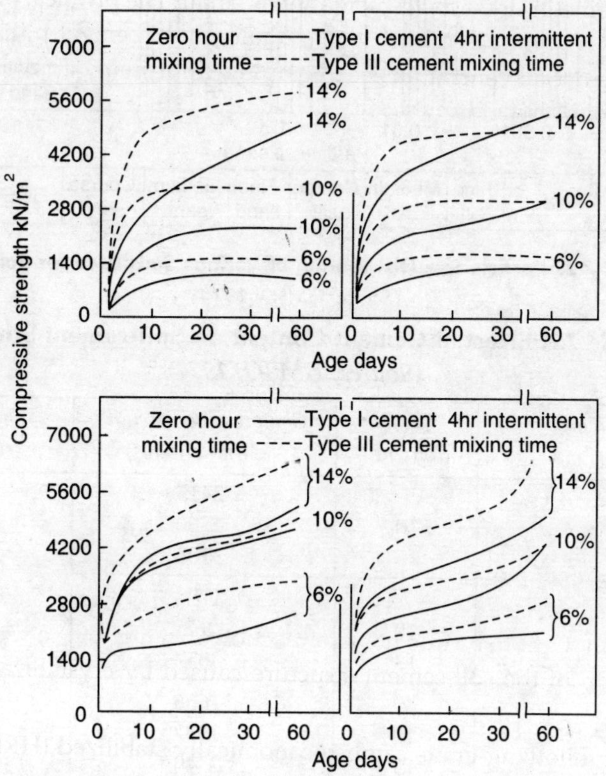

FIGURE 7.1. Comparison of Compressive Strength Obtained with Normal (Type I) Cement and High-early-Strength (Type III) Cement.

(*Source: Felt, 1955*)

compressive strength of specimens made with different proportions of cement. Table 7.1 shows the increase in compressive strength obtained by raising the cement content of various soil-cement mixtures using soils whose particle-size analysis are given in Fig. 7.2. It can be observed that, a given increase in the cement content with the more clayey soils (*e.g.*, soil 1), produced a smaller increase in compressive strength than with sandy soils (*e.g.*, soil 4). It is also observed that the effect of pressure of coarse aggregate (*e.g.*, soil 9) has to give a low compressive strength for lean mixes, but a relatively high compressive strength for rich mixes. Further, high-early-strength cement is (usually) more effective than normal cement (Fig. 7.1).

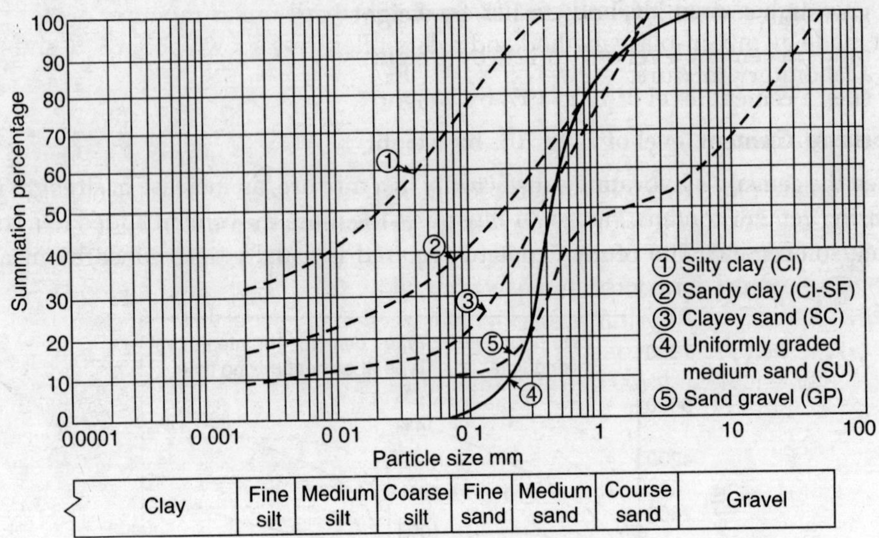

***FIGURE 7.2.* Particle Size Distribution of Various Soils used for Soil-cement.**
(Source: HMSO, 1974)

Table 7.1. Effect of Cement Content on Soil-cement Strength
(Source: HMSO, 1974)

Soil No.	*Soil Type*	*Cement Content %*	*Compressive Strength kN/m² (7 days)*	*Dry Density (kN/m³)*	*Moisture Content %*
1.	Silty Clay	7 10 13	2415 2760 3105	17.43 17.43 17.43	16
2.	Sandy Clay	7 10 13	1794 2622 3657	18.37 18.53 18.53	14
3.	Clayey Sand	7 10 13	1656 1932 2340	17.43 17.74 18.06	12
4.	Clean Sand (uniformly graded)	7 10 13	1449 2829 5934	17.43 18.06 18.53	10
5.	Gravel (poorly graded)	7 10 13	1104 2484 3864	19.47 19.63 19.94	10

7.4.3 Mixing

More uniform soil-cement water mixture, provides strong and durable soil-cement. The intimacy of the mixture is not directly proportional to the mixing energy. As a matter of fact, increase in continued mixing causes a decrease in the degree of mixing and may lead to segregation of components. Thus continued mixing should be only up to the optimal level. Further, mixing after cement hydration has begun can have deleterious effects. It is observed that mixtures made in the laboratory have higher strengths and greater durability than similar mixtures made in the field. Soil-cement made by mix-in-place method and rotary tiller have shown about 50% and 70% of the strength of a laboratory mixture.

7.4.4 Moisture Content

The moisture content plays two roles in soil-cement: (*i*) it influences the compaction characteristics, as with natural soil, and (*ii*) it furnishes water for cement hydration. Of these two, the effect of moisture content on the quality of soil-cement largely arises from its influence on the compaction. The best moisture content for compaction is governed by the soil type and method of compaction. The moisture required for the hydration of cement is adequately provided by the moisture necessary for maximum compaction. The concept of water-cement ratio as used in concrete work is of little value in soil-cement stabilization. Figure 7.3 illustrates the remarkable effect of moulding water on durability and strength.

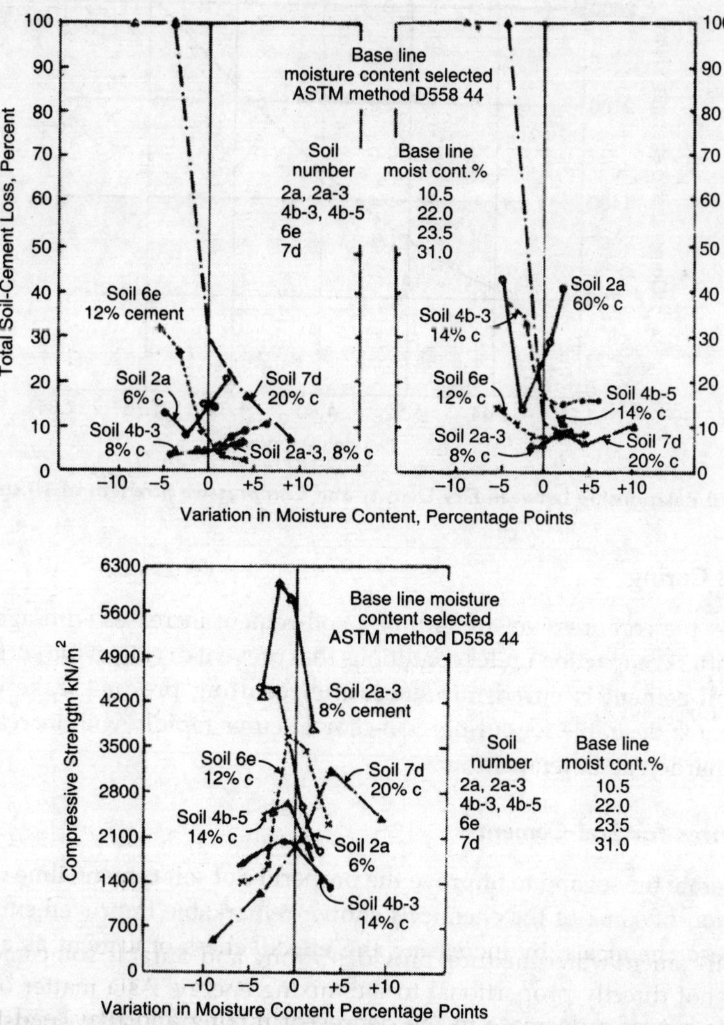

FIGURE 7.3. **Effect of Moisture Content on 28-day Compressive Strength.**
(*Source: Felt, 1955*)

7.4.5 Compaction Conditions

In order to obtain satisfactory soil-cement, adequate compaction is essential. A typical compressive strength/dry density curve is shown in Fig. 7.4, which relates a sandy clay mixed with 10% of cement. As in natural soil it has been observed that for specimens having the same cement content and given the same amount of compaction, but having different moisture contents, the greatest strength is obtained for the one compacted at approximately the optimum moisture content. Published literature suggest, sands should usually be compacted slightly dry, and clays slightly wet, of that moulding-water content which gives maximum density.

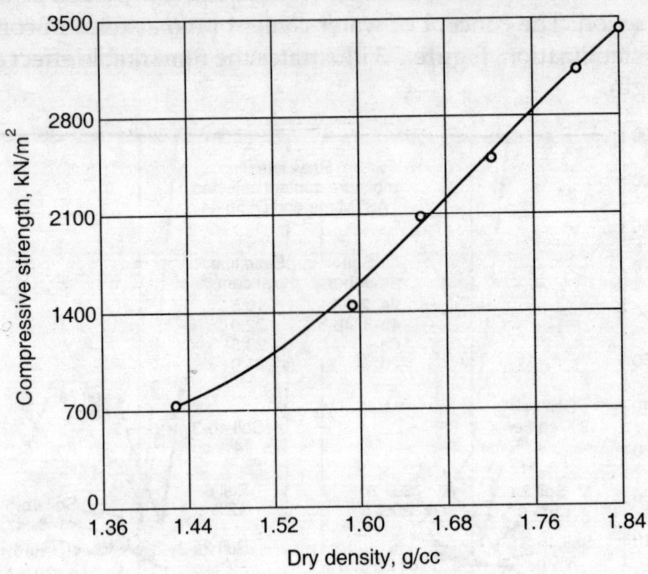

FIGURE 7.4. **Typical Relationship between Dry Density and Compressive Strength of 10 cm Soil-cement Cubes.**

7.4.6 Age and Curing

As with concrete, the compressive strength of a soil-cement increases with age. In practice, soil-cement is cured after compaction under conditions that prevent drying of the surface. The condition under which soil-cement is cured influences the resulting product. Like concrete, a damp environment is most desirable for curing. Soil-cement cures rapidly with increase in temperature although it will harden at all temperatures.

7.4.7 Admixtures for Soil-Cement

In order to accelerate the set and to improve the properties of soil-cement, lime or calcium chloride is added. Addition of some of the chemicals shown remarkable improvement in the strength of soil-cement. These chemicals, by increasing the effectiveness of cement as a stabilizer, permit (Lambe, 1962):

(*i*) A reduction in the amount of cement required to treat a soil responsive to cement.

(*ii*) Stabilization of some of the soils (*e.g.*, certain organic soils) which are not responsive to cement above.

Certain alkalie metal compounds, especially sodium carbonate, suits most of the soils. The type of additive which could be effective should be found by trial. Use of some chemicals along with cement have important advantages such as *(i)* reducing the additive quantity needed to perform a given job, thus simplifying the handling and mixing, and *(ii)* reducing the total stabilizer cost.

7.4.8 Construction of Soil-Cement

Construction of soil-cement usually involves the following operations : (*i*) Shaping the soil to be treated, (*ii*) Pulverizing the soil, (*iii*) Adding water and cement, (*iv*) Mixing, (*v*) Compacting, (*vi*) Finishing, and (*vii*) Curing. Mix-in-place method, Travelling plant method, and Stationary plant method (discussed in Section 7.7) are suitable to produce soil-cement mixtures in the field.

The optimum sequence for adding water and cement and mixing depends on the soil and site conditions. Granular soils are easy to handle compared to plastic soils. Handling of plastic soils can be reduced by adding lime (1 to 3%). The field control of soil-cement stabilization is done by experienced person based on the site visual inspection. Moisture content and dry density have to be checked as per standards.

Construction of soil-cement in frost areas should be done with utmost care. Frost susceptible soils treated with cement are not necessarily made non-frost-susceptible (Lambe, 1962).

Cement is the most successful soil stabilizer and excellent results are guaranteed if used properly.

7.5 BITUMINOUS (CEMENTING) STABILIZATION

Bituminous soil stabilization is an effective method which is being widely used. Bituminous materials are : bitumen, asphalt and tar. **Bitumens** are nonaqueous systems of hydrocarbons which are completely soluble in carbondisulphide. **Asphalts** are materials in which the primary components are natural or refined petroleum bitumens or combinations thereof. **Tars** are bituminous condensates produced by the destructive distillation of organic materials such as coal, oil, lignite, peat and wood.

Bituminous material stabilizes the soil either by binding the particles together or protecting the soil from the deleterious effects of water (*i.e.*, waterproofing) or both these effects may occur together. The first mechanism takes place in cohesionless soils and the second one in cohesive soils.

Among the bituminous materials, most of bitumen stabilization has been with asphalt. Therefore, soil stabilised by asphalt may be referred to as soil-asphalt. Asphalts are produced by three processes : (*i*) Vacuum distillation producing straight-run asphalt, (*ii*) High-temperature pyrolysis of refinery heavies, producing cracked asphalt, and (*iii*) High-temperature air blowing straight-run asphalt, producing blown asphalt.

As the straight-run asphalt has low softening temperature and low melt viscosity, it is commonly used in soil stabilization. Asphalt can not be directly added to the soil because it is too viscous. Its fluidity can be increased by (*i*) heating, (*ii*) emulsifying in water (emulsions), or (*iii*) cut back with some solvent like gasoline (cutbacks). Both emulsions and cutbacks are used in soil stabilization. Although soil-asphalt has varied applications, it is mostly used in bases for highway and airfield pavements.

7.5.1 Nature of Soil

All inorganic soils with which asphalt (emulsion or cutback) can be mixed can be stabilized. Soils satisfying the following requirements yield the best results (Lambe, 1962):

(*i*) Maximum particle size less than one-third the compacted thickness of the treated soil layer.

(*ii*) Greater than 50% finer than 4.76 mm size.

(*iii*) Thirty five to 100% finer than 0.42 mm size.

(*iv*) Greater than 10%, but less than 50% finer than 0.074 mm size.

(*v*) Liquid limit less than 40%.

(*vi*) Plasticity index less than 18%.

Organic matter of acid origin is detrimental to soil-asphalt. Asphalt stabilization can not be effective in fine-grained soils with high *pH* and dissolved salts. It is difficult to handle plastic clays because of mixing problem.

7.5.2 Amount of Asphalt

An increase in asphalt content gives better results. In fine-grained soils addition of asphalt does not increase the strength but tremendously improves the waterproofing property and thereby yielding a better stabilized soil. Asphalt also should be added optimally otherwise results in a gooey mixture which can not be properly compacted.

7.5.3 Mixing

A thorough incorporation of the additive with the soil yields a better stabilized soil. Figure 7.5 shows the effect of mixing on the strength of soil-asphalt.

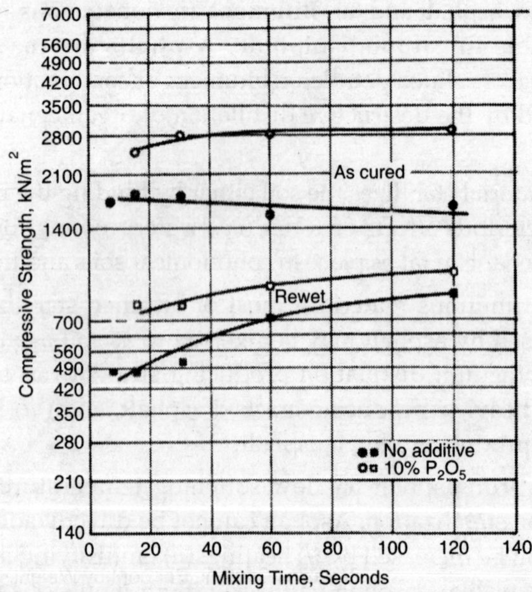

FIGURE 7.5. **Effect of Mixing on Strength of P_2O_5-Modified Cutback Stabilized Clayey Silt.**

(*Source: Michaels and Puzinauskas, 1956*)

7.5.4 Compaction Conditions

The density of a mixture of soil and asphalt is governed by the volatiles content and amount and type of compaction. In general lower the volatiles content, the higher the strength. Further, samples which were cured and then immersed in water showed a maximum strength at a moulding volatiles content near or slightly above that for maximum compacted density and the water pickup and thus strength loss, is least at this moulding volatiles content (Lambe, 1962). In plastic soils the volatiles content which gives maximum cured strength and that which gives optimum density can be quite different and the difference can vary with type of compaction.

7.5.5 Cure Conditions

The following behaviours have been reported to be true (Lambe, 1962) : (*i*) the longer the period of cure and warmer the temperature of cure, the greater the volatiles lost and (*ii*) the longer the period of immersion, the greater the water pickup. The strength of a soil-asphalt is inversely proportional to the volatiles content at the time of test. This valuable property is clearly reflected in Fig. 7.6 in which a general strength-volatiles content relationship was obtained regardless of the formulation employed.

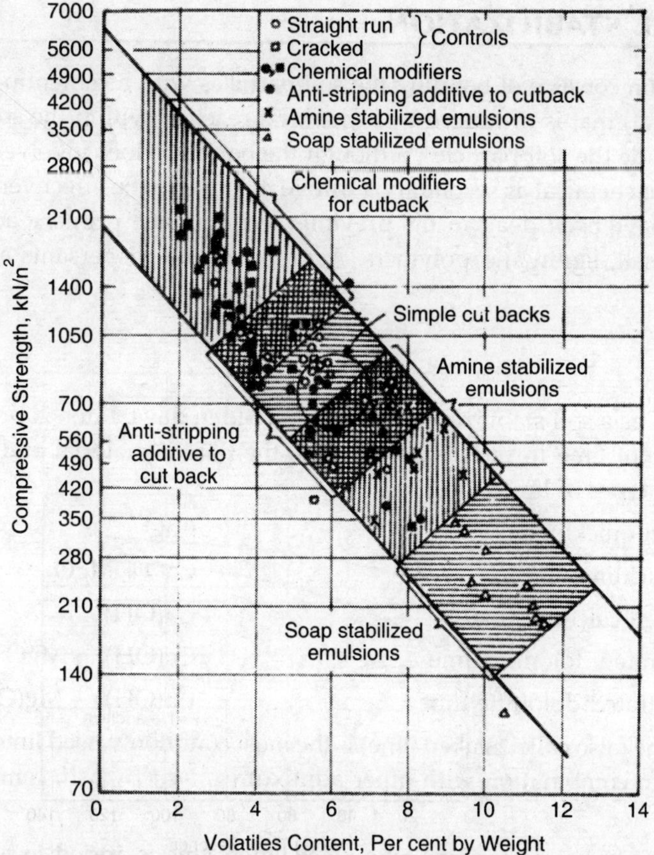

FIGURE 7.6. **Rewet Strength of Asphalt Stabilized Soil V_S Volatiles Content; Generalized Correlation.**

(Source: Michaels and Puzinauskas, 1956)

7.5.6 Construction of Soil-Asphalt

The conventional sequence of construction operations are as follows (Lambe, 1962) : (*i*) Pulverisation of the soil to be treated, (*ii*) Addition of water for proper mixing, (*iii*) Adding and mixing of the bitumen, (*iv*) Aeration to the proper volatiles content for compaction, (*v*) Compaction, (*vi*) Finishing, (*vii*) Aerating and curing, and (*viii*) Application of surface cover. The important items to ensure proper stabilization which need control are mixing, compacting, drying and applying the surface protection. The mixing plants used for soil-cement can be used for soil-asphalt also. The necessary field control tests are moisture content determination before and during processing, bitumen content determination after mixing and density determination after compaction.

The optimum moisture content for stability is usually below that for compaction. As good mixing is generally considered to be most easily obtained at fairly high moisture contents, it is often found necessary, except with sands, to allow a period for the mix to dry between the mixing process and compaction. In practice, treated sands are placed at about 3 to 5% volatiles content whereas cohesive soils are placed at about the optimum volatiles content for compaction. Compared to cutbacks or tars, emulsions provide more latitude in the stabilization of fine-grained soils.

7.6 CHEMICAL STABILIZATION

Chemical stabilization consists of bonding the soil particles with a cementing agent (the primary additive is a chemical) that is produced by a chemical reaction within the soil. The reaction does not necessarily include the soil particles, although the bonding does involve intermolecular force of the soil. The use of chemical as secondary additive to increase the effectiveness of cement and of bitumen (asphalt) have been dealt in the previous sections. The primary additives generally in wide use are lime, salt, lignin and polymers. Aggregants and dispersants are also in use under certain conditions.

7.6.1 Lime

Lime has been used as a soil stabilizer for roads from olden days. Lime is produced from natural limestone. The type of lime formed is based upon the parent material and production process. There are five basic types of lime:

High-calcium quicklime	...	CaO
Dolomite quicklime	...	$CaO + MgO$
Hydrated high-calcium lime	...	$Ca(OH)_2$
Normal hydrated dolomite lime	...	$Ca(OH)_2 + MgO$
Pressure-hydrated dolomite lime	...	$Ca(OH)_2 + Mg(OH)_2$

Hydrated lime (also called slaked lime) is the most commonly used lime for soil stabilization. Lime is also used in combination with other admixtures, *viz.*, fly ash, cement, bitumen for soil stabilization.

Two types of chemical reactions take place when lime is added to wet soil. The first one occurring almost immediately is a colloidal-type of reaction involving any of the following : (*i*) ion exchange of calcium for the ion naturally carried by the soil, (*ii*) a depression of the double layer on the soil colloids because of the increase in cation concentration in the pore water, and (*iii*) an expansion of the double layer of the soil colloids from the high *pH* of the lime. The second reaction takes considerable time in a cementing action. The cementing action, also called pozzolanic action, is not completely understood, but is thought to be a reaction between the calcium from the lime with the available reactive alumina or silica from the soil (Lambe, 1962).

Soil plasticity, density and strength are changed by the addition of lime to soil. Lime generally increases the plasticity index of low plasticity soils and decreases the plasticity index of highly plastic soils. Because of reduction in the plasticity of plastic soils, due to addition of lime, the soil becomes more friable and easy for handling in the field. Addition of lime causes a reduction in the maximum compacted density and an increase in the optimum moulding water content. In general, lime increases the strength of almost all types of soil.

Construction procedure of lime-stabilized soil bases are similar to those employed for soil-cement with a difference that more time is allowed for placement operations for lime. This relaxation is possible as the lime-soil cementation reaction is a relatively slow one. Adequate care should be taken to prevent carbonation of the lime. The normal construction sequence for lime-stabilized bases is as follows : (*i*) Scarify the base, (*ii*) Pulverise the soil, (*iii*) Spread the lime, (*iv*) Mix the lime and soil, (*v*) Add water if necessary to bring to optimum moisture content, (*vi*) Compact the mixture, (*vii*) Shape the stabilized base, (*viii*) Cure-keep moist and traffic-free for at least 5 days, and (*ix*) Add wearing surface.

7.6.2 Calcium and Sodium Chlorides

Both calcium and sodium chlorides as soil stabilizers react in some what similarly. Salt has been used in recent years as additive in the construction of granular stabilized road wearing and base courses. The effect of salt on soils is from (*i*) causing colloidal reactions, and (*ii*) altering the characteristics of soil water. Although calcium and sodium chlorides act as soil flocculants, they are not as effective as other chemicals such as ferric chloride.

Most of the beneficial actions of salt in soil are mainly due to the changes salt makes in the characteristics of the water in the soil pores. These changes reduce the loss of moisture from the soil and are explained by the fact that the salts (especially calcium chloride) are deliquescent and hydroscopic and lower the vapour pressure of water. Frost heave in soil is reduced due to addition of salt by lowering the freezing point of water. As most of the benefits of salt are due to the presence of the salt in the soil pore fluid, any loss of salt concentration may reduce the strength of the stabilized soil. Thus the performance of salt-stabilised soil depends on the amount of ground-water movement. Salt addition shows a slight increase of maximum compacted density and a slight reduction in the optimum moulding water content.

7.6.3 Lignin

Lignin is available both in the powder form and in the form of sulphite liquid. Lignin in both the forms has been in use as an additive to the soil for many years. Lignin is water-soluble, hence its stabilizing effects are not permanent. In an attempt to improve the action of lignin, the chrome-lignin process was developed and studied by Smith (1952). An insoluble gel is formed when sodium bichromate or potassium bichromate is added to the sulphate waste. The effect of lignin on the soil properties is based on the form of lignin and the type of soil treated. Lignin acts as an acid if not neutralised. It is also reported that lignin also reduces from heave.

7.6.4 Waterproofers

Fine-grained soils show considerable strength when they are dry and lose the strength when consumes more water. Waterproofers, *i.e.*, chemicals which prevent or reduce the deleterious attack of water on soils, have proved to be highly useful in stabilization techniques. Waterproofers function as follows (Lambe, 1962) : one end of the waterproofer molecule becomes preferentially adsorbed to, and then reacts with the soil surface; the other end of the molecule, being hydrophobic, repels water and thus makes the soil mineral nonwettable by water. Waterproofers which are recognised as soil stabilizers are alkyl chlorosilanes, siliconates, amines, and quaternary ammonium salts. Waterproofers do not increase the strength of soils but rather help them to retain the natural strength in the presence of water.

7.6.5 Natural and Synthetic Polymers

Polymers are long-chained molecules formed by the linking, *i.e.*, polymerizing of certain organic chemicals called monomers. Natural polymeric materials (like resins) and synthetic polymeric materials (like polyvinyl alcohol (PVA)) have been used as soil stabilizers. There are two ways of chemical incorporation. In the first method, the monomer along with a catalyst (to cause polymerisation) is added which reacts between the soil and monomer. In the second method, performed polymer is added to the soil in the form of a solid, a solution, or an emulsion, causing reactions between the polymer and the soil. Lambe (1962) has presented a few of the resins (both natural and synthetic), which have been used for soil stabilization, along with their applicability and the same is given below:

(A) Natural Resins

(i) **Vinsol Resin.** Imparts almost no strength to soil, but appears to waterproof the fine-grained soils.

(*ii*) **Rosin.** When added to a soil and then reacted with certain metal salts forms insoluble gels which aid stabilization.

(*iii*) **Resin Stabilizer 321.** A rosin derivative; reduces the rate and amount of water adsorption and furnishes sodium ions for ion exchange.

(*iv*) **Stabinol.** A rosin derivative mixed with portland cement; waterproofs the coarser soils.

(*v*) **NSP-121 and NSP-252.** Materials made by National Southern Products Co., have given some water repellency to certain soils.

(*vi*) **NVX.** A product of the Hercules Powder Co., has given some water repellency to certain soils.

(*vii*) **Shellac.** Set obtained by evaporation of a solution of an ester of shellac with maleic acid or by evaporation of solvent from a solution of shellac in alcohol. Sand treated with shellac can have high strength, but practically all this strength is destroyed by water immersion.

(*viii*) **Natural Resin from Tropical Areas.** Number of resins have been investigated for water-proofing. Manila Copal resin and Wallaba resin showed slight effectiveness.

(B) Synthetic Resins

(*i*) **Aniline-furfural.** Waterproofs and imparts strength to soil.

(*ii*) **Polyvinyl Alcohol (PVA).** Forms tough, flexible films when evaporated from aqueous solutions. Films are water-soluble, and to date attempts to make them insoluble have been unsuccessful.

(*iii*) **Polyvinyl Acetate.** Gives very high strength to sand, practically all of which is destroyed by water immersion.

(*iv*) **Resorcinol-formaldehyde.** Gives some strength to sand, but much of it is lost by water immersion.

(*v*) **Others.** Urea-furfural and phenol-furfural resins; Phenol-formaldehyde combinations; Urea-formaldehyde resins; Calcium sulfamate-formaldehyde resins; Ethocel; Methoylol ureas and melamine.

Among the vast number of synthetic polymers, calcium acrylate is of interest to soil engineers as it is the forerunner to stabilization with water-sensitive polymers.

7.6.6 Aggregants and Dispersants

Aggregants and Dispersants are materials that, at low treatment levels, make relatively modest changes in the properties of fine-grained soils. The function of aggregants and dispersants is alteration of electrical forces between soil particles. Like portland cement, aggregants and dispersants do not cement adjacent particles. This structure altering character affect many soil properties such as plasticity, permeability and strength. Although aggregants and dispersants have been widely used in different industries, it is not frequently made use of in soil stabilization. These chemicals have high potentials for use in situations where only a modest improvement in soil behaviour is needed. Since these chemicals are relatively cheap, are effective at low treatment levels, and can be relatively easily incorporated, they can permit comparatively low cost soil stabilization (Lambe, 1962).

Aggregants. These are the materials which are capable of increasing the net electrical attraction between fine-grained soil particles tending to aggregation or flocculation of the soil mass. Aggregants are of two types, *viz.*, in organic salts such as calcium chloride or ferric chloride and polymeric materials such as Krilium. In general the salts are cheaper than the polymeric materials.

When salts are dissolved in the pore water of a soil, they generate cations and anions which can participate in one or more of the following reactions (Lambe, 1962):

(*i*) Exchange with soil ions

(*ii*) Become adsorbed on the particles

(*iii*) Furnish ions that can link particles (*e.g.*, K^+ can fit within the lattices of two adjacent clay sheets upon drying to bond the sheets permanently together).

(*iv*) Increase the ion concentration, thereby reducing the electric repulsion between particles.

Using the above reactions, polymeric materials can also link the adjacent soil particles since the ends of the long chain polymeric molecules can become attached to soil particles. As the action of an aggregant is due to more than one of the possible mechanisms, it is difficult to assess the contribution of each.

If attraction between soil particles is increased, aggregants cause them to stick and form larger particles. The sticking particles tend to arrange themselves in a random loose array. This random arrangement of particles with a loose structure make the aggregants to increase soil permeability. By the aggregant treatment an increase in permeability in the range of two to twenty fold can be obtained. By altering the permeability characteristics of a soil, aggregants can influence the frost-heave behaviour of the soil.

Aggregants improve soil strength by reducing the electric repulsion between soil particles. For example, a leached quick clay can regain strength by the introduction of salt (an aggregant) into the pores of the clay.

Dispersants. These are the materials which are capable of reducing the cohesion between fine-grained soil particles and tend to cause them to disperse by increasing the electric repulsion between adjacent soil particles. A wide variety of materials are available to use as dispersants and the most common ones are the phosphates, sulphonates and versanates. Practically there is no difference in the effectiveness of the various dispersants.

Dispersants reacting with soils adopt the following three mechanisms (Lambe, 1962):

(*i*) **Sequestration.** The polyanionic part of the dispersant removes and insolubilizes any monovalent exchangeable ions.

(*ii*) **Ion exchange.** The dispersant furnishes monovalent ions for exchange reaction with the soil.

(*iii*) **Anion adsorption.** The dispersant furnishes polyanionic groups for adsorption by the soil particles.

Of the above three mechanisms, the anion adsorption is the most important.

Dispersants decrease the particle size, that is, break down aggregates and increase the fluidity of a soil-water system. In a dispersed soil system, adjacent particles do not tend to cohere to form aggregates, but repel each other so they can be easily moved relative to each other. Thus by mechanical work, they can be forced into a mass of high density.

Denser sediments and sediments with particles in a more nearly parallel array, (that is a more tortuous path for flow of water), dispersants cause reduction in soil permeability. The permeability reduction is of the order of one-fifth to one-fiftieth of the original value.

Dispersants can be added to the soil by physically mixing or by injecting. Because of reduction in permeability due to dispersants, there will be a tendency for frost susceptibility also.

7.6.7 Miscellaneous Chemical Stabilizers

There are several chemicals which are used for soil stabilization merely based on the abundant availability or based on some theoretical principles. Among these materials are (Lambe, 1962) : (*i*) Tung oil, (*ii*) Linseed oil, (*iii*) Cotton seed oil, (*iv*) Castor oil, (*v*) Rubber latex, (*vi*) Plasticized sulphur, (*vii*) Molasses, (*viii*) Mineral oil, (*ix*) Sodium carbonate, (*x*) Calcium carbonate, (*xi*) Paraffin, (*xii*) Hydrofluoric acid and (*xiii*) Ionic detergents. These chemicals are nowadays very sparingly used. Two other chemicals, *viz.*, sodium silicate and phosphoric acid have some special properties and they are discussed below.

Alkalic silicates, especially sodium silicate, belongs to a group of inexpensive and versatile compounds. They are used in soil stabilization both as additives to conventional stabilizers (*e.g.*, cement) and as primary stabilizers. Silicates derive their benefits because of their cementitious components. The most widely used form of silicate stabilization is by injection (discussed in Chapter 6).

Phosporic acid has proved as a potent stabilizer. The solidification is attained only by the phosphoric acid. However the reaction can be improved by the addition of fluorine compounds as pure accelerators or inorganic salts as waterproofers. Soils containing strong basic components do not respond to phosphoric acid treatment.

7.7 CONSTRUCTION METHODS

The construction methods and types of equipment which are in use for construction of stabilized roads are discussed below. These methods may also be used for treatment of foundation soils wherever applicable. Different compaction methods are discussed in Chapter 2.

7.7.1 Mix-in Place Method

In this method of construction a "train" of machines is run over the soil to be processed. In order to break up the soil, which is the first process, rippers, cultivators or rotary tillers are used and to loosen the soil to a uniform depth, ploughs or scarifiers are used. Water is then added to the loose soil from a sprayer to bring it to a suitable moisture content for processing. The necessary stabilizer is then added, either by spray wagon, in the case of a fluid stabilizer, or by hand or preferably from a bulk spreader in the case of stabilizer which is in the form of a powder. The mixing of stabilizer into the soil is done by further passes of the rotary tillers or of special soil mixers. Shaping of the loose mixed material is followed by compaction by a suitable roller. No forms are used to define the edge of road construction. It is more economical to construct the stabilized roads slightly wider than the planned width of the road so as to cut back to the derived width when the stabilized construction is completed. Further, the excess stabilized width can be used as foundation for kerbs.

Many stabilizing agents require a period of curing after mixing, before they become fully effective. For necessary curing to attain the surface is covered with a layer of moist soil or straw or alternatively, the surface may be kept damp by frequent application of a light spray of water. Sometimes a bituminous priming coat may be used as a curing agent by applying it soon after compaction. Before the application of a surfacing, the stabilized soil should be sprayed with a priming coat. The priming coat which provides a key for the surfacing, should be appropriate to the type of surfacing material.

The mix-in-place method can be effectively used depending on the type of plant used in the mixing process. The conventional types of plant used are the rotary hoe, the seaman pulvimixer, and single pass stabilizer (HMSO, 1974).

The advantages of the mix-in-place method (applicable to the use of agricultural plant and large single pass machines) are (HMSO, 1974) : (*i*) the plant is simple, cheap and easily transported, (*ii*) the number of machines required can be adjusted to the size of the job, (*iii*) the whole processed section is ready for compaction at the same time, (*iv*) a large average output may be maintained, and (*v*) in a wet climate the loss of water by evaporation may be advantageous, it is the only way of getting rid of excess moisture.

The disadvantages of this method are (HMSO, 1974) : (*i*) it is not easy to obtain a uniform thickness of treatment, because of the difficulty of setting the machines to a given depth, (*ii*) the mixing is not as uniform as with travelling or stationary mixers, (*iii*) heavy rain is liable to spoil a whole section, and (*iv*) in a dry climate the water lost by evaporation is difficult to replace.

7.7.2 Travelling Plant Method

In this method, the procedure for application of the stabilizer is the same as that of mix-in-place method. In the case of cement, the pulverised soil is heaped into a windrow by a specially converted motor-grader or a sizer and the cement is spread on top. Occasionally the cement is spread before windrowing which is unusual. In the case of fluid stabilizers, the soil is windrowed and the stabilizer added by the travelling mixer which, as in cement stabilization, moves along the line of the windrow. Two of the conventional mixers which are the Barber-Greene Mixer and the Gardener Type Mixer (HMSO, 1974).

The advantages of Travelling Plant Method are (HMSO, 1974) : (*i*) accurate proportioning of added water, (*ii*) uniform mixing, (*iii*) short mixing time, (*iv*) a uniform subgrade surface can be obtained, and the depth of treatment can be controlled, and (*v*) it has the highest output for a given expenditure of plant and labour.

The disadvantages are (HMSO, 1974) : (*i*) high initial cost of plant, (*ii*) the consequent need for the plant to work continuously at full capacity, and (*iii*) work may be stopped for a minor breakdown on one piece of plant.

7.7.3 Stationary Plant Method

In the Stationary Mixing plant there are two main types (HMSO, 1974), *viz.*, continuous mixers and batch mixers.

The continuous mixer works under the same principle as that of travelling mixer. In this mixer, an elevating loader supplies material to a hopper with a measuring gate, hence a belt conveyor discharges it to a pug-mill, where water or a fluid stabilizer may be added through spray nozzles, and mixed into the soil. The mixed materials, is then discharged into lorries. Immobilized travelling plant may easily be arranged to serve as a central mixing plant. Depending on the output requirement the size of the central mixing plant is decided.

Batch mixers may be concrete mixers, double-paddle mixers, *etc.*, which are used for small jobs such as for mixing coarse-grained soils with stabilizers. For many of the jobs ordinary tilting-drum concrete mixers have been used. However, the best results are obtained by using double-paddle mixers, pug-mills or roller pair types machines in which soil lumps are easily broken up. The time required for mixing depends on the type of mixer and the type of soil. It is possible to discharge the mixed materials vertically from the mixer into lorries and it can then be transported directly to the site, tipped, spread and compacted in the normal manner.

The advantages of the Stationary Plant Method are (HMSO, 1974) : (*i*) accurate proportioning of the mixer, (*ii*) easy control of depth of treatment, (*iii*) concrete mixers can be used, (*iv*) no additional haulage if soil has to be taken from a borrow-pit, (*v*) small losses of moisture during mixing and transport of material, and (*vi*) the method is suitable for use with form work, for instance when vibrators are required to compact uniform sands or when the stabilized layer is to form a sub-base to machine-laid concrete.

The disadvantages are (HMSO, 1974) : (*i*) expensive if soil in site is processed, and (*ii*) material must be compacted as delivered, and not as a complete section.

7.7.4 Field Control

In order to obtain an efficient result from the above discussed construction methods, it is necessary to check the following items (HMSO, 1974) : (*i*) degree of pulverisation, (*ii*) moisture content, (*iii*) dry density/moisture content relationship, (*iv*) cross-sectional area of windrow, (*v*) depth of treatment or spreading, (*vi*) quality of mixed material, (*vii*) dry density of compacted layer, and (*viii*) stabilizer content. All these items are to be checked except (*i*) which is needed for mix-in-place method only, and (*vi*) which is needed for travelling plant method only. Methods of test for stabilised soils are provided by Indian Standards (IS : 4332, Parts I to X).

Degree of Pulverization. The soil should be pulverized such that 80% of the pulverized material shall pass 4.5 mm sieve. This does not apply to coarse aggregates. The checking is done by sieving representative samples during the pulverisation process.

Moisture Content. In stabilized construction, moisture content determination is the control test most frequently required. Samples are taken for every 100 to 150 of the pulverised soil along the centre-line of the site. After determining the water contents, tentative estimates may be made for the requirement of additional water, if any, so that watering can start as soon as the stabilizer is partly mixed with the soil. After mixing of soil and stabilizer, once again samples are taken at the same locations for further checking. This decides the exact and the final amount of water requirement. Check tests are made at the same locations towards the end of mixing of any water added. Further during compaction, the moisture content is again checked to determine the possibility of any drying of the surface. Stages at which moisture content determinations are desirable during mix-in-place in construction is presented in Table 7.2 (HMSO, 1974). The procedure of checking moisture content is similar in travelling plant method. In stationary plant method it is sufficient to control the moisture content of the incoming soil and the mixed material produced at hourly intervals from a given machine. The field determination of moisture content is discussed in Chapter 2. Also reference may be made to Indian Standard (IS : 4332, Part II-1967) for determination of moisture content of stabilized soil mixtures.

Table 7.2. Moisture Content Determination at Various Stages in Mix-in-Place Construction (*Source: HMSO, 1974*)

Sr. No.	*Stage of Work*	*Object*
1.	Pulverisation of soil	To check whether soil is within 2% of the specified moisture content
2.	Beginning of day's work	To obtain approximate estimate of water requirements
3.	At the completion of mixing of soil and stabilizer	(*a*) To determine the exact amount of water required (*b*) To estimate the rate of evaporation
4.	At the completion of mixing any extra water added after step 3	Check tests
5.	During compaction	Check tests to ensure surface has not dried.

Dry Density/Moisture Content Relationship. This test depends on the soil type and it is essential to check for dry density and moisture content from time to time during the construction to make allowance for any variation from the results of the preliminary tests in the laboratory (IS : 4332, Part III-1967).

Cross-Sectional Area of Windrow. For the travelling plant working on a windrow, it is essential to measure the cross-sectional area of the windrow to ensure that the proportioning of stabilizer is correct and that the correct thickness will be laid.

Depth of Treatment of Spreading. In the mix-in-place construction, the depth of treatment must be controlled from the beginning of the processing, and the pulverising and mixing equipment must be carefully adjusted. If stationary plant is used, the thickness of treated material can be controlled by spreading it with a bulldozer, the blade being supported on runners. The thickness of a compacted layer can be checked while checking for dry density.

Quality of Mixed Material. As the design of a specific work is based on the quality of the stabilized material produced in the laboratory it is essential to check the field plant. Field specimens are made similar to that of the laboratory specimen and appropriate tests are made based on the stabilizer used. For example, a cement-stabilized specimen would be tested for compressive strength after 7 days, and a bituminous-stabilized specimen would be tested for capillary absorption. Samples are usually taken at intervals of 30 to 160 m along the road. In order to check uniform quality, samples should be taken also along the cross-sections of the road.

Density of the Compacted Layer. Dry density of the stabilized soil is determined for every 30 to 160 m and sand replacement method is usually adopted. It is necessary to check the dry density after each day's construction so as to enable adjustments in the compacting procedure to be made. Dry density should not be less than that specified by 0.80 kN/m^3.

Stabilizer Content. In all the methods it is necessary to check the stabilizer content. Running control of this factor enables an excess or a deficiency of the stabilizer to be corrected as soon as possible. As an established field laboratory is needed for accurate check of stabilizer content, it is only set up in medium to large jobs. However, for small jobs it is often useful to take representative samples for testing at a material testing laboratory. Indian Standard has recommended procedures for determining cement, lime and bituminous contents [IS : 4332, Part VII (1973), Part VIII (1969), and Part IX (1970)].

Chapter **8** GEOSYNTHETICS

8.1 INTRODUCTION

Geosynthetics are artificial fabrics used in conjunction with soil or rock as an integral part of a man-made project. The two major groups in geosynthetics are geotextiles and geomembranes. While geotextile is a permeable fabric, geomembrane is an impermeable one. Stressing permeability, or the lack of it, as the distinguishing feature, helps to separate the different roles of these two groups of materials. As more products with slight difference has been manufactured and used for different field applications, they are identified by different terms as listed below (John, 1987):

Geospacers	:	impermeable spacers used within fin drains
Geowebs	:	an American term for cellular geotextiles
Geogrids	:	geotextile related products with large rectangular apertures (more correctly called geotextile grids) or non-rectangular apertures (more correctly called geotextile nets)
Geosynthetics	:	geotextiles, geomembranes and geotextile related products, but excluding those based on natural fibres
Geofabrics	:	planar flat sheet geotextiles and geotextile related products, excluding geotextile mats
Geoproducts	:	geosynthetics, geotextile related products made from natural fibres, geospacers, and metallic soil reinforcement
Geocomposites	:	composites consisting of two or more geoproducts.

Geosynthetics have a wide civil engineering field applications. The main functions which are in use are: (*i*) separation, (*ii*) fluid transmission, (*iii*) reinforcement, (*iv*) filtration, and (*v*) containment and barrier.

Geosynthetics have been applied on geotechnical and construction engineering for the past 25 years. Geosynthetics is no longer a new born technology, but is now entering a more mature stage with constantly expanding applications even outside of the traditional construction field.

8.2 GEOSYNTHETIC TYPES

There are a large number of different geosynthetics are produced, but for the purposes of classification they can be sub-divided generally into the following groups, *viz.*, Wovens, Non-Wovens, Knitted, Biogradable, Nets and Grids, Three dimensional Mats, Composites, and Membranes.

8.2.1 Raw Materials

The raw materials used in the manufacture of geosynthetics are thermoplastics. However, a few specialist geosynthetics may also incorporate either steel wire or natural biodegradable fibres. Examples of biodegradable materials used within geosynthetics are jute, wood shavings, and paper strips. Following is the brief description of the raw materials.

Polyamide. There are two most important types of polyamide (PA). The first one is an aliphatic polyamide obtained by polymerisation of the petroleum derivative ε-caprolactam. Aliphatic polyamides are composed of chains which do not contain ring-shaped rigid structures. The other type is also an aliphatic polyamide. It is obtained by the polymerisation of a salt of adipic acid and hexamethylenediamine and both of them are petroleum products. They are manufactured in the form of thread or tape which are cut into granules.

Polyester. Polyester (PETP) is made by polymerizing ethylene glycol with dimethyl terephthalate or with terephthalic acid. All these three materials are derivatives of petroleum. Polyster is produced discontinuously in two reactors, in series or in a continuous process using more reactors in series.

Polyethylene. Polyethylene (PE) can be produced in a highly crystalline form, which is an extremely important characteristics in fibre-forming polymer. Three main groups of polyethylene are available, *viz.*,

LDPE — low density polyethylene (density 920—930 kg/m^3)

LLDPE — linear low density polyethylene (density 925—945 kg/m^3)

HDPE — high density polyethylene (density 940—960 kg/m^3).

LDPE was the first polyethylene developed followed by HDPE. LDPE is produced at very high pressures whereas HDPE processes are initiated by special catalysts at relatively low pressures and temperatures. HDPE is more rigid, stronger, tougher and has a better chemical resistance than the low density types. However, LDPE is used in applications where its flexibility and water vapour barrier properties can be utilized.

LLDPE is a third group of polyethylenes made also by a low pressure process. It is manufactured by co-polymerizing ethylene with a small amount of alpha-olefins which lowers the density by forming short chain side branches on the linear polymer chain.

HDPE is preferred for the production of PE filaments and tapes for use in geotextiles whereas LDPE and LLDPE are used more for the production of film. Polythelene is availablе in granular or powder form.

Polypropylene. Polypropylene (PP) is a crystalline thermoplastic produced by polymerizing propylene monomers in the presence of a stereo-specific catalyst system. Polypropylene is the highest (900–910 kg/m^3) plastic produced to date. Homopolymers and copolymers are the two types of polypropylene. Homopolymers are used for fibre and yarn applications. Polypropylene is mainly available in granular form.

Polyvinylchloride. Polyvinylchloride (PVC) is mainly used in geomembranes and as a thermoplastic coating material. The basic raw material utilized for the production of PVC is vinylchloride. PVC is a rigid polymeric material but can be converted, when plasticizers are added, into highly flexible products. PVC is available in free-flowing powder form.

Ethylenecopolymer Bitumen. Ethylenecopolymer bitumen (ECB) membranes have been used in civil engineering works as sealing materials. For ECB production, the raw materials utilized are ethylene and butyl acrylate (together forming 50 to 60%) and a special bitumen of 40 to 50%. Stricter specifications are required when they are used as a seal in connection with resistance to hydrocarbons, acids and alkalies.

Chlorinated Polyethylene. Sealing membranes based on chlorinated polyethylene (CPE), are generally manufactured from CPE, the main component, mixed with PVC or sometimes PE. The properties of CPE depend on the quality of the PE and the degree of chlorination. CPE membranes are available in rolls of a maximum width of 2 m and a thickness in the range of 0.6 to 1.5 mm with a roll length of about 20 m.

Apart from the raw materials explained above there are other polymers also available with less market value.

8.2.2 Wovens

There are a large number of geosynthetics produced which can be sub-divided into several different categories based upon their method of manufacture and is shown in Fig. 8.1 on next page. They may be basically belong to natural fabrics or synthetic fabrics. Under synthetic fibres they may be broadly classified as conventional geotextiles, geotextile related products and geomembranes.

Wovens were the first to be developed from synthetic fibres. The type accounts for about 25% of the geotextiles market in terms of volume. As their name implies, they are manufactured adopting techniques similar to weave clothing textiles. This type has the characteristic appearance of two sets of parallel threads or yarns. The yarn running along the length is known as a warp and the one perpendicular to it is called a waft (Fig. 8.2 on next page).

The majority of low to medium strength woven geosynthetics are manufactured from polypropylene which can be in the form of : extruded tape, slit film, monofilament, or multifilament. Often a combination of yarn types is used in warp and waft directions to optimise performance/ cost. Higher permeability are obtained with monofilament and multifilament than flat tape construction only.

8.2.3 Non-wovens

Non-woven geosynthetics can be manufactured from either short staple fibre or continuous filament yarn. The fibres can be bonded together by adopting thermal, chemical or mechanical (needle punched) techniques or a combination of techniques. Whether at staple fibre or a continuous fibre is used has very little influence on the properties of the non-woven geosynthetics.

- Geosynthetics
  - Natural fibres
    - Jute grids, paper strips, wood shavings
  - Synthetic fibres
    - Conventional geotextiles
      - Non-woven
        - Mechanically bonded
          - Needle-punched
        - Chemically bonded
        - Thermally bonded
          - Calendered single polymer
          - Melded
          - Calendered multi-polymer
      - Knitted
      - Woven
        - Mono-filament
        - Multi-filament
        - Slit film
    - Geotextile related products
      - One-dimensional
        - Strips ties
      - Two-dimensional
        - Open mesh
          - Grid nets
        - Close mesh
          - Webs
      - Three-dimensional
        - Mats
    - GEO membranes
      - Liner
      - Barrier

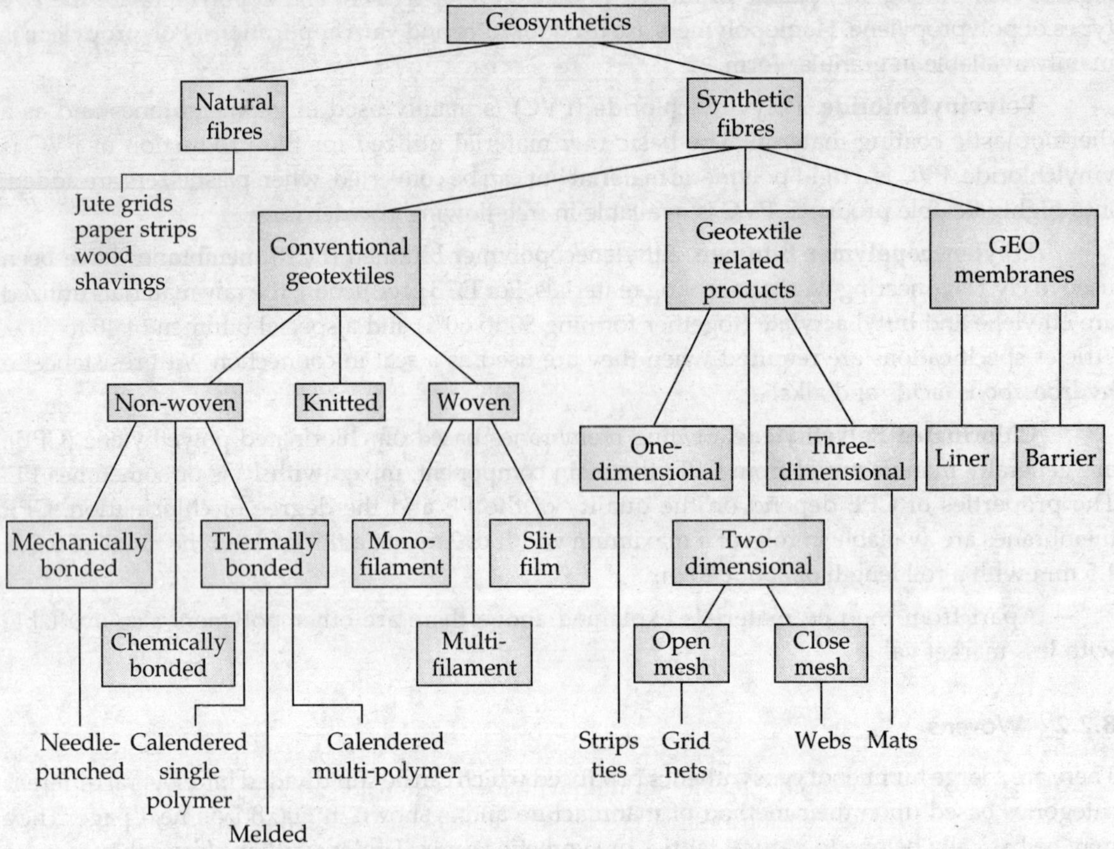

FIGURE 8.1. **Geosynthetics Classification Groups.**
(Source: John, 1987)

The thermally bonded non-wovens contains a wide range of opening sizes than is found in a woven geosynthetics and are relatively thin with a typical thickness of about 0.5 to 1 mm. Chemically bonded non-wovens are comparatively thick usually in the order of 3 mm thick. Chemical bonding is the least common method for forming non-wovens. Mechanically bonded non-wovens have a typical thickness in the range of 2 to 5 mm and also tend to be comparatively heavy because a larger quantity of polymer filaments is needed in order to provide a

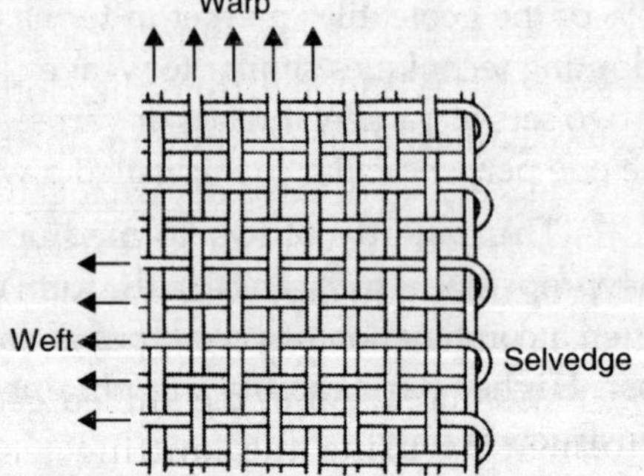

FIGURE 8.2. **Woven Fabric with Selvedge.**
(Source: Zanten, 1986)

sufficient number of entangled filament cross-overs for adequate bonding. Figure 8.3 illustrates a thermally bonded non-woven geosynthetics.

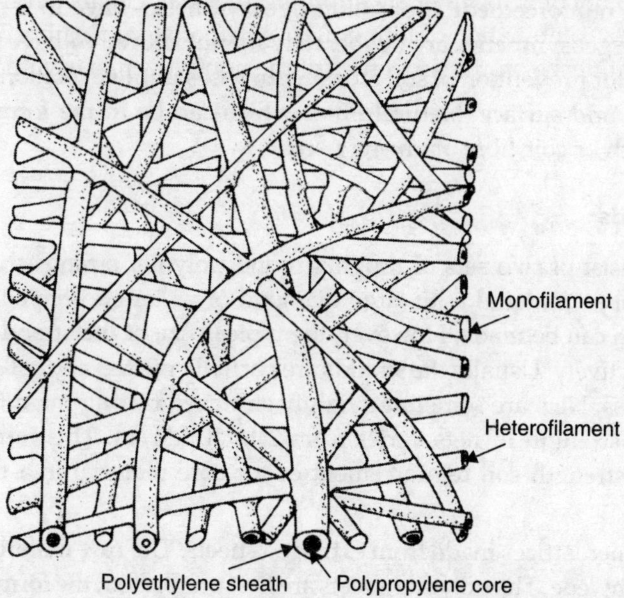

FIGURE 8.3. **Highly Magnified View of a Melded Geotextile.**
(*Source: John, 1987*)

Non-wovens are not load resisting geosynthetics since their tensile strength is limited. Non-wovens derive their benefit by the mechanical inter-locking and or the chemical bonding with the polymer type used. The significant difference between wovens and non-wovens is that the polymer filaments are aligned in the directions of the warp and waft in the weaving process of the wovens. The result of this alignment is that the polymer type directly effects the stress-strain relationship and as a result higher tensile strength with lower extension are obtained. It is reported that for high strength, low extension and low creep geosynthetics for soil reinforcement, the polyester is the optimum choice (Rawes, 1989).

8.2.4 Knitted

Knitted geosynthetics are manufactured using another process which is adopted from the clothing textile industry, namely that of knitting. In this process interlocking a series of loops of yarn together is made. There are different types of knit used within the clothing textile industry. An example of a knitted fabric is illustrated in Fig. 8.4. Only a very few knitted types are produced. All of the knitted geosynthetics are formed by using the knitting technique in conjunction with some other method of geosynthetic manufacture, such as weaving.

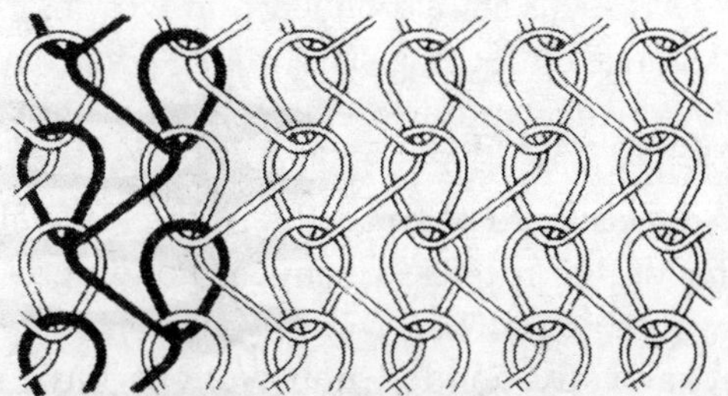

FIGURE 8.4. **Wrap Knitted Fabric.**
(*Source: Zanten, 1986*)

8.2.5 Biogradable

The main intension of every geosynthetic application is long life apart from the function it is intended to play. In certain soil reinforcement applications, geosynthetics have to serve for more than 100 years. But biogradable geosynthetics are deliberately manufactured to have a relatively short life. This is generally used for prevention of soil erosion purposes until vegetation can become properly established on the ground surface. Natural fibres which can be in the form of paper strips, jute nets, wood wool mulch or coir fibre are being used.

8.2.6 Nets and Grids

Geosynthetic nets consist of two sets of roughly round polymer strands that cross at a constant angle to give a very open material with large diamond or rectangle shaped apertures. Nets with apertures up to 75 mm can be made. However, the typical size of the strands and the apertures is 2 mm and 7 mm respectively. Usually the strands are partially melted and rolled to produce thermal bonds where they cross. Nets are sometimes lightly stretched during manufacture to increase the elastic modulus. The strength of nets varies from 2 to 10 kN/m. This form of geosynthetics is mainly used for low strength soil reinforcement or for core material in a fin drain geosynthetic composite.

Grids are polymer lattices made from extruded sheets. The raw materials are polypropylene or high density polyethylene. The polymer sheets are first perforated, the form, size and distribution of holes being determined by the end product. The perforated sheets are then stretched in one direction while it is gently heated. The action of stretching the sheet align's the polymer's long chain molecules in the direction of stretch, giving the grid a high tensile stiffness in this direction. A uniaxial lattice, that is a grid stretched in one direction, is thus produced (Fig. 8.5). The term uniaxial arises from the alignment of the stretched polymer ribs and the greatest strength properties in one direction.

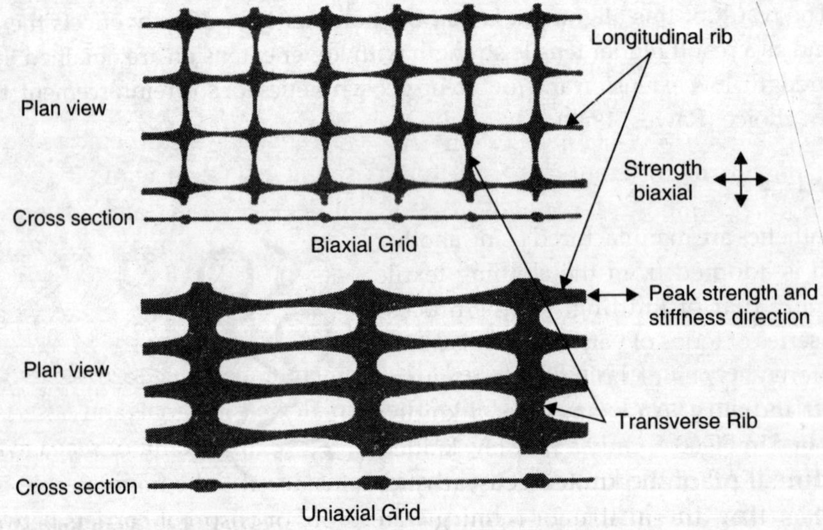

FIGURE 8.5. **Uniaxial and Biaxial Grids.**
(*Source: John, 1987*)

An alternative form of grid may be produced by clamping the uniaxial lattice on the stretched side and applying a second stretching on the transverse direction. This gives a biaxial grid with a square aperture shape (Fig. 8.5). The biaxial grid is used for gabions while the uniaxial one is used for reinforcing soil structures.

8.2.7 Three-dimensional Mats

These mats have large openings, similar to that found in some of the finer geosynthetics nets. The distinguishing feature between mats, nets and grids is that the mats have a distinct three-dimensional structure, whereas the form of the nets and grids is nearer to that of a two-dimensional planar structure.

Three-dimensional mats are produced by extruding for example monofilaments on to a rotating profile roller followed by cooling. As a result the threads fuse together at crossings. The structure contains typically less than 10% volume of the polymer filament. Mats can be produced from any extrudable polymers, but the generally used one is polyamide. The mats are available in size of 6 m width with 5 to 25 mm thick. These three-dimensional mats are often used in the manufacture of composites for drainage.

8.2.8 Composites

This is an area which has particularly developed in recent years. Composites can be produced from two or more of the geosynthetics described in the previous sections. Also, it is possible to form composites from one or more of the geosynthetics used together with geomembrane or geospacer. Evidently the composites have better properties to meet the needs of a specific application. For example, a combination of woven and non-woven geosynthetics are used for soil reinforcement or as separator depending on which one is used as a prime constituent. If non-woven is the backing material and woven is a prime constituent, then the woven is manufactured directly on top and then stitched to the non-woven. This can produce a composite with an aligned warp, useful for soil reinforcement applications. If non-woven is a prime constituent, then the woven sheet acts as a thin strong backing sheet to which the non-woven can be needled to produce a thick dense material. This form of composite is useful as a separater in difficult and dynamic load conditions. Other forms of composites are: non-woven sheet and waffled core composites, non-woven and mat composites, non-woven sheet and net composites and woven sheet and net core composites.

8.2.9 Membranes

Membrane geosynthetics are generally referred to as geomembranes which are thin two-dimensional sheets of materials with a very low permeability. These are flexible materials and can be strengthened with a fabric or film. Geomembranes are only subject to a small amount of seepage as a result of permeation. For all practical purposes they may be considered to be impermeable to both gases and fluids. Thus, they are suitable for forming waterproof or gasproof barriers between adjacent bodies of soil or soil and fluid.

Geomembranes are manufactured from synthetic (thermoplastic) materials such as HDPE, LDPE, PVC, CPE, *etc.*, or bitumen products such as ECB. Generally they are either produced as extruded sheet polymer or as a composite.

There is a distinction between membranes manufactured in the factory or fabricated on the site. In on-site fabrication of thermoplastic membrane a warm or cold viscous material is applied directly on to the surface which is to be sealed. The membranes may be reinforced by spraying the viscous material on to a membrane or fabric backing. The following techniques are commonly used in the factory manufacture of thermoplastic geomembranes (Zanten, 1986):

(*i*) non-reinforced geomembranes, extruded or rolled from a polymer or by spreading the polymer on a sheet of paper which is removed at the end of the process.

(*ii*) reinforced membranes, fabric or membrane is impregnated or covered with the bitumen-polymer mixture; rolling may also be used to produce the final product.

(*iii*) reinforced laminated geomembranes, a geomembrane is combined with a fabric or membrane by rolling or coating.

Viscous manufacturing techniques can produce materials with thickness up to 15 mm. The width is usually not more than 2 m with a thickness of 0.5 to 2.5 mm. Membranes reinforced with bitumen may be of 4 to 5 m wide with thickness of 1.5 to 6 mm.

Reinforced bituminous membranes are built up from different elements, each with its own specific function, either in the production process or in use. The basis of the reinforced membrane is a subtrate of polyamide fabric or polyester membrane, soaked in a bituminous mixture. Based on the production process employed, a glass membrane and/or a polyester film either as a substrate or as a filler can be included. The polyester film also acts as a barrier to the penetration of plants and roots through the membrane. Usually a thick layer of bituminous mixture is coated on either side of the reinforced membrane.

8.3 PROPERTIES OF GEOSYNTHETICS

The properties which are to be satisfied by a geosynthetics depend on the function it has to fulfil. Geosynthetics can have a variety of functions such as reinforcement, filter, drainage, separation layer, *etc*. For example, for reinforcement purposes the mechanical properties such as modulus of elasticity and strength have to be emphasized. On the other hand for filters the emphasis is on hydraulic properties such as hydraulic conductivity. Apart from these, the durability and resistance against mechanical wear and tear have to be guaranteed. Considering all the aspects the properties of geosynthetics can be grouped under the following categories:

(*i*) material and fibre properties

(*ii*) geometrical aspects

(*iii*) mechanical properties

(*iv*) hydraulic properties

(*v*) durability or chemical properties

These properties are discussed briefly subsequently.

8.3.1 Material and Fibre Properties

A description of various raw materials which are used for geosynthetics manufacturing is presented in Section 8.2.1. Standard methods of testing are needed for determining the material and fibre properties and geosynthetics as a whole. A large number of test methods have been standardised in a few countries. In India a complete test standardisation for geosynthetics is not yet available.

Temperature and water content of the materials of geosynthetics are the two aspects which affect the various properties of geosynthetics. Because of this, conditioning of the material over a certain period in air at a certain temperature and relative humidity is prescribed in various test standards.

The fibre used in the manufacture of geosynthetics are built up from linear macromolecules, whose properties are based on the bonding forces between the atoms. Further the fibre properties also depend on the structure in which the macromolecules are arranged. As the basic materials are polymers, they behave like a visco-elastic material.

8.3.2 Geometrical Aspects

Different types of geosynthetics are described in Section 8.2. The choice of particular type and construction depends on the properties required. Other aspects to be considered for the selection are: field boundary conditions and methods of execution. Width and length, thickness, mass per unit area, and the available prefabrication techniques all play a role in the definite choice of geosynthetics and construction (Zanten, 1986).

In principle the length of geosynthetics is unlimited but it is limited by the transport facilities involved and the ease of handling on site. Depending on the mass per unit area the length varies from 50 to 200 m. For wovens and non-wovens the width is 5 to 5.5 m. Thickness of a geotextile is defined as the distance between the upper and lower surface of the material, measured under a specified pressure. The thickness of most geotextiles lies between 0.2 to 10 mm. The thickness of a geomembrane is of more importance than geotextiles for the following reasons (Zanten, 1986): (*i*) to ensure sufficient impermeability to liquids and gases, (*ii*) to ensure that the geomembrane can adequately resist mechanical forces, especially in the construction phase, (*iii*) to ensure reliable techniques for welding the membrane sheets together, (*iv*) to allow the possibility of embossing the geomembrane, whereby the friction between membrane and ground on which it is applied is increased. Variation in thickness of geomembranes may have extreme effects on stress-strain relationships. Increase in thickness at spots in the membrane also lead to concentration of tension. Increases in membrane thickness is not of concern in reinforced membranes.

In general the mass per unit area of non-wovens and wovens are of the range 100 to 1000 g/m^2 and 100 to 2000 g/m^2 respectively. Lighter grades in the range of 100 to 200 g/m^2 are most commonly used.

There is a difference in length between straightened yarns and the yarns in the wavy form in the fabric. This difference in length is referred to as crimp, a parameter which is significant for the structure of the fabric and also for the pore shape and size. Crimp affects the mechanical properties of the fabric.

8.3.3 Mechanical Properties

In the general sense mechanical properties are the load-deformation characteristics of geosynthetics. In turn the mechanical properties of geosynthetics depend on the mechanical property of the fibre material, fibre structure, the yarn structure and the structure of the geosynthetics. The effective performance of geosynthetics in civil engineering applications in general and as soil reinforcement in particular depends exclusively on the tensile stress-strain characteristics of the soil-geosynthetics interface friction behaviour. In addition to tensile tests in which the geosynthetics is subjected to loads and/or displacements, tests have to be conducted in which the load is applied perpendicular to the geosynthetics as occurs in many practical cases. The relevant properties are burst and puncture strength. For example, a burst could occur due to a gas or liquid pressure acting on a membrane. The interface friction, fatigue resistance, creep resistance, tear-strength, abrasion resistance and seam strength are the other important mechanical properties. For details of tests procedures reference may be made to Zanten (1986). A reference may also be made to Christopher (1989) for necessary test standard to determine a particular property of a geosynthetics. For easy reference the geotextile properties and testing methods selection (Christopher, 1989) is presented in Table 8.1.

Role of properties and characteristics as related to function of geosynthetics is illustrated in Fig. 8.6 (Gicot and Perfetti, 1982).

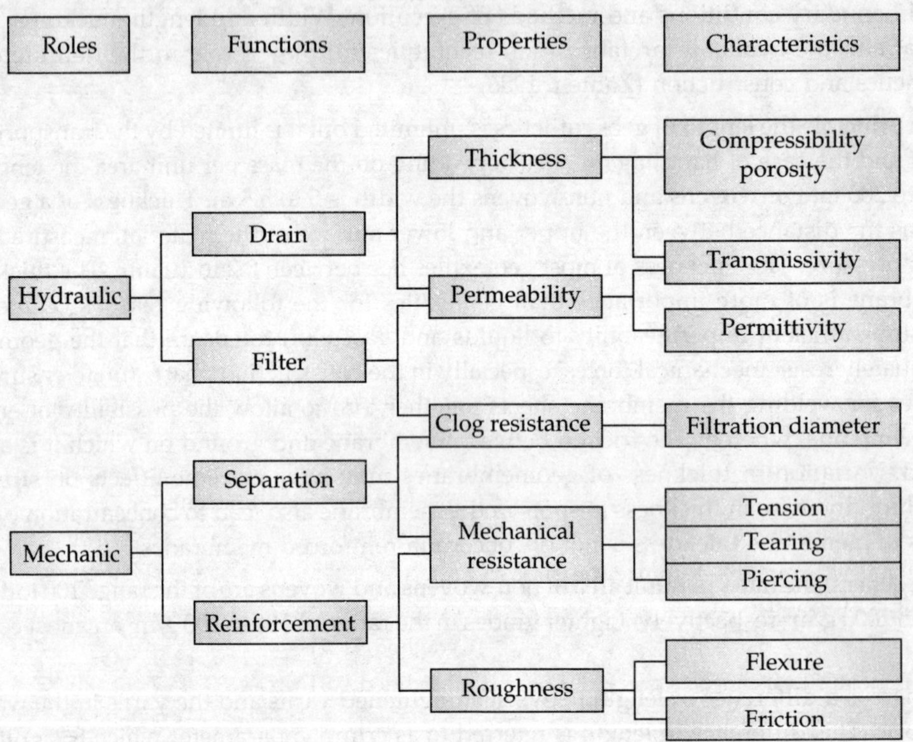

FIGURE 8.6. **Role of Properties/Characteristics as Related to Function.**
(*Source: Gicot and Perfetti, 1982*)

Table 8.1. Geosynthetics Properties and Testing Methods for Geosynthetics Selection (*Source: Christopher, 1989*)

Property	*Test Method*
I. General Properties (from manufactures)	
Type and Construction	N/A
Polymer	N/A
Weight	ASTM proposed (FHWA Manual)
Thickness	ASTM proposed (FHWA Manual)
Roll Length	Measure
Roll Widths	Measure
Roll Weight	Measure
Roll Diameter	Measure
Specific Gravity and Density	FHWA Manual
Absorption	FHWA Manual
Surface Characteristics	N/A
Geotextile Isotropy	N/A
II. Index Properties	
Mechanical Strength-Uniaxial Loading	
(*a*) Tensile Strength	
1. Grab Strength	ASTM D-4632
2. Strip Tensile Strength	ASTM D-1682—Sections 18 and 20 using a CRE of 12-inch/min.
3. Wide Width Strength	ASTM D-4595 (FHWA Manual)
(*b*) Poisson's Ratio	No Test
(*c*) Stress-Strain Characteristics	ASTM D-4595 (FHWA Manual)
(*d*) Dynamic Loading	No standard
(*e*) Creep Resistance	FHWA Manual
(*f*) Friction/Adhesion	Modified Corps of Engineers EM 1110-2-1906, using Ottawa 20-30 sand
(*g*) Seam Strength	See II (*a*) Tensile Strength above; use (1), (2) or (3) depending on requirements
(*h*) Tear Strength	ASTM D-4533
Mechanical Strength—Rupture Resistance	
(*a*) Burst Strength	Mullen Burst-ASTM D-3786, (Fed. Std: 191 A, Method 5122)
(*b*) Puncture Resistance	Modified ASTM D-4833
(*c*) Penetration Resistance	No standard (Dimensional Stability)
(*d*) Fabric Cutting Resistance	No standard
(*e*) Flexibility (Stiffness)	Modified ASTM D-1388-Option A using 2-in.× 12-in, sample (FHWA Manual Fed. Std. 191 A, Method 5206)

(Table Contd...)

Table 8.1. *(Contd...)*

Endurance Properties	
(*a*) Abrasion Resistance	ASTM D-4886
(*b*) Ultraviolet (UV) Radiation Stability	ASTM D-4355
(*c*) Chemical and Biological Resistance	No standard for geotextiles (For textiles; Fed. Std. 191 A, Methods 5760, 5762, 2015, 2016 and 2053)
(*d*) Wet and Dry Stability	No standard
(*e*) Temperature Stability	No standard
Hydraulic	
(*a*) Opening Characteristics	
1. Apparent Opening Size (AOS)	ASTM D-475 (FHWA Manual)
2. Porimetry (pore size distribution)	Use AOS for O_{95} O_{85} O_{50} O_{15} and O_5
3. Percent Open Area (POA)	FHWA Manual
4. Porosity (*n*)	No standard
(*b*) Permeability (*k*) and Permitivity	ASTM D-4491 (FHWA Manual)
(*c*) Soil Retention Ability	Empirical Relations to Opening Characteristics
(*d*) Clogging Resistance	No standard—See Soil-Fabric Tests
(*e*) In-Plane Flow Capacity (Transmissivity)	ASTM D-4716
III. Performance Properties For Soil-Geotextile Systems	
Stress-Strain Characteristics	Tension Test in Soil-McGown, *et. al.*, (1982), Triaxial Test Method-Holtz, *et. al.*, (1982) California Bearing Ratio on Soil Fabric System-Christopher (1983) Tension Test in Shear Box-FHWA Manual
Creep Tests	Extension Test in Soil-McGown, *et. al.*, (1982) Triaxial Test Method-Holtz, *et. al.*, (1982) Extension Test in Shear Box-Christopher (1983)
Friction/Adhesion	Direct Shear Method-Modified, Corps of Engineers Procedure EM 1110-2-1906; Bell-2-1906; FHWA Manual Pull-Out Method-Holtz (1977)
Dynamic and Cyclic	No standard procedure
Loading Resistance	Soil-Fabric Permeameter Gradient Ratio-(FHWA Manual)
Soil Retention and Filtration Properties	Slurry method-CALTRANS-Hover (1982) Application Motiel, *e.g.*, Virginia HTRC Silt Fence Method (FHWA Manual)

8.3.4 Hydraulic Properties

Geosynthetics are used as filters in bank protection works, downstream culverts and in bed protection works along canals. In these situations the filter is laid directly on the subsoil and guarantee the soil tightness of the structure. Further, to avoid unnecessary building up of excess pore water pressures under the protection the geosynthetics must also be sufficiently water permeable. Geosynthetics used in road construction are installed in the substructure or as a layer separating the substructure from the subsoil. Geosynthetics are also used in land drainage, in this respect providing a water permeable layer or a water drainage. In all these fields geosynthetics must satisfy certain requirements for water permeability and soil tightness. Thus to satisfy the filtration and drainage requirements, geosynthetics need to fulfil two important conditions, *viz.*,

(*i*) It should provide adequate hydraulic conductivity to allow the free flow of water out of the surrounding soil into the drain.

(*ii*) It should prevent particle movement from the parent soil into the drainage media. These conditions signify that the geosynthetics must have a required opening size as well as a tight fabric structure. The hydraulic properties of geosynthetics are controlled by: (*i*) fabric opening characteristics, (*ii*) water permeability, and (*iii*) clogging resistance.

Fabric opening characteristics include apparent opening size, opening size distribution, percent open area, and porosity. The ability of a fabric to conduct flow is dependent on the opening characteristics of the geosynthetics. In order to assess the performance of geosynthetics accurately, these should be used in combination with measured permeability values. The apparent opening size (AOS) is a measure of the largest effective opening in a geosynthetics. The pore size distribution of fabric is determined by extending the AOS technique to several other bead sizes. The percent open area is defined as the ratio of the total of individual open areas in a given fabric specimen to the total area of the fabric specimen. Porosity is related to the ability of water to flow through the fabric and defined as the ratio of volume of voids to total volume.

In general permeability of a geosynthetics must be substantially greater than that of the protected soil. High fabric permeability also infer that partial clogging will not reduce fabric permeability. Permeability and transmissivity are the two parameters which govern the hydraulic conductivity. Transmissivity of a geosynthetics is defined as the volumetric flow rate per unit thickness under laminar flow conditions, in the in-place direction of the fabric.

Clogging is defined as the movement by mechanical action or hydraulic flow of soil particles into the voids of a fabric and retention therein. Under study state and reversing flow conditions, fabric clogging or blinding causes a decrease in water flow rate and corresponding increase in hydraulic head loss through the geosynthetics. The mechanisms responsible for fabric clogging is not fully understood yet. The pertinent tests for determining hydraulic properties are given in detail by Rao *et. al.*, (1990) and Zanten (1986).

8.3.5 Durability

In addition to the physico-mechanical properties the chemical nature of the polymer from which the geosynthetics is manufactured is also of great importance principally in connection with the desired durability. Polymers have a significant characteristics that they are relatively insensitive to the action of a great number of chemicals and to the environmental effects. Evidently this is an added advantage compared to conventional construction materials. Nonetheless each plastic has a number of weaknesses which should be considered in the design and application. Table 8.2 illustrates the resistance of various geosynthetics (Zanten, 1986). Many of the synthetic polymers are sensitive to oxidation because of oxidation, the mechanical properties such as strength, elasticity and strain absorption capacity deteriorate and the geosynthetics ultimately becomes brittle and cracks. Special additives are available to counteract these processes. Polymers used in geotextiles and geomembranes differ considerably in their intrinsic resistance to oxidation. Currently no method exists to evaluate the chemical stability of geosynthetics.

As a matter of fact it is very difficult to predict durability through direct testing. However, it is possible to assess durability through index testing which helps to provide a relative comparison of the products. Abrasion resistance and Ultra Violet (UV) resistance are other two factors which affect the durability of geosynthetics.

Abrasion Resistance. Geosynthetics may be subjected to abrasion in many ways. Abrasion could occur as a result of friction produced by various types of movement of rock and soil against the surface of the fabric such as wave action or riprap, sand scour sediment in a stream, or the aggregate cover in fabric reinforcement applications such as roadways and railways and may also occur due to installation (Hodge, 1987). Standard tests are available to evaluate the degree of abrasion.

Table 8.2. The Resistance of Various Geotextiles
(Source: Zanten, 1986)

Polymer	*PA*		*PETP*		*PP*		*PE*		*Soft PVC*	
Loading duration	*Short*	*Long*	*Short*	*Long*	*Short*	*Long*	*Short*	*Long*	*Short*	*Long*
Resistant against:										
Dilute acids	+	0	++	+	++	++	++	++	+	0
Concentrated acids	0	–	0	–	++	+	++	+	0	–
Dilute alkali	++	+	++	0	++	++	++	++	++	+
Concentrated acids	0	–	0	–	++	++	++	++	+	0
Salt (prine)	++	++	++	++	++	++	++	++	++	++
Oil (mineral)	++	++	++	++	+	0	+	0	+	0
Glycol	+	0	++	0	++	++	++	++	++	++
Micro organisms	++	+	++	++	++	++	++	++	+	0
UV Light	+	0	+	0	0	–	0	–	+	–
UV Light (stabilized)	++	+	++	+	++	+	++	+	++	+
Heat dry (up to 100°C)	++	+	++	++	++	+	++	0	+	0
Steam (up to 100°C)	++	+	0	–	0	–	0	–	0	–
Moisture absorption	++	++	++	++	++	++	++	++	+	+
Detergents	++	++	++	++	++	++	++	++	++	++
Tendency to creep	++	+	++	++	+	0	+	0	+	0

Degree of resistant: – = not resistant; 0 = moderate ; + = passable; ++ = good. This assessment of resistance is valid under normal conditions and temperatures:

1. During execution
2. During usage
3. Depending on type of softener at high relative humidities.

Ultra Violet Resistance. In order to minimise UV degradation, geosynthetics in most applications is embedded in soil. During the manufacturing process, additives such as carbon black or UV stabiliser are added to increase their UV resistance. However, geosynthetics may be exposed to sunlight in applications like reinforced soil walls and erosion control and during storing before installation. Thus it is essential to evaluate the UV resistance of geosynthetics. Standard tests are available to determine the UV resistance.

The relative importance of the properties of geosynthetics as a function of the type of structure is presented in Fig. 8.7 (Gicot and Perfetti, 1982).

Type of structure	Functions				Durability					Hydraulic resistance		Clog resistance	Mechanical resistance							Roughness		
										Permeability			Tension		Tearing		Piercing			Flexure	Friction	
	Drainage	Filtration	Separation	Reinforcement	Chemical	Thermal	Photochemical	Biological	Thickness	Permittivity	Transmissivity	Filtration diameter	Low speed	Creep	Low speed	Dynamic	Perforation	Puncturing	Bursting	Flexibility	Low speed	Abrasion
Access road	•	•	⬤	⬤			•	●	•	●	•	●	⬤	⬤	●	⬤	⬤	⬤	●	⬤	⬤	•
Subgrade	•	•	⬤	●			•	●	•	●	•	●	●	●	●	⬤	⬤	⬤	●	●	●	•
Light traffic road	•	•	⬤	•			•	●	•	●	•	●	●	•	●	●	●	●	●	●	•	•
Stroage area	•⬤	•	⬤	•	•⬤	•⬤	•⬤	●	•	●	•⬤	●	•⬤	•	●	●	•⬤	•⬤	•⬤	•	•	•●
Fill on compressible soil	●	●	⬤	●			•	●	●	●	●	●	⬤	⬤	⬤	⬤	⬤	⬤	⬤	⬤	⬤	
Rail road	⬤	●	⬤	•	•⬤		•	●	⬤	•	⬤	⬤	●	●	⬤	⬤	⬤	●	●	●	●	
Drainage trench	●	⬤	●		•		•	⬤	⬤	⬤	●	⬤	•		●	●	●	●	●	⬤		
Vertical drainage	⬤	⬤					•	●	⬤	⬤	⬤	⬤	●		●	•	•	•	•	●		
Erosion control	•	⬤	●	●	•		•⬤	⬤	•	⬤	•	⬤	⬤	⬤	⬤	⬤	⬤	⬤	⬤	⬤	⬤	⬤
Sport ground	●	●	⬤				•	●	●	●	●	●	•		•	•	•	•	•	•		
Impermeable liner	⬤	●	⬤	•⬤	•⬤	•⬤	•⬤	⬤	⬤	•	⬤	⬤	⬤	•	●	●	●	⬤	⬤	⬤	⬤	
Road revetment			⬤	⬤	⬤	⬤	•	●	●				⬤	⬤						⬤	⬤	

• Not Very Important ● Important ⬤ Very Important

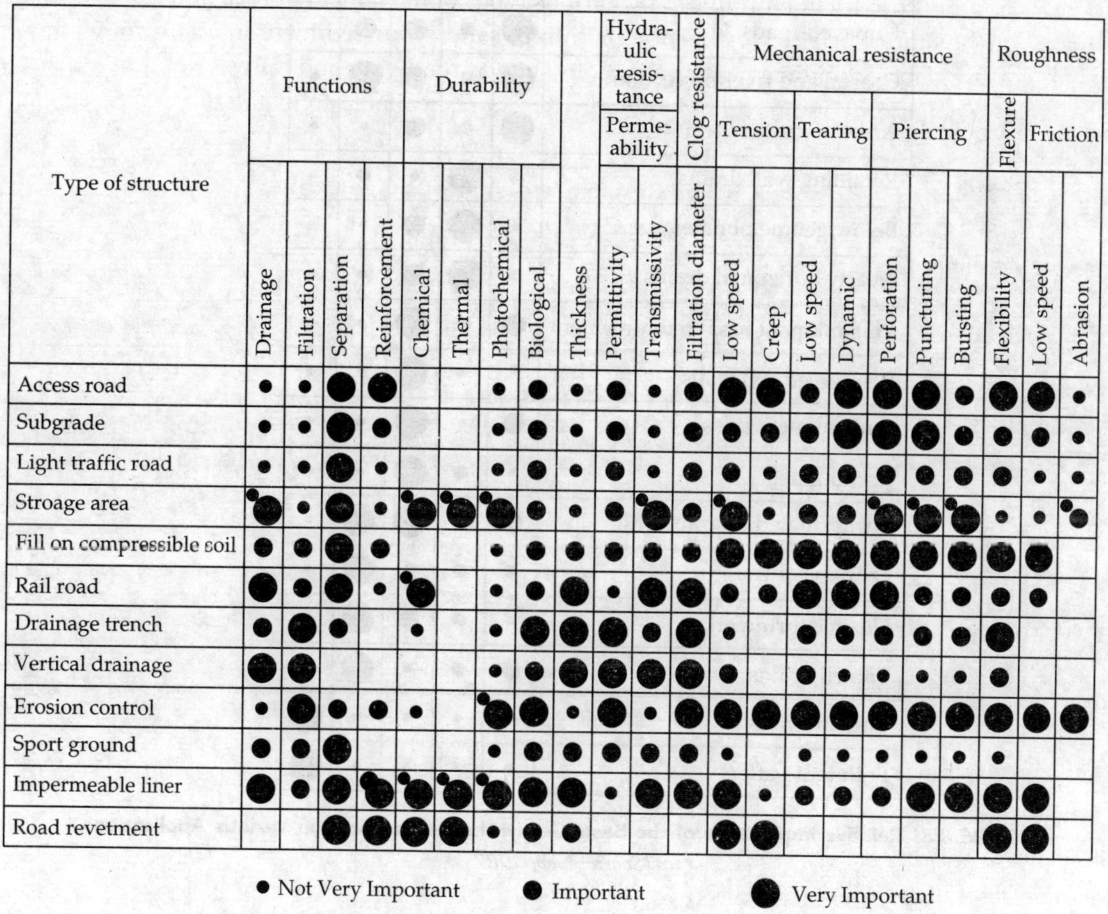

FIGURE 8.7. **Importance of Properties as a Function of Type of Structure.**

(*Source: Gicot and Perfetti, 1982*)

8.4 APPLICATIONS OF GEOSYNTHETICS

It has been discussed earlier the five basic functions of a geosynthetics, *viz.*, separation, fluid transmission, reinforcement, filtration and containment and barrier. These basic functions are accomplished in various field applications which are subsequently presented. Figure 8.8 illustrates the relative importance of the basic geosynthetic functions in various applications. A few applications of geosynthetics in India for various civil engineering works have been reviewed by Veeraraghavan (1987) and Rao and Raju (1990).

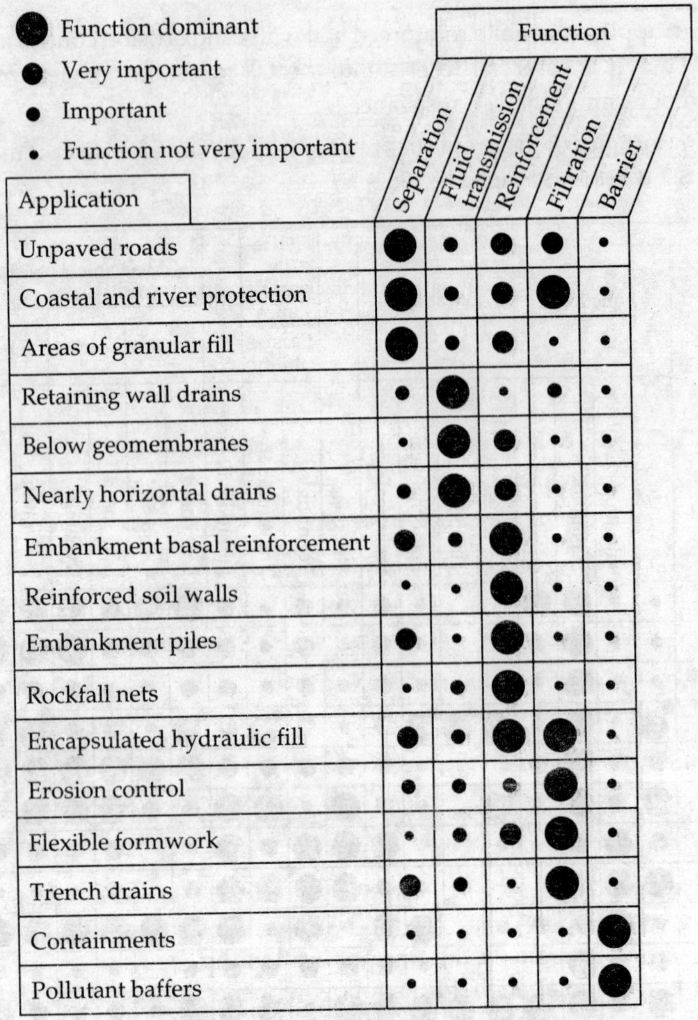

⬤ Function dominant
● Very important
• Important
· Function not very important

Application	Separation	Fluid transmission	Reinforcement	Filtration	Barrier
Unpaved roads	⬤	•	●	●	·
Coastal and river protection	⬤	•	●	⬤	·
Areas of granular fill	⬤	•	●	·	·
Retaining wall drains	•	⬤	·	•	·
Below geomembranes	·	⬤	●	·	·
Nearly horizontal drains	•	⬤	●	·	·
Embankment basal reinforcement	●	•	⬤	·	·
Reinforced soil walls	·	·	⬤	·	·
Embankment piles	●	·	⬤	·	·
Rockfall nets	·	•	⬤	·	·
Encapsulated hydraulic fill	●	•	⬤	⬤	·
Erosion control	•	•	•	⬤	·
Flexible formwork	·	•	●	⬤	·
Trench drains	●	•	·	⬤	·
Containments	·	·	·	·	⬤
Pollutant baffers	·	·	·	·	⬤

FIGURE 8.8. **Relative Importance of the Basic Geosynthetics Functions in Various Applications.**
(*Source: John, 1987*)

8.4.1 Separation

In principle a geosynthetics, when used as a separator, must prevent the inter-mixing of particles from two soil layers with different properties. This prevents contamination which may impair the intended behaviour of granular soil layers. The field application of geosynthetics as separators commonly adopted are: unpaved roads, paved roads, railways and protection of geomembranes.

Unpaved Roads. Use of geosynthetics for unpaved roads are a well-established and a most common one. Although the prime function of a geosynthetics in unpaved road construction is separation, the secondary functions of reinforcement and filtration remain essential. Providing a geosynthetics sheet between a granular sub-base and a weak subgrade helps to stabilize an unpaved road in a number of ways as shown in Fig. 8.9. The geosynthetics while maintaining separation (John, 1987):

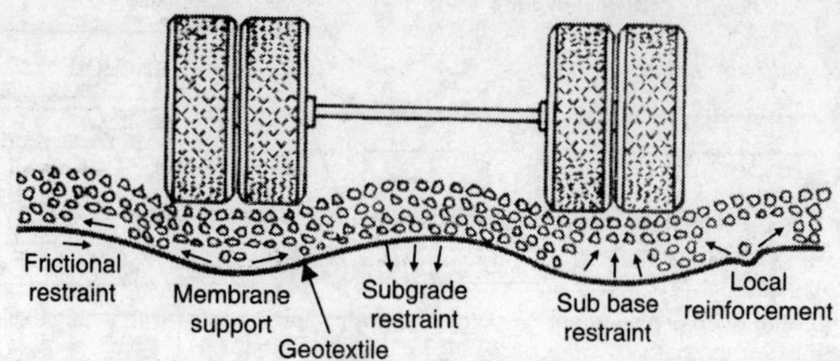

FIGURE 8.9. **Geotextile Stabilization of an Unpaved Road.**
(*Source: John, 1987*)

(*i*) Provides local reinforcement.
(*ii*) Restrains the aggregate from downward and lateral movement in the ruts.
(*iii*) Restrains the subgrade soils from upward and lateral movement between the ruts.
(*iv*) Acts as a support membrane.
(*v*) Provides sufficient friction to limit lateral sliding of the aggregate.

Most of the actions explained above belong to reinforcement category, showing the essential nature of the secondary role. In the absence of a geosynthetics, mixing up of the granular sub-base of soft mud from the underlying soil would rapidly reduce its strength to that of the soft mud alone. Thus a geosynthetics sheet placed beneath the granular sub-base can greatly reduce maintenance, enable better compaction during construction and reduce the thickness of aggregate layer.

Problematic subgrade soils are highly compressible soils such as peat and saturated soft to medium stiff clays. In peats the problem is mainly settlement and in clays development of high pore water pressure due to vehicle movement.

Paved Roads. Rutting type of deflections of the surface is unacceptable in the case of paved roads. In this case, geosynthetics can be provided at three different locations in a permanent road, *viz.*, at the interface between the aggregate sub-base and the subgrade soil, within the pavement structure, or with a surface overlay. In the first application the geosynthetics acts in a similar way as that in the unpaved road and can yield the following benefits (John, 1987):

(*i*) Prevents pavement sub-base aggregate from penetrating the subgrade soil
(*ii*) Prevents fine soil particles from the subgrade soil entering the sub-base aggregate
(*iii*) Reduces the need for excavation of soft fine subgrade soils
(*iv*) Speeds placement of the sub-base aggregate during construction
(*v*) Reduces rutting of the sub-base aggregate while it is being used as a haul road
(*vi*) Evens out settlement of the sub-base aggregate over any pockets of soft material that may have been overlooked.

In the second application, a high geosynthetics elastic stiffness is required to bring in some reinforcing effect. For this application, the most effective location for the geosynthetics is within the base course or between the base course and the weaving course at a depth of not less than 40 mm (John, 1987). The presence of the geosynthetics improves the tensile strength and gives the road a greater resistance to cracking and helps to provide a longer fatigue life.

In the third application, the geosynthetics is placed on the surface of an existing pavement prior to laying an asphaltic overlay. Presence of the geosynthetics restricts propagation of reflection cracks (Fig. 8.10) and thereby increasing the life of the overlay (John, 1987).

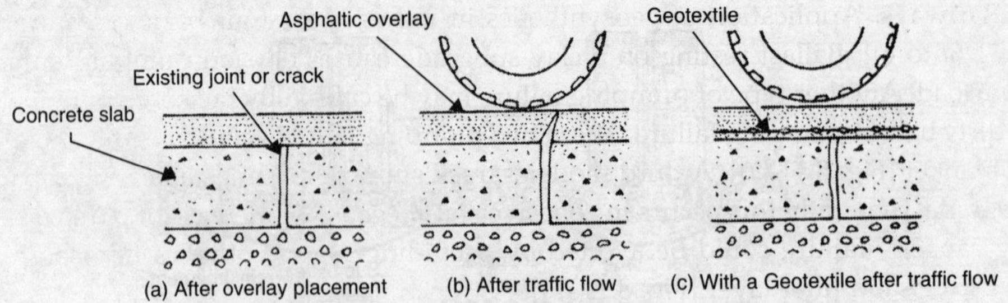

***FIGURE 8.10.* Control of Reflection Cracking with a Geotextile.**

(*Source: John, 1987*)

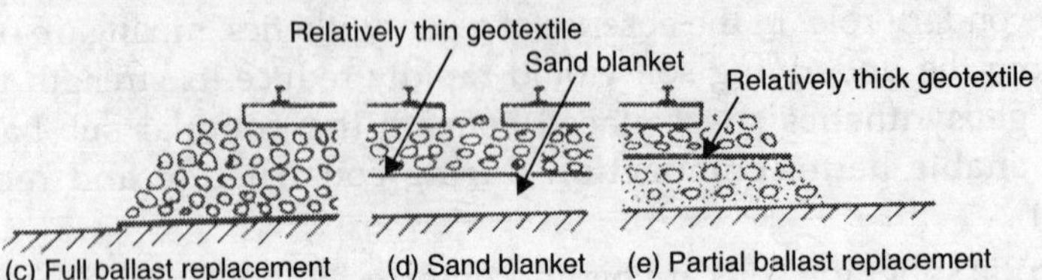

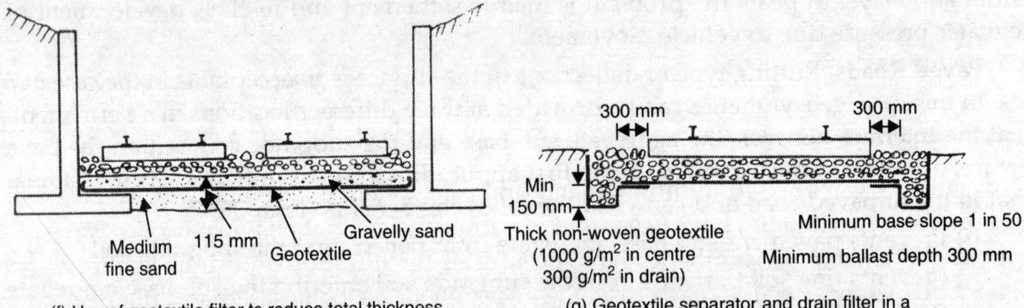

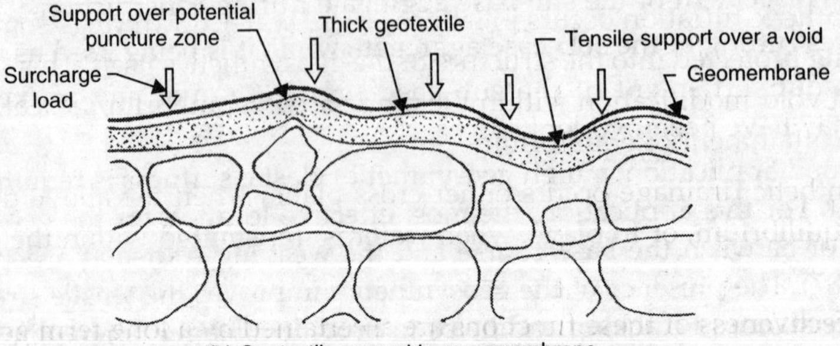

***FIGURE 8.11.* Applications of Geotextiles as Separator.**

(*Source: John, 1987*)

Railways. Application of geosynthetics in railways for four basic cases are illustrated in [Fig. 8.11 (*a* to *g*)]. Ballast resting on a clay subgrade causes erosion pumping failure due to the dynamic load. Another type of pumping failure may be caused by the contamination of the ballast by the dirty ballast pumping failure. These two pumping failures could be prevented by providing a sand blanket [Fig. 8.11 (*a*)]. A third mode of track support failure is a bearing capacity failure of the subgrade. Generally this occurs in cohesive soils because of increase of pore water pressure. A bearing capacity failure could be averted by providing a geosynthetics film sandwiched in the middle of the sand filter layer [Fig. 8.11 (*b*)].

Figures 8.11 (*c*) to (*e*) show three alternative ways of incorporating geosynthetics within the track-bed so that they may act as a filter. For such use the geosynthetics should have higher puncture resistance and strength and also should be capable of dissipating the excess porewater pressure. Otherwise the excess development of pore water may cause liquefaction failure.

In another situation use of a geosynthetics in conjunction with fine granular materials could reduce the thickness of filter layer [Fig. 8.11 (*f*)]. For example, this can occur where the track runs alongside a retaining wall with shallow foundations.

Geosynthetics are also used in railway track surface water drainage systems. In such situations, the application of geosynthetics is similar to that in highway drains [Fig. 8.11 (*g*)]. In this case a heavier grade geosynthetics has to be used.

Protection of Geomembranes. Geomembranes are used in connection with containing open areas of water, such as in canal or dam linings, or alternatively containing pollutants, such as around industrial waste storages. Geomembranes have to resist tensile forces, puncture and wear from objects. It is now the established practice to use geotextiles to protect geomembranes from the hazards. Geotextiles often used for this purpose are usually thick, mechanically-bonded non-wovens. Fig. 8.11 (*h*) illustrates the functions of a geotextile as a sepator in the protection of a geomembrane.

8.4.2 Filtration and Fluid Transmission

Geosynthetics can be used in filtration and drainage system in a similar way as that of soil filters and drains. Geosynthetics have been used in various field situation with success. Some of the field conditions wherein geosynthetics have been used for filtration and drainage are illustrated in Fig. 8.12 (Koerner, 1985). In each case, the geosynthetics is being used as a filter (for cross plane flow), a drain (for in place flow), or both. The two functions are distinguished as given below:

Geosynthetic filtration occurs in fabrics where water flow brings some of the finer particles of the soil being protected into the structure of the geosynthetics itself. This soil modification above the fabric and void modification within the fabric attain equilibrium. Clear water passes through only after the attainment of equilibrium.

Geosynthetic drainage occurs either cross plane, when the above described process have reached the equilibrium, or in plane, when water is transmitted within the geosynthetic structure itself.

The effectiveness of these functions are ascertained by a long term geosynthetics flow tests. One such a test, as reported by Koerner and Ko (1982), is presented in Fig. 8.13. In this illustration, the final portion of the curve is a critical one. When slope m_f reaches zero, it signifies that equilibrium has been established within the soil and the geosynthetics. On the other hand if the slope m_f continues to decrease, the geosynthetics or upstream soil is probably clogging and ultimately result in a complete cutoff flow of water.

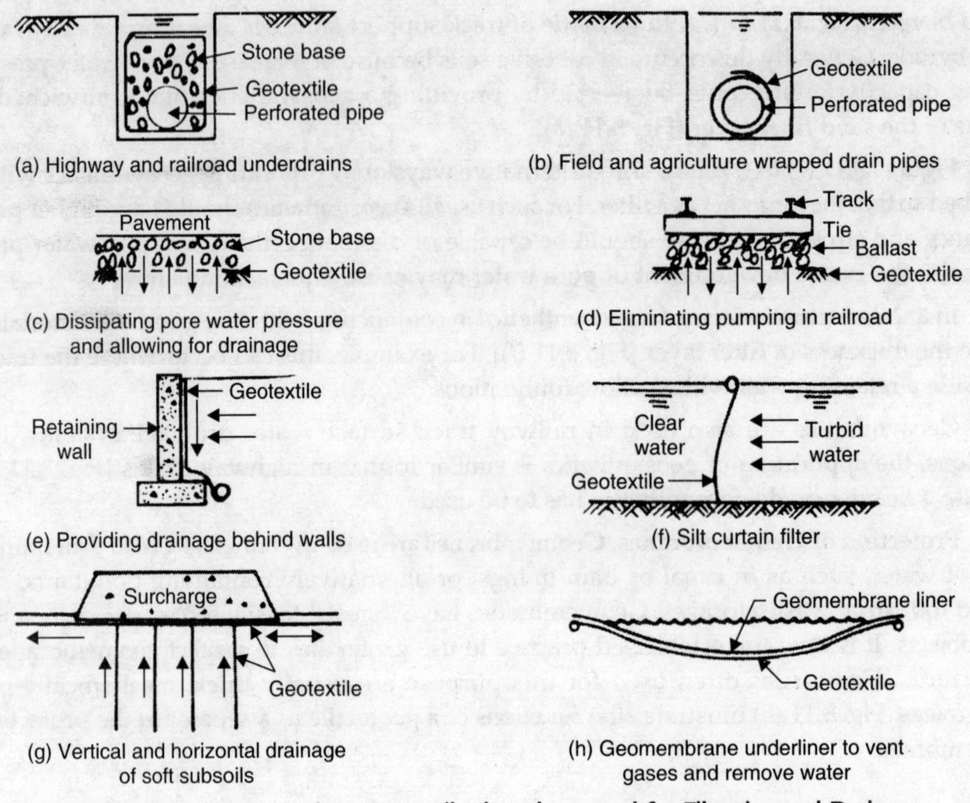

***FIGURE 8.12.* Situations where Geotextiles have been used for Filtration and Drainage.**
(*Source: Koerner, 1985*)

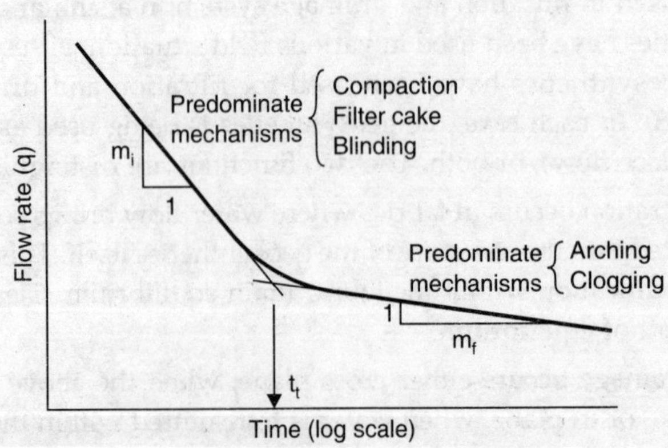

t_t = Transition time (Minutes for sands, hours for silts, days for clays)
m_i = Initial slope = Function of soil compaction filter cake buildup and Blinding over fabric voids
m_f = Final slope = Function of soil arching over fabric voids and clogging within fabric

***FIGURE 8.13.* Generalized Long-term Flow Response of Soil-geotextile Systems.**
(*Source: Koerner and Ko., 1982*)

Various drainage applications are illustrated in Fig. 8.14. For details of drainage applications reference may be made to Zanten (1986) and John (1987).

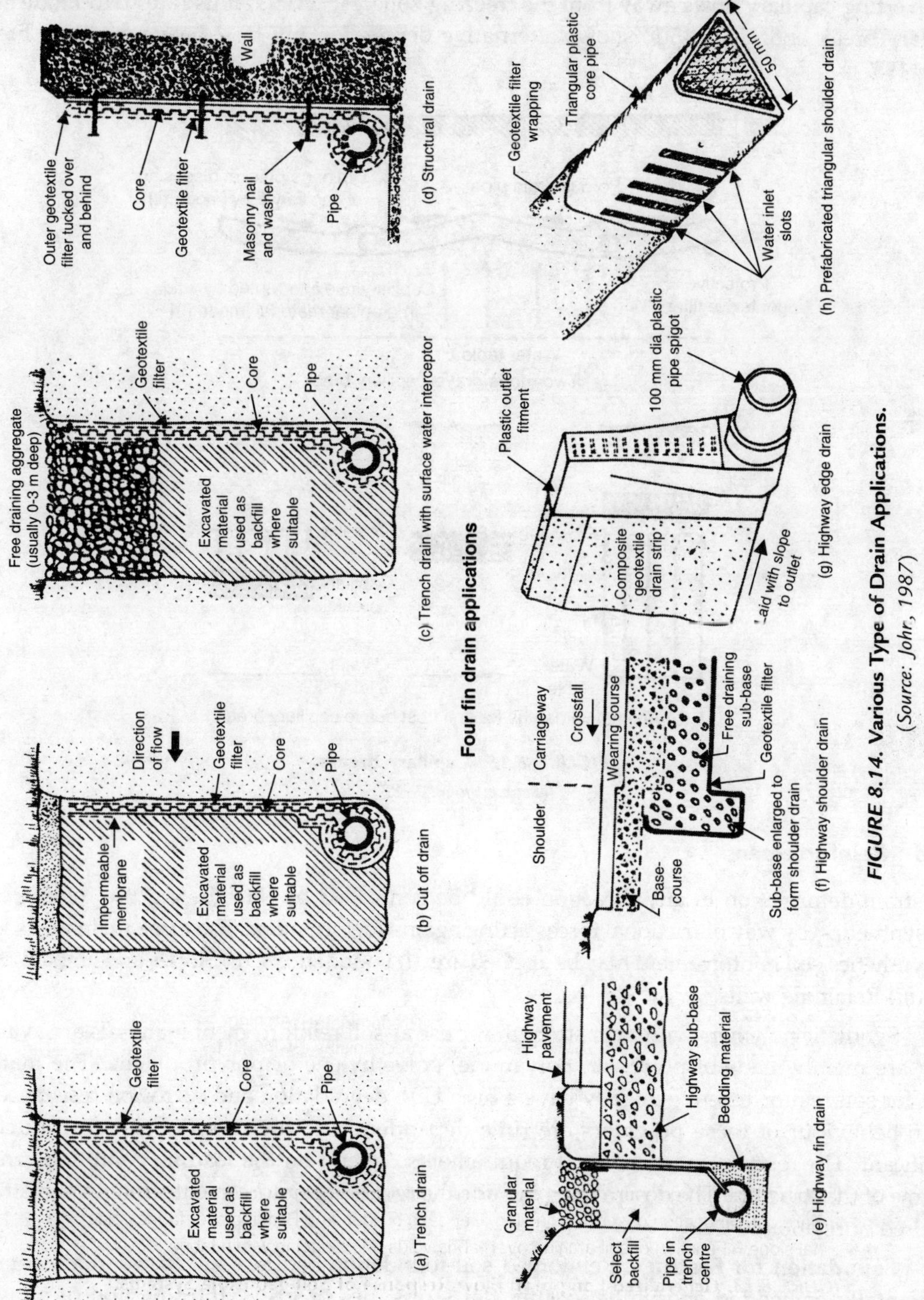

***FIGURE 8.14.* Various Type of Drain Applications.**
(Source: Johr, 1987)

At low temperatures, very small ice crystals form in the soil water and viscosity also increases. Under the influence of these factors the capillary rise can be high. A capillary break can function in two different ways to limit frost heave: (*i*) Interrupting the capillary rise by large air-filled voids, (*ii*) Diverting capillary flows away from the freezing zone. Fig. 8.15(*a*) illustrates two-mode gravel capillary break and Fig. 8.15(*b*) shows alternative depths for anti-frost heave capillary breakes (John, 1987).

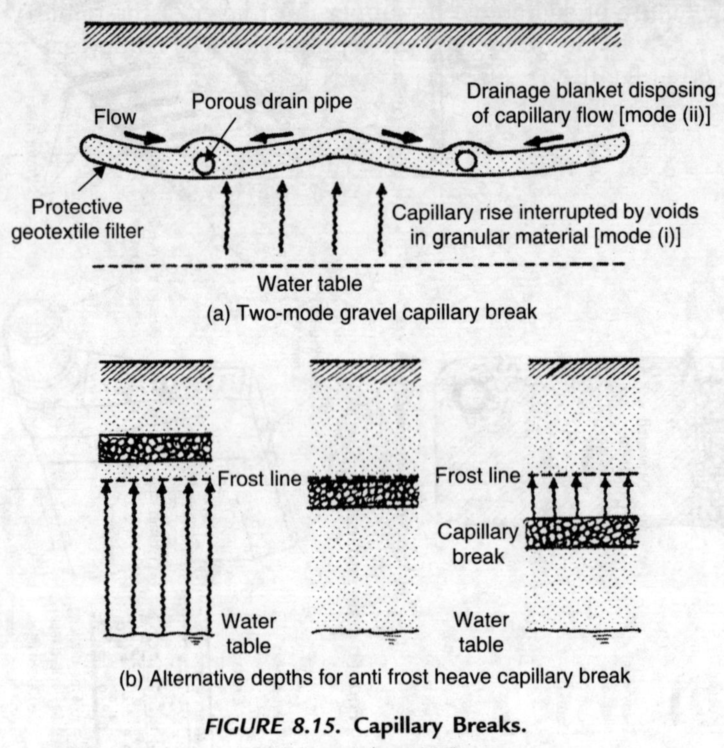

FIGURE 8.15. **Capillary Breaks.**
(*Source: John, 1987*)

8.4.3 Reinforcement

The strain/deformation in any direction could be controlled by introducing the reinforcement (geosynthetics) by way of frictional forces acting against deformation. The three main areas where geosynthetics-soil reinforcement may be applied are: (*i*) Foundations, (*ii*) Slopes and embankments, and (*iii*) Retaining walls.

Synthetic polymers with high strength for use as soil reinforcement materials are available. They are mainly made of polyester, polyamide, polyethylene or polypropylene. The materials used for soil reinforcement generally have a high U.V. degradation and corrosion resistance. The strain behaviour of these polymers are time dependent, *i.e.*, the creep of these products are significant. The reinforcement material requirements depend on the loading conditions and the lifetime of the structure. The design must include the required reinforcement strength and strength required in relation to time.

Foundation for Footings. Reinforced soil foundation bed is a soil foundation containing horizontally embedded geosynthetics/thin flat strips, ties or grids. Introduction of a layer of reinforced soil between a footing and weak subsoil can increase the bearing capacity substantially, thus obviating the necessity of combined footing or a raft foundation. Since geosynthetics are considered to be non-biogradable and are quite cost effective, they are used in the form of geotextiles/geogrids and geocells.

Binquet and Lee (1975) have identified three possible modes of failures for the reinforced soil for a footing, *viz.*,

(*i*) Shear failure of soil above the upper-most layer of the reinforcement: The mode of failure is possible if depth to the topmost layer of reinforcement is sufficiently large [Fig. 8.16 (*ii*) (*a*)].

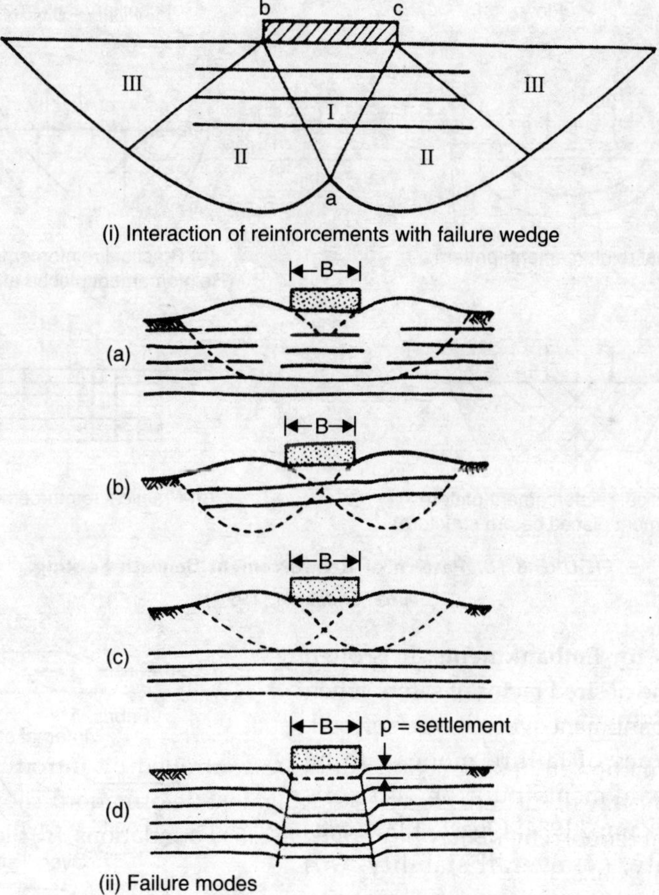

FIGURE 8.16. **Failure Wedges and Modes.**

(*Source: Binquet and Lee, 1985, (a) to (c); Koerner, 1985, (d)*)

(*ii*) Reinforcement pull-out failure: This type of failure occurs for reinforcement placed at shallow depths beneath the footing with insufficient anchorage [Fig. 8.16 (*ii*) (*b*)].

(*iii*) Reinforcement tension failure: This type of failure occurs in the case of long and shallow reinforcement for which the frictional pull-out resistance is more than the tensile strength [Fig. 8.16 (*ii*) (*c*)].

Another mode of failure due to long term (creep) settlement under sustained surface loads and subsequent stress relaxation [Fig. 8.16 (*ii*) (*d*)] has been identified by Koerner (1985).

Ideal pattern of reinforcement for footings should be in accordance with the directions of principle tensile strain (Fig. 8.17). In order to satisfy this condition, the reinforcement should lie along the lines shown in [Fig. 8.18 (*a*)], starting with horizontal reinforcement below the footing and becoming progressively more vertical at some distance away on either side. It is impractical to provide a continuous system of reinforcement. The practical feasibility of providing reinforcement is illustrated (Sridharan, 1990) in Fig. 8.18 (*a*) to (*d*).

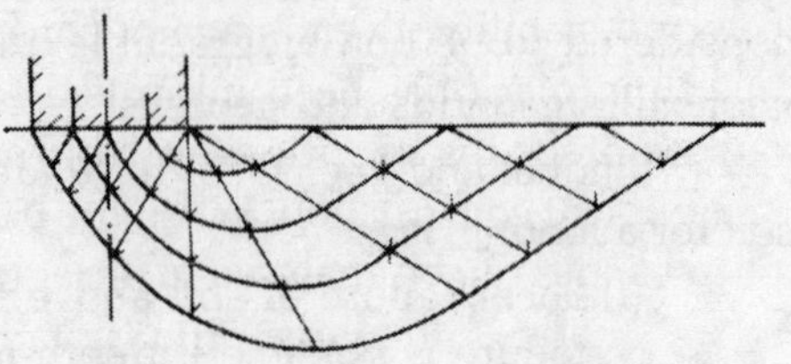

FIGURE 8.17. **Zero Extension Characteristics.** (*Source: Bassett and Last, 1978*)

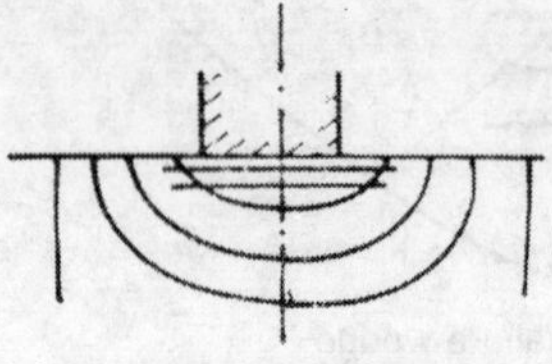

(a) Ideal reinforcement pattern

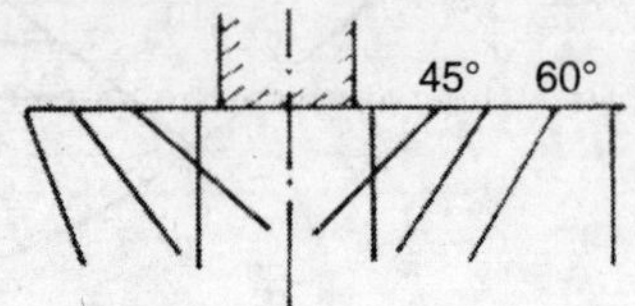

(b) Practical reinforcement pattern (Reinforcement placed after structure)

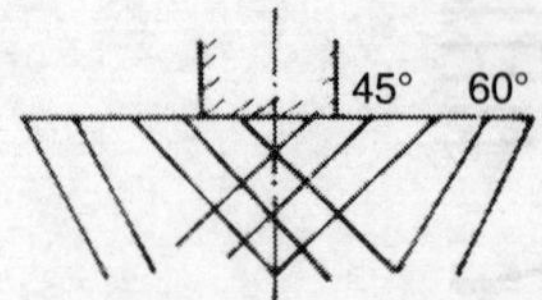

(c) Practical reinforcement pattern (Reinforcement placed before structure)

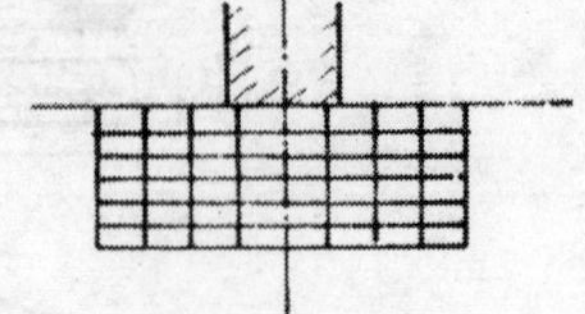

(d) Practical reinforcement pattern

FIGURE 8.18. **Pattern of Reinforcement Beneath Footing.** (*Source: Sridharan, 1990*)

Foundations for Embankments. It is often difficult to obtain the desired factor of safety when constructing an embankment over a weak/soft soil. In general four types of failure modes can be identified for embankments built on soft soils (Risseeuw and Voskamp, 1984; Quast, 1983): *viz.*, (*i*) internal stability, (*ii*) overall stability, (*iii*) foundation stability, and (*iv*) bearing capacity failure. The failure modes are illustrated in Fig. 8.19.

Normally the slope of the embankment is chosen in such a way that the internal stability of the embankment is assured assuming that the bearing capacity of the subsoil itself is sufficient to carry the load of the fill and prevent the fill pressing through into the subsoil. Generally the bearing capacity of the subsoil will be adequate to carry the load of the embankment.

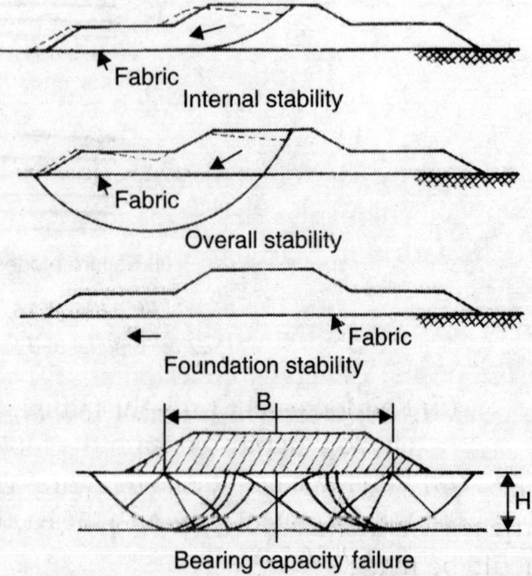

FIGURE 8.19. **Failure Modes for Embankments Built on Soft Soils.** (*Source: Zanten, 1986*)

In case of inadequacy, the embankment has to be constructed in stages so as to prevent loss of overall stability. Every stage of construction has to be taken up only after ascertaining the soil already in place has been improved to the extent necessary. Alternatively the subsoil may be consolidated, by adopting appropriate soil improvement technique, before taking up the embankment construction. When the bearing capacity of the subsoil is sufficient, the three types of failure modes can be analysed in order to determine the strength required in the fabric. A schematic representation of the three primary failure modes for a fill on a low-bearing capacity soil with a reinforcing fabric at the interface is shown in Fig. 8.19 (Zanten, 1986).

The shearing resistance between the reinforcing mat and subsoil limits the maximum restoring force that can be obtained from the fabric. If the force is more than the maximum shearing force along the anchor length of the fabric it will be pulled out of the soil and result in the failure of the embankment.

In order to obtain the full benefit of the geosynthetics tensile strength, it is important that the construction sequence is so planned such that tension on geosynthetics is applied gently. There can be no tensile force mobilised in the geosynthetics unless it is subjected to strain. The geosynthetics sheet should not slack or folded beneath the embankment after construction. The following construction sequence is recommended in order to avoid such problems (John, 1987, Fig. 8.20): (*a*) Place the geosynthetics, (*b*) Place fill at the embankment edges, (*c*) Fold over the geosynthetics ends, (*d*) Place more fill at the embankment edges, (*e*) Place fill in the central region, (*f*) Raise the height of the embankment edges, and (*g*) Complete embankment construction.

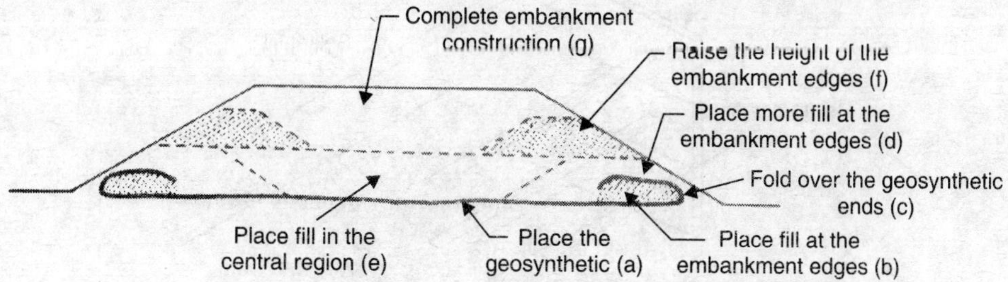

FIGURE 8.20. **Recommended Embankment Construction Sequence.**

(*Source: John, 1987*)

Steep-faced Embankments. On restricted sites it can be advantageous to reinforce the face of the embankments. This helps to reduce the area of land required and the volume of fill. The presence of reinforcement at the face of the embankment allows the use of less expensive fill materials.

Because of high stress concentration steep-faced embankments are not recommended over soft deposits. Generally cohesionless fill within the embankment is favoured for the reasons of high soil to reinforcement surface friction, absence of pore water pressures and ease of compaction. In case of non-availability of cohesionless fill, geotextile grids may be employed with cohesive or partly cohesive soils, as the soil to reinforcement bond for a grid does not rely solely upon the surface friction. In such cases the construction should be rigid and adequate drainage provisions should be made.

In order to get the best performance of reinforced embankment slopes, careful thought must be given to the surface finish. Without the aid of a facing, effective embankment can not be constructed. The most widely used method of restraining the face material is the wrap-around method (Fig. 8.21). Other facing system in use are sandbag facing support system and Textomur facing system.

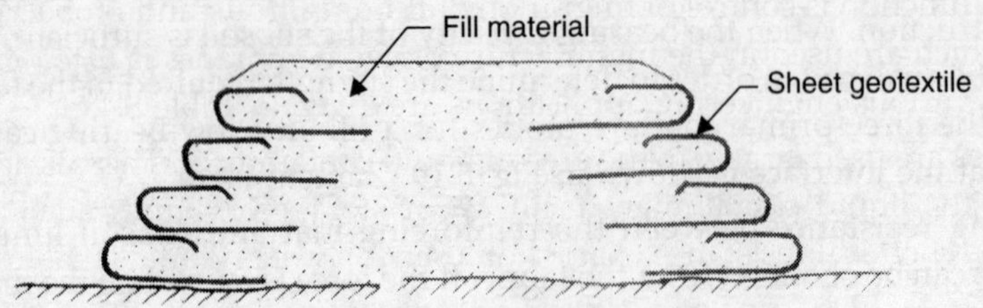

FIGURE 8.21. **Wrap Around Facing Method.**
(*Source: John, 1987*)

Reinforced Soil Walls. Reinforced soil retaining walls made use of tensile reinforcing elements in a fill that is normally largely cohesionless. Under favourable conditions, significant economic advantage can be derived. The advantages claimed by this technique are: flexibility of the reinforced soil mass and large base widths of structures. The three main components of a reinforced soil wall are: (*i*) The facing, (*ii*) The reinforcement, and (*iii*) The select fill (Fig. 8.22).

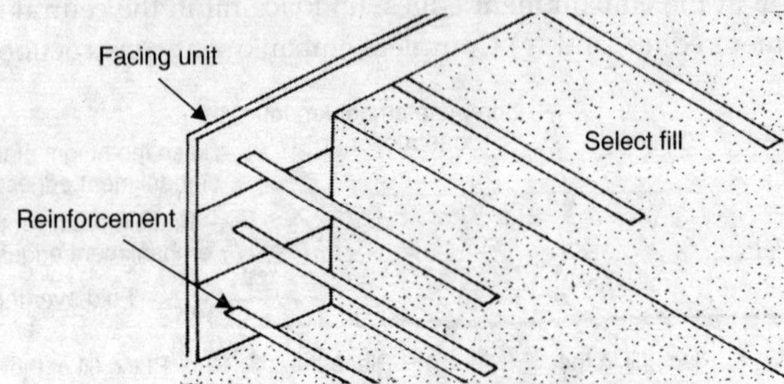

FIGURE 8.22. **Reinforcement Soil Wall Components.**

The facing units are usually interlocking concrete panels. Joints provided in the panels and the counter-balancing pull from the soil reinforcement, reduces the bending moment to a lower value than that of reinforced concrete cantilever wall. As the bending moment is low, there is a large saving in concrete which increases with height. Further the joints provide a greater tolerance to differential settlements. Easy of assembly, stability when partially complete and absence of major temporary works can also be advantageous on congested or tidal silts. Reinforced soil wall demands a select fill because it relies upon friction between the select fill and the reinforcement. The cost of fill obviously depends on its availability and whether the wall is above or below the existing ground surface. For further details of reinforced walls, such as short-term walls, long-term walls and anchored walls, reference may be made to John (1987).

8.4.4 Containments and Barriers

An unique type of geosynthetics is geomembranes. Geomembranes are used as flexible impermeable liners and barriers. The term liner is applied when the geomembrane is used as an interface or a surface revetment. When the geomembrane is used inside an earth mass it is termed as barrier.

Geomembranes are employed in a wide range of civil engineering works, such as potable water reservoirs, distribution canals, municipal and hazardous solid waste landfills, liquid waste impoundment, cut-off walls, dam facing, final closure landfill covers, spill containment systems, wherein the primary function is control of migration of fluids (Giroud and Frobel, 1983). Common types of polymers which are used in the manufacture of geomembranes is listed in Table 8.3. Use of geomembranes in various engineering applications are given in Table 8.4.

Geomembranes are used largely in the area of liquid containment. There is no unique design of lining system for the liquid containment facility. For storage of liquids such as acids, bases, heavy metals, *etc.*, the chemical resistance chart should be followed. Liquids that are combinations of different chemicals, the aggressive one should be considered. For storing an unidentifiable or unknown variety of liquid utmost conservatism should be used and in such a situation high density polyethylene (HDPE) is employed because of its rivertness to various chemicals. For canals and other water storage works unreinforced polyvinyl chloride (PVC) is used. In India low density polyethylene (LDPE) is successfully used in the canal lining systems (Shobha, 1990).

Table 8.3. Common Types of Polymers used in Manufacturing Geomembranes
(*Source: Giroud and Frobel, 1983*)

Thermoplastic Polymers	Polyvinyl chloride (PVC); Oil Resistance PVC (PVC-OR); Thermoplastic Nitrile-PVC (TN-PVC); Ethylene Intepolymer Alloy (EIA).
Crystalline Thermoplastics	Polyethylene (VLDPE, LDPE, LLDPE, MDPE, HDPE, referring to very low, low linear low, medium and high density); High density Polyethylene-Alloy (HDPE-A); Poly-propylene; Elasticized Polyolefin.
Thermoplastic Elastomers	Chlorinated Polyethylene (CPE); Chlorinated Polyethylene-Alloy (CPE-A); Chlorinated Polyethylene (CSPE, Common name is Hypalon); Thermoplastic Ethylene Propylene Diene Monomer (T-EPDM).
Elastomers	Isobutylene Rubber (IIR, common name is Butyl rubber); Ethylene Propylene Diene Monomer (EPDM); Polychloroprene (CR, Common name as Neoprene); Epichlorohydrin Rubber (CO).

Table 8.4. Use of Geomembranes in Various Engineering Applications
(*Source: Shobha, 1990*)

Treatment Basins (liquid containment)	Many private generators of toxic waste are required to design treatment basin with geomembrane lining systems. In this application, chemical resistance and impermeability of the lining are important.
Canals (liquid containment)	Canals are built for transportation of water for agricultural or industrial usage. Geomembrane lining systems for irrigation water reduces the water loss through seepage.
Disposal basins (solid containment)	Disposal basin are for the permanent storage of waste that can not be treated adequately or economically. These wastes could be municipal solid waste, industrial solid waste or chemical waste. Geomembranes with drainage system are used at the bottom side and top of the basin to 'minimise' the flow of leachate from the landfill.
Upstream face of dams (impermeable barrier)	Geomembranes are used on the upstream face for protection of embankment from wind and wave action.
Curtain walls (impermeable barrier)	Geomembranes are used in a narrow trench to construct a cut-off wall system.

Table 8.5. Types of Waste Stored in Landfills
(*Source: Koerner, 1985*)

Type of Waste	*Quantity*	*Quality*
Municipal	Moderate	Least dangerous
Residual Building demolition: Treated or untreated waste, Non-treated, non toxic waste, Near hazardous waste	Largest	Increasing order
Hazardous	Small	Dangerous
Radioactive	Small	Most dangerous

Three basic types of wastes are:

(*i*) Solid Domestic—Typical house-hold refuse and general solid matter.

(*ii*) Solid Industrial—Furnance residues, metals, wood, ashes.

(*iii*) Chemical—Oil sludge, powder compound by-products, organic and inorganic materials are the lanfills which are to be stored using geomembranes.

In earlier days clay liners were used for soil containments. Because of some drawback of clay liners, use of geomembranes for solid containment had been initiated. While storing wastes below ground level adequate care has to be taken for leachate. Although some of the landfills are toxic, hazardous or radio active, most of them are harmless. Types of wastes, stored, in landfills are listed with ranking of danger and quantity of waste stored in Table 8.5.

Geomembranes as an impermeable barrier has been in use in various geotechnical engineering structures, *e.g.*, in dams, embankments and cut-off walls. Generally a high strength geotextile is placed between the geomembrane and crushed stones, in dams and foundations, for puncture protection of the geomembranes. The selection of geomembranes is made based on the following criteria (Shobha, 1990):

(*i*) Withstand sufficient settlement

(*ii*) Chemical resistant to the railing material contained

(*iii*) Satisfactory performance at various temperatures

(*iv*) Good quality of seams at low temperatures

(*v*) Cost competitive.

Geomembranes have been used in cut-off trenches individually or in conjunction with other materials depending on individual project and geological details.

Potential applications of geomembranes, can be summarised as under (Zanten, 1986):

(*a*) Sealing against fluid percolation

— coast (from the sea)

— river banks

— shipping canals and locks (a seepage-reducing slope or a screen)

— reservoirs, water storage (*e.g.*, pools for recreational purposes or to contain water for fire-fighting)

— trenches below the groundwater level

— sites for dumping sand for constructional purposes, terrain bunding

— irrigation canals
— storage weirs and dams
— road sections below the groundwater level (viaducts, tunnels, cuttings).

The function of the geomembrane in these cases is to form a barrier between the water and the surroundings, and to ensure that water transport through the geomembrane is reduced to a minimum (liner).

(*b*) As buffers against pollutants

— permanent or temporary storage of waste products, for instance domestic, chemical or solid wastes, sludge from harbours, fly ash, gypsum, manure, polluted soil *etc*.
— wastewater treatment plants
— top cover
— basins for use in emergencies with, for example, storage tanks
— roads in areas used for extraction of groundwater for domestic use
— areas such as petrol station, parking areas.

In these cases the function of the geomembrane is to create a barrier between two media and to prevent any mixing of these media.

Recent applications of Geomembrane and Geonet in the liner systems of Landfills are explained below.

Landfills in continuation of discussion made in Chapter 1, waste materials may be classified in general into four major categories as (*i*) municipal waste, (*ii*) industrial waste, (*iii*) hazardous waste and (*iv*) low-level radioactive waste. Waste materials are generally placed in landfills.

Leachate as discussed in Chapter 1, materials collected in landfills, interact with moisture received from rainfall and snow to form a liquid called leachate. The Chemical composition of leachates varies widely depending on the waste materials involved. As these leachates may pollute the groundwater they are required to be contained in landfills with some type of liner system. Clay layer and Geomembrane are the two single-liner systems.

Single Liner System important aspect of geomembrane to be used as a landfill liner is the seaming of geomembranes. Geomembrane sheets are seamed together in factories to get large sheets. These sheets are seamed finally at field to fit in the final position. Some of the types of seams are given in Fig. 8.23:

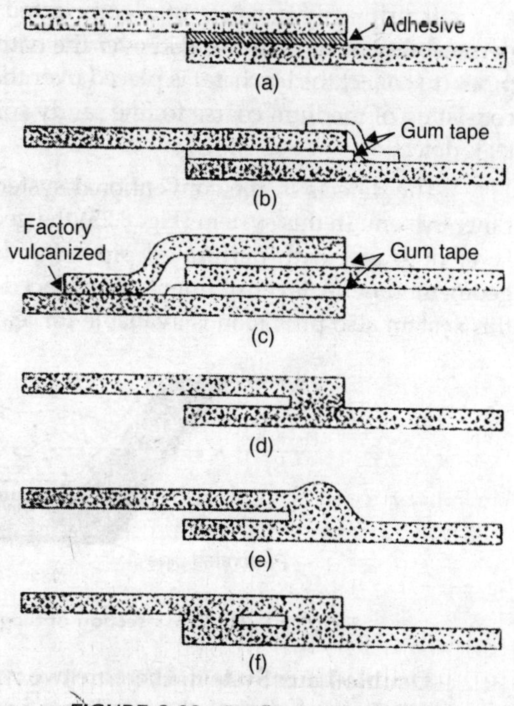

***FIGURE 8.23.* Configuration of Field Geomembrane Seams.**
(*Source: US Environmental Protection Agency, 1989*)

Lap Seam with Adhesive is one in which a solvent adhesive is used [Fig. 8.23 (*a*)]. After the application of the solvent, the two sheets of geomembrane are overlapped and pressure applied by roller. Lap seam with Gum Tape is one which is used in thermoset material [Fig. 8.23 (*b*)]. Tongue-and-Groove Splice is the one which is used for the splice [Fig. 8.23 (*c*)] extrusion. Weld Lap Seam is got by extrusion or fusion welding done on geomembranes made from polyethylene. Here a ribbon of molten polymer is extruded between the two surfaces to be joined [Fig. 8.23 (*d*)]. Fillet Weld Lap Seam is similar to an extrusion weld lap seam. Here for fillet welding the extrudate is placed over the edge of the seam [Fig. 8.23 (*e*)]. Double Plot Air or Wedge Seam is the one in which hot air is blown to melt the two opposing surfaces. After the opposite surfaces are melted, pressure is applied to form the seam [Fig. 8.23 (*f*)].

In earlier days, till 1982, single clay liner system was in use for landfill. A typical cross-section of single clay linear system for a landfill is shown in Fig. 8.24 (Das, 2000).

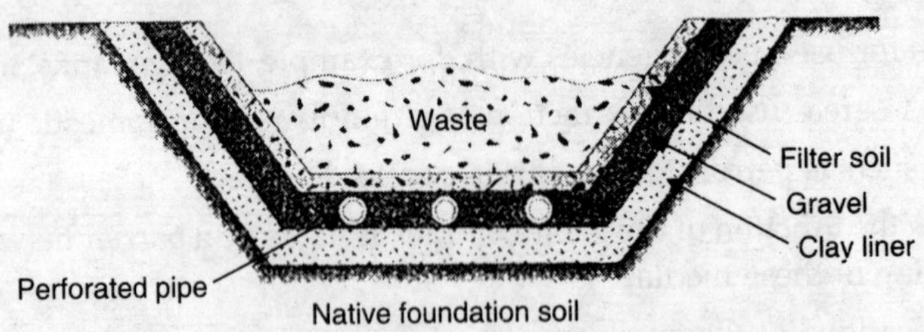

FIGURE 8.24. **Cross-section of a Single Clay Liner System for a Landfill.**

It primarily consists of a compacted clay liner of 0.9 to 1.8 m thick with a maximum permeability of 10^{-7} cm/sec laid over the native foundation soil. A layer of gravel with perforated pipes (to collect the leachate) is placed over the clay liner. Over the gravel layer, a layer of filter soil consisting of medium coarse to fine sandy soil is provided. This liner system has no provision for leak detection.

The defects in the conventional system have been partly overcome in the Geomembrane Liner System. In this system (Fig. 8.25) the geomembrane is laid over the native foundation soil. A layer of gravel with perforated pipes (for lechates collection and removal) is placed over the geomembrane. A layer of filter soil is placed between the solid waste material and the gravel. In this system also provision is available for leak detection (Das, 2000).

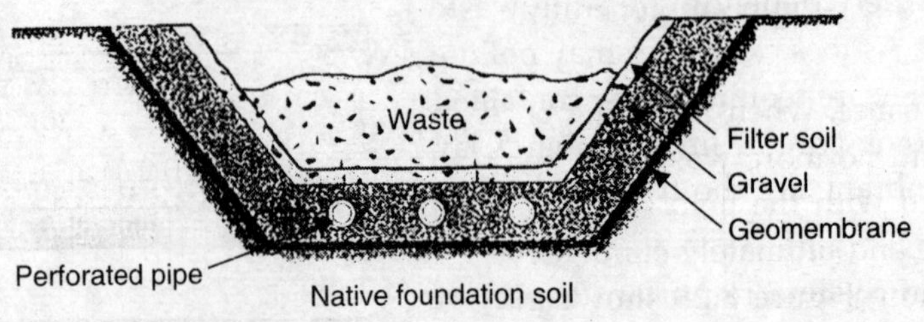

FIGURE 8.25. **Cross-section of Single Geomembrane Liner System for a Landfill.**

Double-Liner System. There are two types of Double-Liner System in use in different countries since 1990. In the first type the two liners comprise of Geomembrane and Geonet (Fig. 8.26). Here two geomembrane liners are provided. One is a primary liner and the other one is a secondary liner. Above the top liner is a primary leachate collection and removal system (Das, 2000).

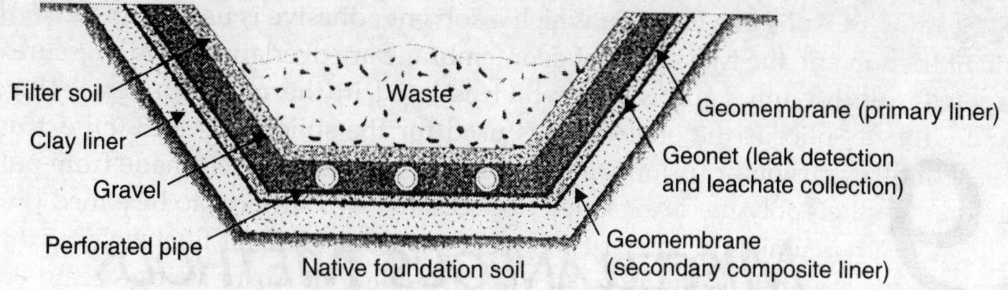

FIGURE 8.26. **Cross-section of Double-liner System–Type-I.**

A leak detection, collection and removal (LDCR) of leachate provision is made between the primary and secondary liners. The compacted clay liner should be of atleast 1 m thick with $k \leq 10^{-7}$ cm/sec. In this type the primary lechate collection is made of a granular material with perforated pipes and a filter system. The primary layer is a geomembrane whereas the secondary liner is a composite layer of geomembrane with a compacted clay layer below it. The LDCR is made of a geonet.

The Type-II Double-liner system is shown in Fig. 8.27. Here the primary leachate collection system is similar to that of Type-I (Das, 2000).

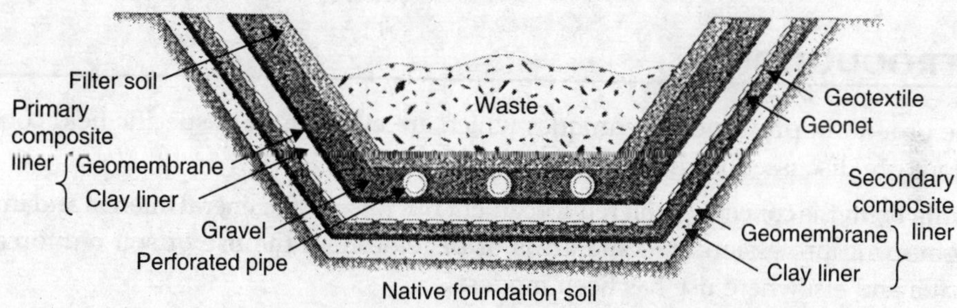

FIGURE 8.27. **Cross-section of Double-Linear System–Type-II.**

But the primary and secondary liners are both composite liners (geomembrane-clay). The LDCR system is a geonet with a layer of geotextile over it. The geotextile layer acts as a filter and separator. Gomembranes with thickness ranging from 1.8 to 2.54 mm are generally used.

Landfill-closure. When the filling of landfill is over and no more waste can be placed, it is closed with the help of a cap which will reduce and ultimately eliminate leachate generation. Figure 8.28 shows a schematic arrangement of layering system recommended as a cap. It primarily comprises of a compacted clay layer cap over the solid waste. Over which a geomembrane liner, followed by a drainage layer and a top soil cover are provided.

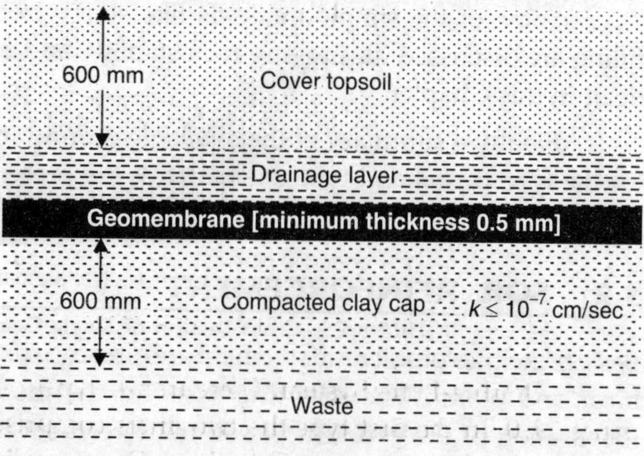

FIGURE 8.28. **Schematic Diagram of Landfill Closure.**
(Source: Koerner, 1994)

Chapter **9** MISCELLANEOUS METHODS

9.1 INTRODUCTION

Some of the ground improvement techniques which are either used in specific field conditions or rare situations are discussed in this Chapter.

Among them the concept of soil reinforcement has raised the general interest and imagination of civil engineering profession. In some areas of development the use of soil reinforcement has been dramatic and elsewhere use has been modest.

The another method of concern is ground improvement by thermal stabilization. The principles of thermal effects on soils have been well recognised. In principle both heating and cooling can be employed for stabilization. But because of economy the cooling methods have become very popular during recent years.

Some of the simple methods like slurry trenches, moisture barriers, prewetting and void filling are discussed. Improving rock stability and quality for foundation purposes have also been dealt in this Chapter.

Finally a summary of the available ground improvement techniques and their applicability for the type of soil and type of structures are presented.

9.2 SOIL REINFORCEMENT

Basic principles of soil reinforcement are already existing in nature and are demonstrated by animals, birds and plants. The scientific basis for the modern concept of soil reinforcement lies in the idealisation of the problem of soil reinforcement in the form of a weak soil reinforced by high-strength thin horizontal membranes (Westergaard, 1938). The modern form of soil reinforcement was first applied by Vidal (1969). Based on the Vidal's concept the interaction between the soil and the reinforcing horizontal members is solely by friction generated by gravity. Applying this concept, retaining walls were built in France in 1986. Nowadays this technique is widely used in Europe and U.S.A. This technique is yet to become popular in India, and the constraining factor being identified as the non-availability and cost of reinforcing materials.

Reinforced soil is somewhat analogous, to reinforced concrete. But direct comparison between the functions of reinforcement in the two cases is not valid. The mode of action of reinforcement in soil is not one of carrying developed tensile stresses as in reinforced concrete but of anisotropic reduction of normal strain rate (Jones, 1985).

The supporting capacity of soft, compressible ground may be increased and the settlement may be reduced through use of tensile reinforcement in planes normal to the direction of applied stress of compression reinforcement in the direction parallel to the applied stress. Tensile reinforcement used for soil reinforcement are wider in range which are discussed subsequently. Commonly used compression reinforcement elements include columns and walls formed in-situ using gravel, sand, soil-cement, soil-lime or admixtures.

In recent years theoretical developments associated with reinforced soil have been extensive. However, and complete understanding of every aspect and use of the subject has not yet been fully developed. Nonetheless, an understanding of the fundamental principles was established by the early developers and users and these have been since confirmed or further expanded. For details of theories, analysis and design of soil reinforcement the reader may refer to Jones (1985). The type of materials and field applications pertaining to ground improvement are discussed subsequently. Rigid procedures for design should be avoided and adequate engineering judgement is needed when using earth reinforcing techniques.

9.2.1 Materials

There are three basic materials or material composites required in the construction of any reinforced soil structure. They are:

(*i*) soil or fill matrix
(*ii*) reinforcement or anchor system
(*iii*) a facing if necessary.

Other materials which are required, in addition to the above, to cover associated elements such as foundations, drainage connecting elements and capping units and to act as barriers and fencing. There used to be adequate inter-relationship between the materials used. Based on design considerations and availability, the materials are selected.

Soil/Fill. The shear properties of soil can be improved as theoretically any soil could be used to form earth reinforced structure. In long term conventional structures the soil used is usually well-graded cohesionless fill or a good cohesive frictional fill although pure cohesive soils have been used with success. The advantages of cohesionless fills are that they are stable, free draining, not susceptible to frost and relatively non-corrosive to reinforcing elements. The only disadvantage is its cost. In case of cohesive soils the main advantage is availability but there may be long-term durability problem together with distortion of the structure.

As a convenient compromise between the technical benefits from cohesionless fill and the economic benefits from cohesive fill, cohesive-frictional fill may be preferred. Sometimes the use of waste materials as fill for reinforced soil structures is attractive from an environmental as well as economic view point. Mine wastes and pulverised fuel ash are the wastes usually employed.

Reinforcement. A variety of materials including steel, concrete, glass, fibre, wood, rubber, aluminium and thermoplastics can be used as reinforcing material. Reinforcements may take the form of strips, grids, anchors and sheet material, chain, planks, rope, vegetation and combinations of these or other material forms.

Strips are flexible linear elements having their breadth greater than their thickness. Strips are formed from aluminium, copper, polymers and glass fibre reinforced plastics (GRP) and bamboos. The form of stainless galvanised or coated steel strips are either plain or with projections such as ribs to increase the friction between the reinforcement and the fill. Similar to strips, planks formed from timber, reinforced concrete or prestressed concrete can also be used.

Grids and geogrids are also used as reinforcement. Grids are formed from steel in the form of plain or galvanised weldmesh or from expanded metal. Grids formed using polymers are referred to as geogrids (explained in Chapter 8) and are normally in the form of an expanded proprietary plastic product.

Sheet reinforcement may be formed from metals such as galvanized steel sheet, fabric (textile) or expanded metal not meeting the criteria for a grid.

Flexible linear elements having one or more pronounced protrusions or distortions which act as abutments or anchors in the fill or soil. They may be made from materials like steel, rope, plastic (textile) or combinations of materials such as webbing and tyres, steel and tyres, or steel and concrete.

Composite reinforcements can be formed by combining different materials and material forms such as sheets and strips, grids and strips or strips and anchors, depending on the field problem requirement.

The principal requirements of reinforcing materials are strength, the stability (low tendency to creep), durability, case of handling, a high coefficient of friction, and/or adherence with the soil, together with low cost and ready availability.

9.2.2 Applications

Application areas for the use of earth reinforcements are many. Jones (1985) identifies several soil reinforcement field applications, *viz.*, bridge works, dams, embankments, foundations, highways, housing, industry, military, railways, root pile systems, pipe works, waterway structures, and underground structures. The applications relevant to ground improvement are illustrated in Fig. 9.1 (Jones, 1985). For the illustrations shown in Fig. 9.1, the materials given in Table 9.1 could be used as soil reinforcements and their advantage and limitations are also given in Table 9.1.

The applications illustrated by Jones (1985) should not be taken as being the only effective or rational solution to any problem. In practice, applications and techniques may often be combined and the introduction of new construction materials enable other applications to be considered. As a matter of fact, the variety and wide range of the areas of application for these techniques is unlimited.

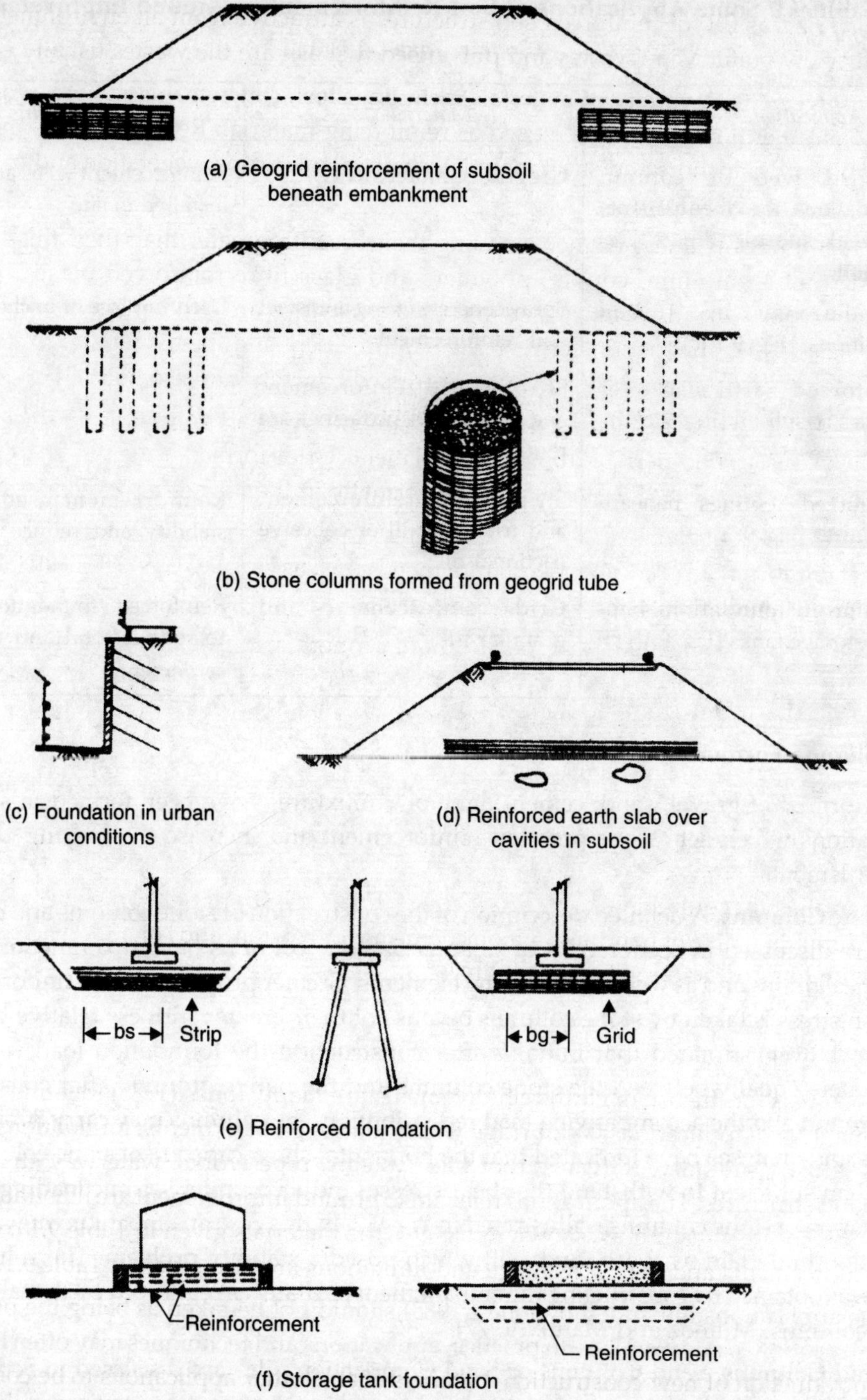

***FIGURE* 9.1. Applications of Soil Reinforcement for Ground Improvement.**

(*Source: Jones, 1985*)

Table 9.1. Some Applications of Soil Reinforcement in Ground Improvement

(*Source: Jones, 1985*)

Application	*Materials*	*Comments*
1. Geogrid web or column foundations for embankments on weak subsoil [Fig. 9.1 (*a*) and (*b*)]	Geogrid reinforcement	Reinforcement webs and columns installed in site.
2. Foundations in urban conditions. [Fig. 9.1 (*c*)]	Spray concrete facing and steel nail reinforcement	Used in place of anchor systems.
3. Reinforced Earth slab over cavities in subsoil [Fig. 9.1 (*d*)]	Strip or grid reinforcement and selected fill, no facings are required	
4. Reinforced footings beneath structures [Fig. 9. 1 (*e*)]	Strip or grid reinforcement and frictional fill or cohesive frictional fill.	Reinforcement used to ensure stability and reduce settlement.
5. Reinforced foundations beneath storage tanks [Fig. 9.1 (*f*)]	Grid reinforcement and granular fill.	Reinforced foundations are used to reduce total and differential settlement.

9.2.3 Columns Formed In-situ

Columns formed of gravel, sand, cement, lime or admixtures have been formed in-situ in weak soil formations which act as compression reinforcement and increase the bearing capacity and reduce settlement.

Stone Columns. A detailed description of the construction of stone columns and other salient features are discussed in Section 5.3.4 of Chapter 5. Stone columns have two functions, *viz.*, they act as vertical drains and as well as reinforcing elements. Greater proportion of an uniformly applied foundation stress is taken by stone columns because of their greater stiffness relative to the native soft ground. It is estimated that initially after construction the foundation load is distributed approximately equally between the stone columns and the native ground. After consolidation of the soft ground and the accompanying load redistribution, the columns may carry 80% of the load. Field tests and analyses have indicated that the horizontal shear capacity of stone columns should be more than sufficient to withstand the shear stresses induced under seismic loading conditions. An application of stone columns is illustrated in Fig. 9.2. In this case an embankment was proposed on a location underlain by loose sandy silt which posed a stability problem. The solution to this problem was obtained by constructing the embankment partially as reinforced earth wall, supported on stone columns (Mulnoz and Mattox, 1977).

Sand Columns. Sand Columns or Sand Compaction piles are discussed in Section 5.3.3 in Chapter 5. The required relative density can be obtained by varying the spacing and the size of the compaction piles. For moderate depths up to 15 m, this method is economical.

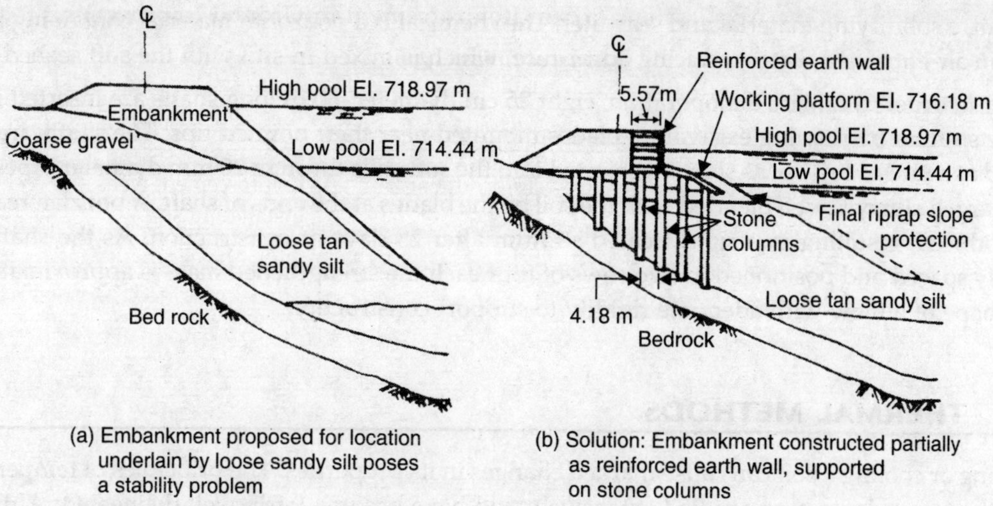

***FIGURE* 9.2. An Example of the Application of the Stone Column Technique Combined with a Reinforced Earthwall to Support a Railroad Embankment Along a Lakefront.**
(*Source: Munoz and Mattox, 1977*)

Soil-cement or Soil-lime Columns. Several procedures are available for the construction of soil-cement or soil-lime column or pile. A special hollow rod with rotating vanes is augered into the ground to the desired depth and the stabilizing admixture is simultaneously introduced. By this process piles up to 600 mm diameter can be constructed.

Cement, in the order of 5 to 10% of the dry soil weight, is best for use in sandy soils. In these materials compressive strengths more than 10 MN/m^2 can be achieved. Lime is effective in both expansive plastic clays and in saturated clays. In these materials compressive strengths of the order of 1 to 2 MN/m^2 can be obtained. Installation is comparable to those for installation of sand columns. If over-lapping columns are constructed, then a soil-cement or soil-lime wall will result. Broms and Bowman (1979) have used soil-lime columns beneath light structures and road embankments to reduce settlement resulting from consolidation of thick deposits of soft glacial clays. Unslaked or quick lime was mixed with clay and columns up to 10 m long and 50 cm in diameter were constructed. It has been reported that lime clay mixture has substantially increased the strength of the natural clay and the columns transferred surface loads to deeper, stronger strata. Brown and Bowman's (1979) study has revealed that the increase in shear strength from lime mixing was most-significant in fresh and brackish water-deposited glacial clays, but in soft organic clays and salt water-deposited clays the strength increase was low.

The soil-lime columns have been widely used for varied applications, *viz.*, to decrease the negative skin friction on piles, to prevent the lateral displacement of soil around pile foundations from creep, to increase the stability of clay slopes, to reduce the lateral earth pressure on retaining structures and a replacement for sheet piles in deep excavations to prevent bottom heave (Hunt, 1986).

Deep Chemical Mixing Method. This method has been in use in Japan since 1976 for treating very soft seabed alluvial soils for support of water front retaining structures (ENR, 1983). In this method a plant, operating from a barage, produces a slurry consisting of conventional Portland cement, a solidifying material and seawater. The chemical is a pozzolith material containing lignin and an air-entrained water-reducing admixture, which is mixed in-situ with the soft seabed soils.

In order to initiate the operation, eight 25 cm diameter, 60 m long shafts are inserted into a stable stratum by rotating excavating blades mounted near their pointed tips. The shafts are then raised at a rate of 1 m/sec as slurry is injected into the soft soils through 20 mm diameter pipes. The discharged slurry is well blended with the soil by the blades at the ends of shaft. A pozzlan reaction is produced, resulting in a rigid treated stratum after 28 days of construction. As the shafts are closely spaced and positioned in two rows of four each the strengthened mass is approximately in the shape of a blow with adequate rigidity to support construction.

9.3 THERMAL METHODS

Heating or cooling a soil can cause marked changes in its properties. The principles of temperature effects on soils have been studied extensively and have become fairly well delineated. Although thermal stabilization appears to be very effective, it has several inherent undesirable features which have totally limited its use. The cost is the main disadvantage since cost of heating or cooling to effect stabilization is no way competitive with the cost of other techniques.

9.3.1 Stabilization by Heating

In general, the higher the heat input per mass of soil being treated, the greater the effect. Even a small increase in temperature may cause a strength increase in fine-grained soils by reducing the electric repulsion between the particles, a flow of pore water due to thermal gradient and a reduction in moisture content because of increased evaporation rate. Hence, it is technically feasible to stabilize fine-grained soils by heating. The effect of increase in temperature and the corresponding possible change in soil properties are the following:

Temperatures ~ 100°C—Cause drying and significant increase in the strength of clays, along with decrease in their compressibility.

Temperatures ~ 500°C—Cause permanent changes in the structure of clays resulting in decrease of plasticity and moisture adsorption capacity.

Temperatures ~ 1000°C—Cause fusion of the clay particles into a solid substance much like brick.

It is reported that heat has changed an expansive clay into an essentially non-expansive material. Burning of liquid or gas fuels in boreholes or injection of hot air into 0.15 m to 0.20 m diameter boreholes can produce 1.3 to 2.5 m diameter stabilized zones after continuous treatment for about 10 days. Soviet engineers have used this technique exhaustively for strengthening partially saturated loessial soils. This technique was found to be economical than pile foundations in loessial soils. It should be noted that the injection of hot gases is applicable only to nonsaturated soils. Beles and Stanculescu (1958) have used thermal stabilization to increase the strength and to decrease the compressibility of cohesive soils. This technique can be favourably used only when a site is located near a large and inexpensive heat source.

9.3.2 Stabilization by Cooling

A reduction of heat (*i.e.*, cooling) in a clayey soil increases interparticle repulsion resulting in a small loss in strength and moves the pore water because of the imposed thermal gradient. Freezing of pore water in soil is the most effective method of thermal stabilization. Ground freezing technique has gained popularity during recent years.

Water in a soil freezes at or below 0°C with the initiation of freezing, the soil strength increases rapidly with decreasing temperature as more soil is frozen and the strength of ice increases. Frozen soil is far stronger and less pervious than unfrozen ground. It also forms non-vibration-sensitive barrier to seepage flow or soil deformation. This technique can stabilize a wide range of soil types.

Ground freezing is accomplished by bringing a refrigerent into the proximity of soil pore water. The pore water may be stationary or moving at a rate less than 2 meters per day. Once freezing is initiated the pore water around the refrigerant pipes begins to freeze, and with continued exposure the ice layer expands until it comes into contact with the ice spreading out from adjacent refrigerant pipes. Thus a continuous wall is formed.

A number of schemes are possible to provide refrigerant to the soil. Different schemes are reviewed by Shuster (1972) and the same is illustrated in Fig. 9.3. Two most common schemes are the use of expandable liquid refrigerants, like liquid nitrogen, liquid carbon dioxide, or liquid propane, and the in-situ pumped-loop method *via* a secondary coolant.

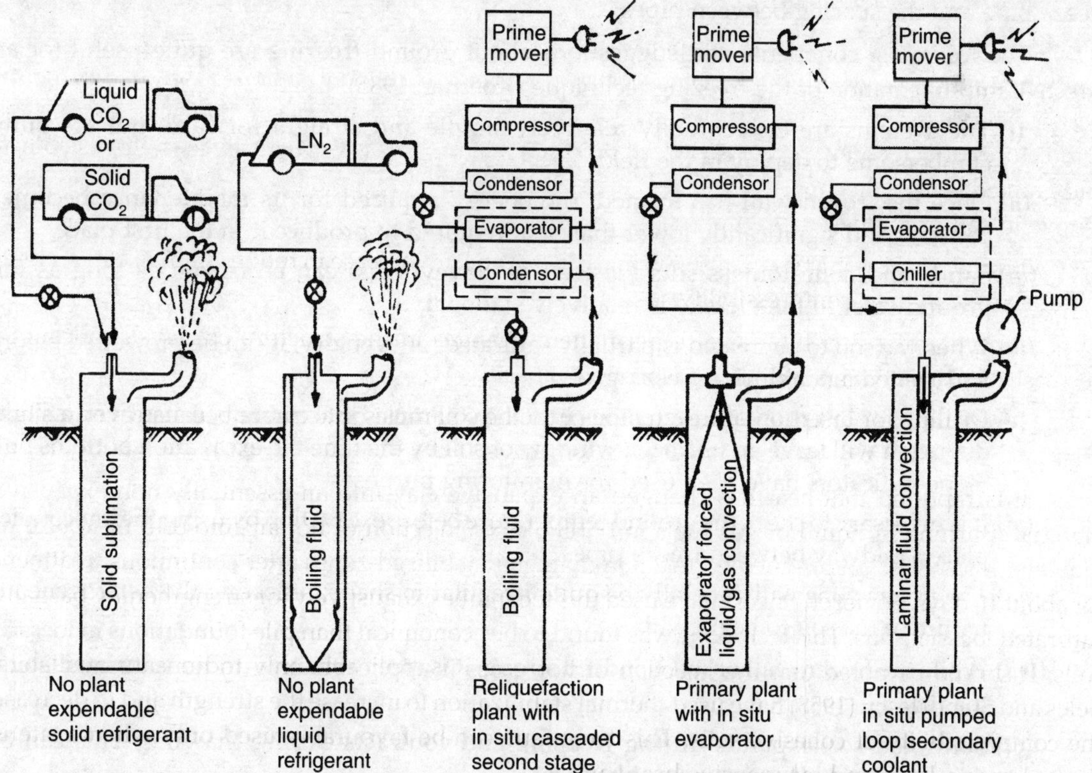

***FIGURE* 9.3. Alternative Refrigeration Method for Ground Freezing.**
(*Source: Shuster, 1972*)

Of these two methods, the first one is relatively simple. Freeze pipes are installed at one metre centres, sufficiently vented and the liquid refrigerant is injected and allowed to boil. Freezing takes place very rapidly but the frozen zone is often very irregular and energy consumption by this process is very high. However, for a small project of short duration freezing technique is quite useful.

The second technique is a much more popular one which uses a primary plant and a pumped loop secondary circulating coolant. This method is well suited to single installations of freeze pipe elements intended to provide maintenance freezing over a long period of time. The primary source of the refrigeration is one- or two-stage ammonia or freon refrigeration plant. The coolant distribution system consists of a closed-loop supply manifold connected to a number of parallel-connected freeze pipes placed in the ground and to a return manifold. Diameter of ground freeze pipes is typically 100 to 200 mm with sealed lower ends. Feed pipes of 25 to 75 mm diameter are placed inside the larger capped pipe. Coolant is circulated into the feed pipes and returns through the annular space between the two. The proximity of the return coolant to the soil is what produces the freezing. The conventional coolant is brine, a mixture of sodium chloride and water with 10 to 23% sodium chloride, although diesel oil, propane, and glycol-water mixtures have also been used.

Most of the design of ground freezing systems is done empirically on the basis of past practice. Generally several weeks are required for a ground freezing system to be effective. The time required for complete freezing depends on the type of coolant used, the temperature achieved, the size of freeze pipe and the spacing between pipes.

The following comments and suggestions about ground freezing are quite useful for an effective implimentation of the freezing technique (Koerner, 1985):

(*i*) The systems are conceptually relatively simple and straight forward, but are quite cumbersome to deploy in the field.

(*ii*) Once the frozen wall has formed, the energy required for its maintenance becomes constant and significantly lower than that required to produce it in the first place.

(*iii*) Sands and cohesionless silts (as well as clayey soils) can be frozen as long as the groundwater in the vicinity is relatively stationary.

(*iv*) When the soil to be treated is partially saturated, or even dry, it can be prewetted before and wetted again during freezing.

(*v*) Drilling for insertion of freeze pipes into the soil must be accurate because even a slight deviation will leave an unfrozen window of soil by the time the excavation bottoms out. Slope indicators have been used for monitoring purposes.

(*vi*) It is necessary to check the ground temperature before excavation by using thermocouples placed midway between freeze pipes.

(*vii*) The frozen zone will generally be quite irregular in shape, reflecting the heterogeneous nature of the soil to begin with.

(*viii*) While the frozen wall of soil is indeed strong, it is creep-sensitive under sustained lateral loading.

(*ix*) Saturated soil expands during freezing and contracts during thawing. This can be important in underpinning applications.

(*x*) A competent speciality contractor should perform the work. If the situation is critical, it is no place for amateurs!

9.4 OTHER SIMPLE METHODS

There are different simple methods of ground improvement which are in practice adopted under varied field conditions. Such methods are given in the following sections.

9.4.1 Slurry Trench

In order to form a deep impervious membrane in a pervious formations, a trench filled with a mixture of sand-gravel and semifluid plastic clay is formed. The trench is made using a dragline, clamshell bucket or rotary paddle. The width of trench may vary from 1 to 4 m. A viscous mixture of highly plastic clay and water, termed slurry, is filled in holes to support the excavation walls. The slurry is usually made from commercial betonite similar to the drilling mud used in oil wells. The permanent membrane is formed by a mixture of the same slurry with sand and gravel that displaces the lighter slurry to form a void-bound structure. The coarse grained particles contact one another to form a reasonably rigid mass with enough slurry in the voids to provide imperviousness. The slurry material either stiffens thixotropically or hardens with the aid of a cementing agent. It has been reported that membranes deeper than 60 m have been constructed by this process (Sowers, 1979).

9.4.2 Moisture Barriers

Expansive clays such as black cotton soils and partly saturated lightly cemented soils such as loess can provide excellent foundation support for light structures provided they are maintained relatively dry. Thus critical zones under foundations should be protected from water. This is accomplished by providing moisture barriers which keep away water entering the critical zones. Membranes such as polyethylene sheeting and asphalt treated fabrics are effective for this purpose. Sand or gravel blankets and trenches may also be effective moisture barriers, provided they can be maintained at a low degree of saturation. Unsaturated coarse layers which have low hydraulic conductivity effectively prevent significant water transfer across the layer.

9.4.3 Prewetting

One technique for stabilization of expansive soils that can be effective under light structures such as residential houses is to flood the area prior to construction. As in natural expansive soils extensive network of fissures and cracks present initially, ponding process is easily facilitated. Because of prewetting the water content will be closer to be attained after construction, hence volume changes will be small subsequently. It is the usual practice to treat the surface with a layer of lime to a depth of 0.3 to 0.5 m after ponding. This treatment provides a working platform for construction and an impermeable moisture barrier to retard subsequent dessication of the prewetted soil.

9.4.4 Void Filling

Void filling is a simple technique which could be applied for granular soils. In cohesionless soils the filling of voids reduces the permeability and reduces the water penetration resulting in the maintenance of soil strength. A variety of materials are used for void filling, *viz.*, Portland cement, soluble silicate gels, organic monomer-polymers. Fine-grained materials such as silt, fly ash and clay can also be used in void filling. Swelling materials with fine particles can lodge in the voids and later expand to fill the voids. Emulsified asphalt or latex can be treated so that the emulsion breaks in the soil voids and creates an impervious mass. Most of these materials are placed by injection.

9.5 IMPROVING ROCK STABILITY AND QUALITY

Foundations on rock usually are an economical and troublefree solutions as long as the rock is reasonably shallow, relatively unweathered and sound. If rock conditions are unfavourable careful evaluation have to be made in particular if structures imposing heavy loads such as tall buildings, towers, chimneys, power plants, bridges and concrete dams are to be constructed on it. Type of foundation, shallow or deep, depends on the rock formation (shallow or deep), the type of proposed structure, and the rock quality. Where rock is exposed for examination, the engineer or geologist will have much more confidence in judging rock quality than he will for the case of deep foundations where examination is either impossible, not practical or limited in area. Typical critical problems one has to face are : (*i*) heavy column loads on unsound rock, (*ii*) bridge foundations constructed in canyon walls, and (*iii*) foundations for concrete, buttress or arch dams with high seepage pressure. In order to overcome such problems, rock conditions must be evaluated carefully with respect to both rupture and deformation as affected by rock quality, rock surface irregularity and groundwater conditions. The necessary treatments required for unfavourable rock-mass are discussed subsequently.

9.5.1 Treatment and Purpose

There are five main categories of techniques for treating rock masses (Hunt, 1986):

(*i*) Excavation and the removal of soft, weathered, or highly fractured or otherwise undesirable materials to provide for support on rock of adequate quality, to provide for a stable slope or to remove permeable material from beneath dams.

(*ii*) Injection with grout to increase rock strength and improve support characteristics for foundations and tunnels and to decrease permeability and cleft-water pressures, thereby reducing seepage and improving the stability of dam foundations, tunnels, and other underground openings.

(*iii*) Rock bolts or cable anchors to reinforce slopes and underground excavations and to improve uplift resistance for concrete dams.

(*iv*) Surface treatments with shotcrete or gunite to reinforce loose masses of small rock blocks, to deter rock decomposition in slopes, or to reinforce the exposed surfaces in tunnels and other underground excavations.

(*v*) Drains to relieve cleft-water pressures, including subhorizontal drains, galleries, and deep pumped or gravity wells for slopes or simple drain holes for dam foundations.

9.5.2 Uplift Pressure and Seepage Control

The case of dam foundation seepage has two major aspects : first flow quantity and reservoir loss and second uplift pressures. If the depth of fractured foundation rock is shallow it can be removed and a core trench or a cut-off wall may be constructed. If the fractured rock is too deep, then a grout curtain connected to a grout blanket may be provided. In earth dams the major concern usually is with flow and reservoir loss through the abutment and foundation rock.

Grout curtains are capable of reducing uplift pressures beneath the dam and reducing pore pressures in the abutments. As complete cut-off is seldom achieved with a grout curtain, it is always desirable to install drain holes to ensure that pore pressures are relieved. Drain spacing for concrete dams depends on rock conditions but often ranges from 3 to 10 m on centres. In earth dams, drainage holes may be drilled beneath the dam at locations where the intercepted water can be discharged into the internal drainage system.

9.5.3 Rock Reinforcement

Rock Bolts and Cable Anchors. Apart from grouting the rock-mass strength can be increased with rock bolts and cable anchors. Rock bolts are used in slope stabilization, in open exacavations, in tunnels, caverns and mines and in concrete dam foundations to provide resistance to, uplift and sliding. The latter conditions may prevail where weak foundations are interbedded with strong layers. They are used to sustain high residual stresses causing floor uplift or wall deflection in open or closed excavations. Also applicable areas where there is a concern is that high grouting pressures will cause displacement of shallow rock. Cable anchors are mostly used for slope stabilization.

Rock bolts are tensile units employed to keep rock mass in compression. It is installed as nearly perpendicular to joints as practicable. Ordinary types of rock bolts are illustrated in Fig. 9.4

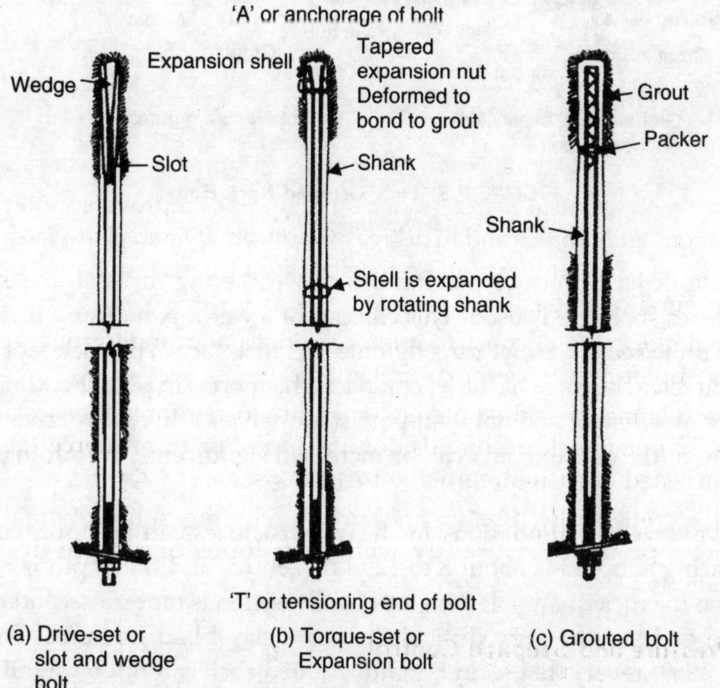

***FIGURE* 9.4. Types of Ordinary Rock Bolts.**

(Source: Lang, 1972)

(Lang, 1972). The ordinary types consist of rods installed in drill holes by driving and wedging, by driving and expanding, or by grouting with mortar or resins. Bolt heads are then attached to the rod and torqued against a metal plate to impose the compressive force on the mass. Fully grouted rock bolts, Fig. 9.5 (Lang, 1972), provide more permanent bolts than the ordinary types. Grouting with resins are becoming popular because of easy installation.

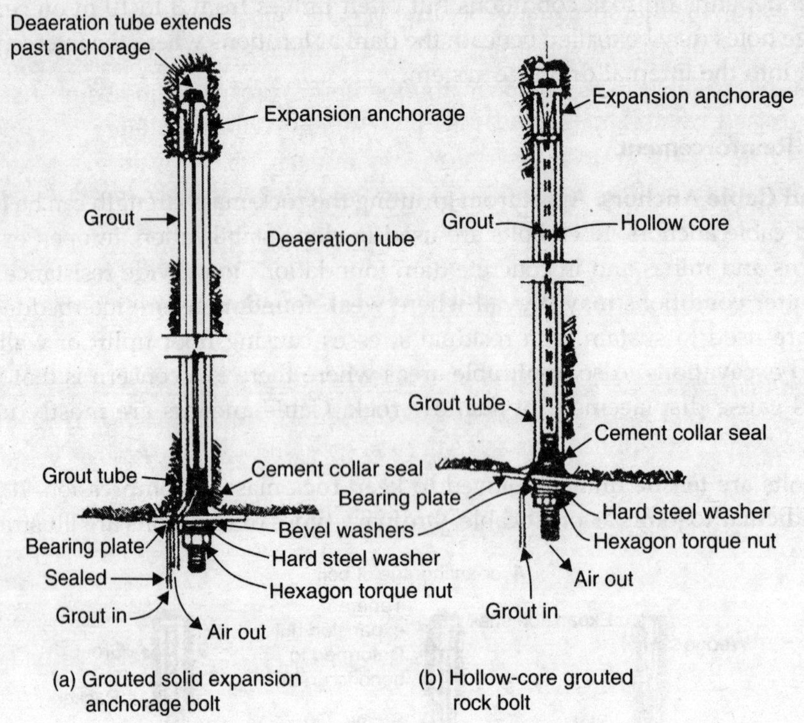

***FIGURE* 9.5. Fully Grouted Rock Bolts.**
(*Source: Lang, 1972*)

Shotcrete. In order to reduce possibilities of weathering in weak rocks and to reinforce fractured rock masses shotcrete is used. This consists of a wet-mix mortar with aggregate as large as 2 cm which is projected by air jet directly onto the rock face. The rock face is cleaned before application of shotcrete. The force of the jet compacts the mortar in place bonding it to the rock. As shotcrete acts as reinforcement and not a support, it is often used in place of rock bolts. If necessary the tensile strength of the concrete mix can be increased significantly by adding 25 mm long wire fibres.

Rock Improvement. Foundations for heavy structures can be improved by compaction grouting. The spacing adopted is about 3 to 12 m on centres and the depth is ranging from 10 to 30 m depending on the rock quality. If the grout consumption is more, a secondary grid is adopted with spacing 2 m or less. It is very difficult to treate clay-filled fractures. The published data (Lancaster-Jones, 1969) reveal that sound, tightly jointed rock can not be significantly improved. However, in sound to moderately sound rock with open joints the best results are obtained. Further by this process even poor quality rock can be improved significantly.

9.5.4 Rock Deterioration Prevention

It is of concern to consider the possible deterioration of rock with time beneath a foundation. Rocks that are highly soluble (such as limestone; sandstone that is cemented with gypsum, calcite, or other carbonates; or even dikes of calcite in metamorphic or igneous rocks) are susceptible to attack by moving water. Grout curtains are employed to prevent or impede deterioration of soluble materials. Cavity growth and roof collapse because of water-table lowering caused by groundwater withdrawl is also of concern which is also prevented or impeded by grouting.

Many shales are subjected to rapid weathering upon exposure. Thus exposure to weathering should be minimised by making excavations in smaller sections. Foliated metamorphic rocks (gneisses and schists) may be deteriorated due to contaminated groundwater resulting from leaking sewage pipe.

9.6 REMEDIATION OF CONTAMINATED SOILS

As an unanticipated side effect of industrial development, nations are experiencing events of soil, groundwater, and surface water contamination on a unprecedented scale. Further many of these contaminants pose an immediate threat to public health and the environment. Many sites in industrialized nations are contaminated with potentially active chemicals and metals such as arsenic, barium, cadmium, chromium, mercury, nickel, selenium, zinc and lead. Such metals, when concentrated in soil and sediments, pose a threat to health and environment.

For planning and implementation of a remediation program on a contaminated soil there is a need to determine the potential for contaminant migration, including its distribution and the toxicity of the contaminants. For this an environmental site assessment has to be made before remediation.

The Environmental Site Assessment (ESA) are effected in two phases. In the Phase-I ESA the following aspects are to be considered (Hess, 1993):

(*i*) Potential environmental liability on a property scheduled for a transaction has to be identified.

(*ii*) Hazardous materials that might be disturbed and released during subsequent construction activities have to be identified. In general, hazardous substances pose a reduced risk if left undisturbed. Early identification allows alternative plans for excavation.

(*iii*) Information about environmental conditions during planning stages is to be provided so that alternative design can be formulated to avoid complications and reduce cost.

(*iv*) Worker and public safety via exposure to soil, water and air contamination are to be ensured.

Many of the physical and chemical properties of the containment will dictate the over mobility and toxicity of the plume. These are done under the Phase-II ESA Program so as to ascertain the influence on the selected remedial program. Some of the more critical properties to assess in the Phase-II ESA are (Pichtel, 2007):

(*a*) Water solubility
(*b*) Viscosity
(*c*) Specific gravity
(*d*) Soil sorption coefficient
(*e*) Biodegradability index
(*f*) Vapour pressure
(*g*) Vapor density.

Depending on the situation various methods are used as remediation which are explained below (Pichtel, 2007).

9.6.1 Isolation of Containment Plume

Containments from a site may enter groundwater through several avenues such as (*i*) infiltration of precapitation or run-on from the surface, (*ii*) groundwater migration into the contaminated zone, and (*iii*) migration of contaminants as free liquids through soil.

In order to control the containment migration it may be necessary to contain the plume and prevent additional dissolution of hazardous constituants by water entering the contamination zone. During such an isolation a detailed remediation can be programmed.

The isolation process is achieved by any one of the following methods, *viz.*, (*i*) Diverting groundwater flow, (*ii*) Providing subsurface barriers and (*iii*) Capping.

Diverting Groundwater Flow: Vertical or horizontal movement of groundwater can be altered by the injection of water into a substratium or by the extraction of groundwater from selected locations. This process shall help to divert, contain or remove a contaminant plume.

In case of removal and containment of a hazardous material, series of extraction wells are sunk with a down gradient from a contaminated zone. The arrangement functions to remove groundwater and contaminants.

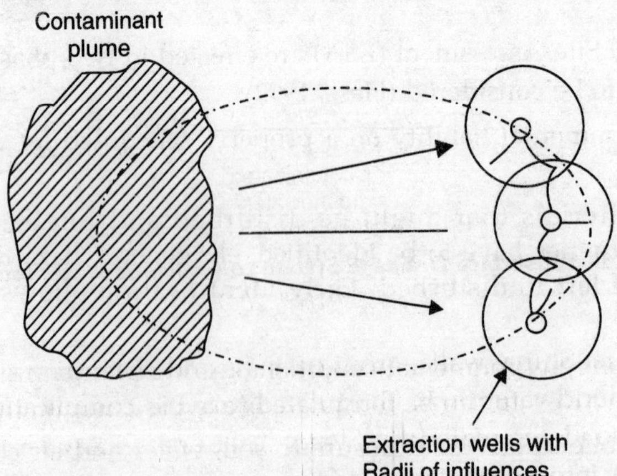

***FIGURE* 9.6. Use of Extraction Wells to Remove and Containment of Hazardous Plume.**
(Source: *Pitchel, 2007*)

An alternate approach is to inject water directly into the subsurface to completely push the plume away from an area requiring protection (Fig. 9.7). A major disadvantage of this method is that continuous operation and ongoing maintenance are required for the entire period of the remedial activity which will be costlier over a long period.

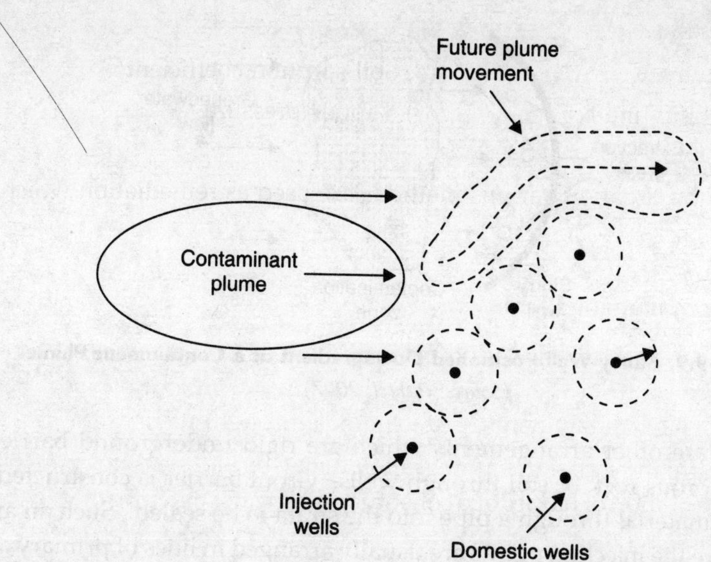

***FIGURE* 9.7. Use of Water Injection into the Subsurface for Diversion of Containment Plume.**
(*Source: Pichtel, 2007*)

Compared to the pumping systems explained above, a subsurface drainage arrangement is a passive system which does not require substantial maintenance. Subsurface drains can also be installed (Fig. 9.8) to isolate a contaminated area by intercepting uncontaminated groundwater before it comes in contact with the site.

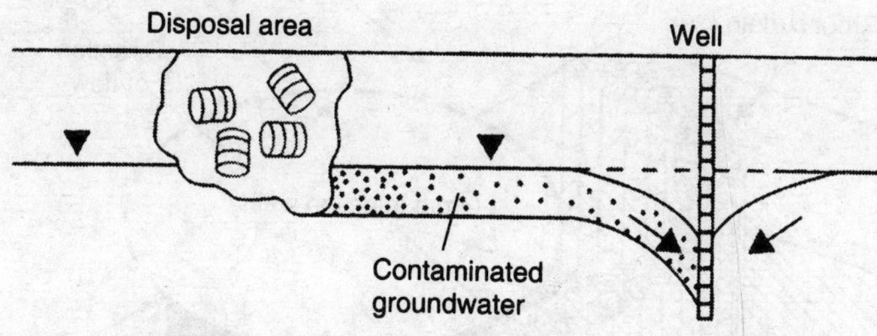

***FIGURE* 9.8. Subsurface Drainage System to Contain a Hazardous Plume.**
(*Source: Pichtel, 2007*)

Subsurface Barriers: Slurry walls, grout curtains and sheet piling are the surface barriers adopted diverting the groundwater flow. Slurry walls are the common installations for reducing groundwater flow in consolidated materials. Slurry walls are rigid underground physical barriers formed by pumping slurry into a deep vertical trench during excavation. The slurry is then allowed to set and then back filling with native soil or some other suitable material. One such arrangement of slurry wall is shown in Fig. 9.9. Slurry Walls may be placed upgradient or downgradient of the containment site or it my encircled. Figure 9.9 shows the slurry wall positioned downgradient of a containment plume. Site topography affects the choice of slurry material. A soil-bentonite mixture is suitable for a nearly level site as it can flow easily. For sites of irregular topography, concrete-bentonite mixtures is recommended as it sets rather quickly (Smith *et.al.*, 1995).

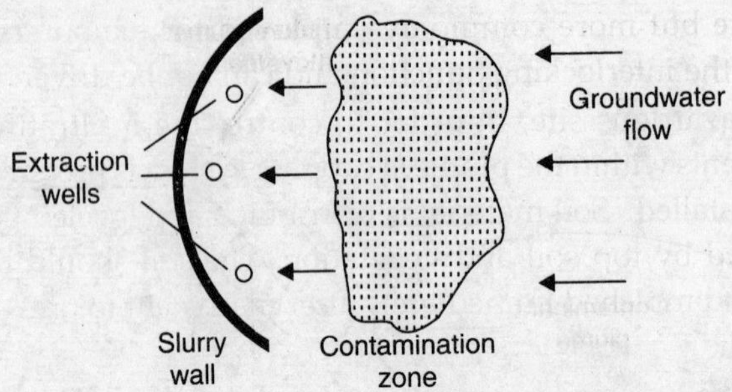

FIGURE **9.9. Slurry Wall Positioned Downgradient of a Containment Plume.**
(*Source: Pichtel, 2007*)

Grout curtains are other arrangements which are rigid underground barriers formed by injecting grouts into porous rock or soil through wells. Grout barrier is constructed by pressure-injecting the grouting material through a pipe into the strata to be sealed. Such an arrangement is shown in Fig. 9.10. Here the injection points are usually arranged in lines of primary and secondary grout holes. The spacing of grouting points is calculated based on soil permeability, grout viscosity and set-time of the grout. The spacing should be such that to allow grout injection radii from adjacent injection points to overlap (Smith *et.al.*, 1995). Clays have been widely used in grouts because of their relatively low cost.

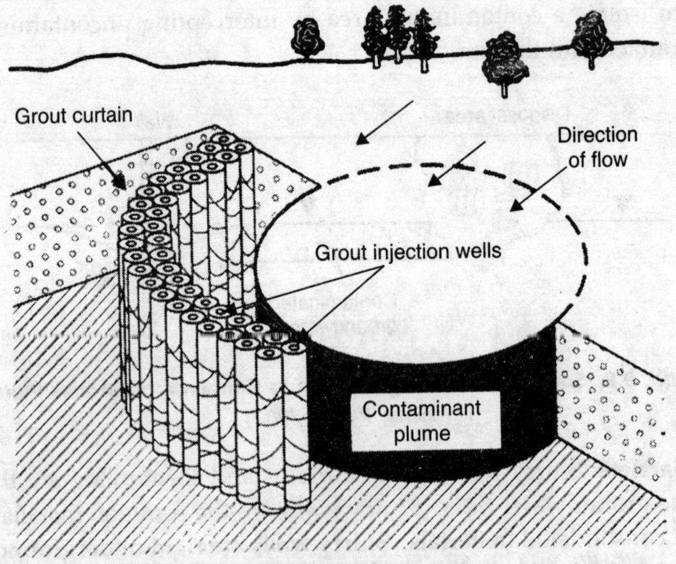

FIGURE **9.10. Configuration of Grout Curtain.**
(*Source: Pichtel, 2007*)

Sheet-pile barrier is another type of arrangement made to divert ground water flow. This involves driving rigid sheet piles into the ground to form a physical barrier to groundwater movement. Steel or concrete sheet piles are used. Steel sheet piling is more commonly used as it can resist shocks, physical stress and easy for extension laterally through interlocking. This can not be a permanent measure but more commonly employed for temporary dewatering. There is a possibility of leakage at the interlocking junction and can not be driven in rocky soils.

Capping: At a hazardous site, in order to control the infiltrating water and to prevent solubilitation of constituents within the plume, a capping system is preferred. After the preparation of the site a soil cap is installed. Soil material is distributed and leveled then the cap is covered by a clear soil later followed by top soil and vegetation. The soil should be lined and fertilited, if necessary, then sealed and mulched immediately after placement to prevent erosion. Deep routed crops should be avoided.

9.6.2 Extraction Processes

Extraction processes involve the removal of inorganic and/organic containments from soil for necessary recovery, treatment and disposal. This process is also called as soil flushing, soil washing, chemical leaching or solution mining. In principle in all these processes a chemical separation is conducted in which containments are solubilized or similarly desorbed from solid forms and recovered. In situ extraction processes are applicable for either the vadose zone or the saturated zone. Two of the processes are explained below.

Soil Flushing Process: At the affected site, the flushing solution is applied through sprinklers or irrigation or by subsurface injection. A sufficient period is allowed for the reagents to percolate downward and react with contaminants. The contaminants are subsequently mobilized by solubilization or formation of emulsions (Smith *et. al.*, 1995). Wells or subsurface drains are provided to collect the elutriate. They are then removed, treated or recycled back to the site (Fig. 9.11).

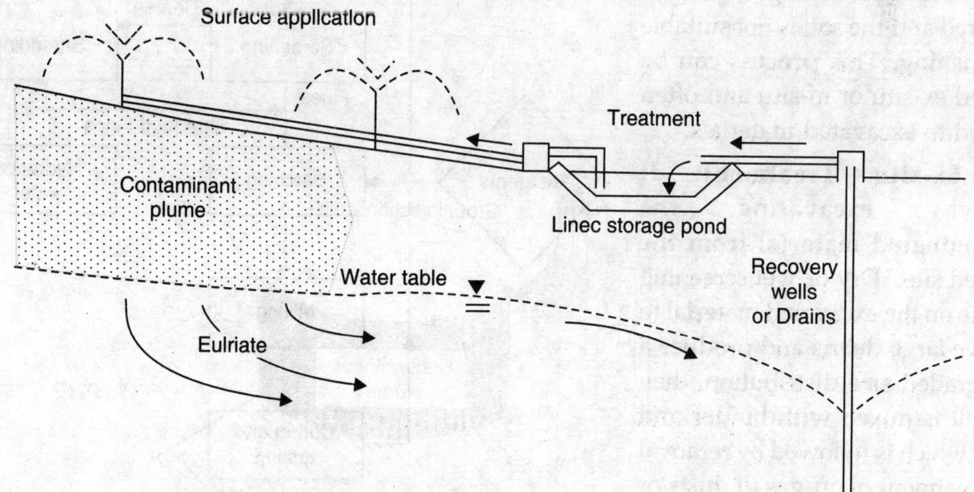

***FIGURE* 9.11. In Situ Flushing Process in a Contaminated Soil.**
(*Source: Pichtel, 2007*)

Soil Washing Process: It is an ex-situ process in which the contaminated soil is physically removed from the affected area and placed in a reactor vessel and vigorously reacted with the washing solution. There may be repeated cycles of washing, centrifugation, and separation of soil material (Fig. 9.12). Eventually the residual extractant is rinsed out of the cleaned soil with water and the soil is returned to the ground. The recovered extractant can be reused for additional washings.

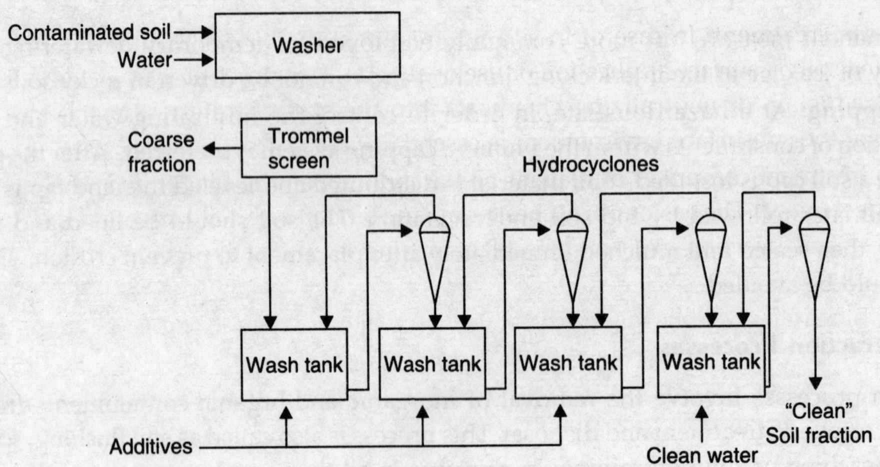

***FIGURE* 9.12. Schematic Diagram of an Ex-situ Washing Process for a Contaminated Soil.**
(*Source: Griffiths, 1995*)

9.6.3 Solidification/Stabilization (S/S)

This technique is employed in situations where large quantities of toxic/or relatively immobile contaminants occur, large debris are scattered and the soil is not suitable for flushing. This process can be applied ex-situ or in-situ and often applied to excavated materials.

Ex-situ Treatment: It involves excavating the contaminated material from the affected site. Dry or wet screening is done on the excavated material to remove large debris and produce a well-graded size distribution, then the soil is mixed with binder and water which is followed by removal and treatment of off-gas (if dusts or volatile gases are produced) and then transfer of the treated materials to a disposed area. A flow diagram of the process is shown in Fig. 9.13. Ex-situ treatment can be accomplished in a fixed facility or in a mobile treatment plant transported to the site.

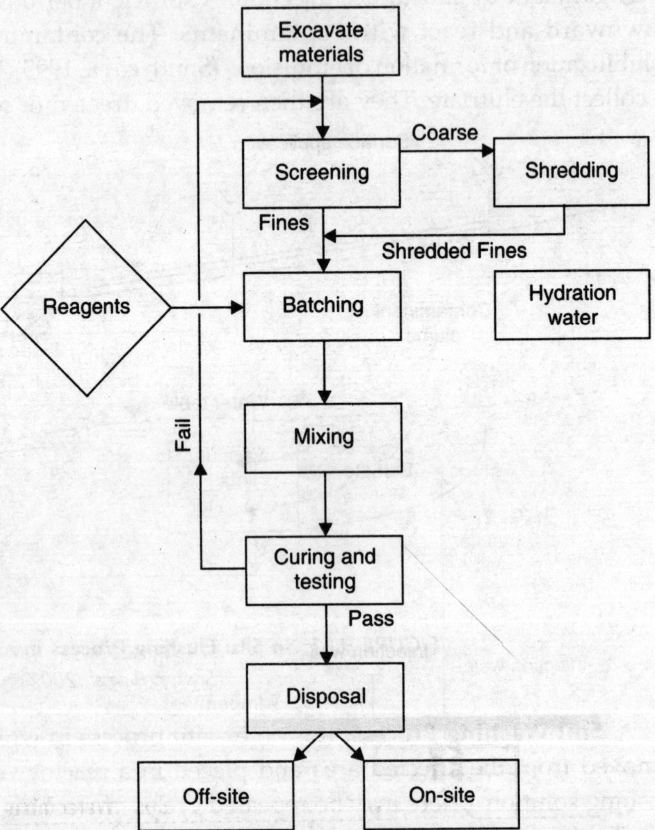

***FIGURE* 9.13. Flow-diagram of Ex-situ of S/S of Contaminated Soil.**
(*Source: Pichtel, 2007*)

In-situ Treatment: In case of in-situ S/S treatment the binder is introduced directly as a contaminated soil by surface application or use of augers. This processing involves in introducing oxidizing, reducing, or neutralizing chemicals into the groundwater system. Thus adequate knowledge of the subsurface environment is essential.

This treatment comprises of reagent preparation, mixing of binder and contaminated soil, and off-gas treatment. The basic approaches for mixing binder with soil are in-place mixing, vertical auger mixing and injection grouting. In-place mixing is applicable only to shallow or surface contamination.

In-situ treatment claims many advantages. A major advantage of in-situ treatment is that it eliminates labour, energy and costs involved. It is applicable at limited-space sites. Further dust generation is remarkably reduced. The success of the treatment lies in the complete and uniform mixing of the binder with the contaminated soil. Disadvantage of in-situ treatment is not feasible in the presence of clays, oily sands and cohesive soils.

9.6.4 Soil Vapour Extraction

This extraction process involves in the removal of votalite organic compounds from a contaminated subsurface by drawing air currents through a network of wells. This treatment is specially applicable in vadose zone wherein the interstitial spaces where hydrocarbon vapours get entrapped in the flow of extracted air. In the saturated zone a different technique called air spacing is used to treat groundwater. Low-molecular weight, volatile contaminants are easily extracted than the heavier contaminants. Lighter contaminants include butane, pentane, and hexane; the aromatic benzene, toluene, xylenes, ethylbenzene and other alkylbenzenes (Pichtel, 2007).

A schematic diagram of the soil vapour extraction system is shown in Fig. 9.14. The operation involves the following steps (EPRI, 1988 and Pichtel, 2007).

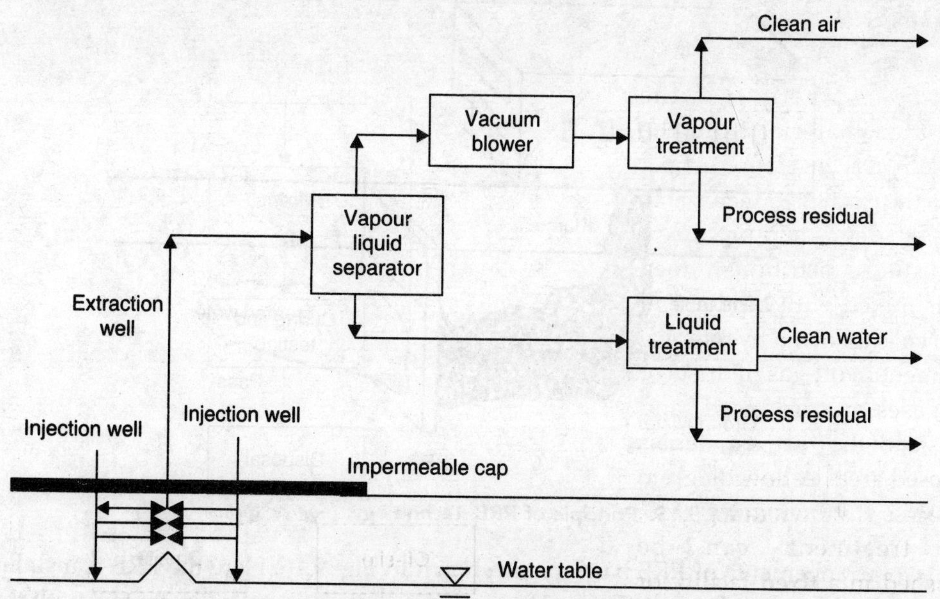

FIGURE **9.14. Schematic Diagram of Soil Vapour Extraction System.**
(*Source: Pichtel, 2007*)

(i) The influent air is heated by an air heater located upgradient of the injection post. Subsurface temperature causes warming of air which serves two purposes (*a*) to increase the volatilization rate of hydrocarbon directly and (*b*) to stimulate indigenous microbiological activity. The later activity further works to decompose organics.

(ii) A grid network of slotted pipe allows air to flow through the system.

(iii) Induced draft fans establish the airflow through the vadose zone.

(iv) A vapour treatment unit recovers and treats volatilized hydrocarbons. This process minimizes emissions to the atmosphere.

(v) Provision of air flow meters, flow control volves and sampling ports are included in the design (EPRI, 1988).

9.6.5 Permeable Reactive Barriers

Isolation of the contaminant plume (also called as pump-and-treat systems) explained in Section 9.6.1. has been found to have widespread application and has become a standard technique. Several drawbacks are found in this application including more cost and long period of operation (Keely, 1989). Such drawbacks are overcome by the installation of Permeable Reactive Barriers (PRBs) at the downgradient of a containment plume. The basic principle of PRB technology is shown in Fig. 9.15.

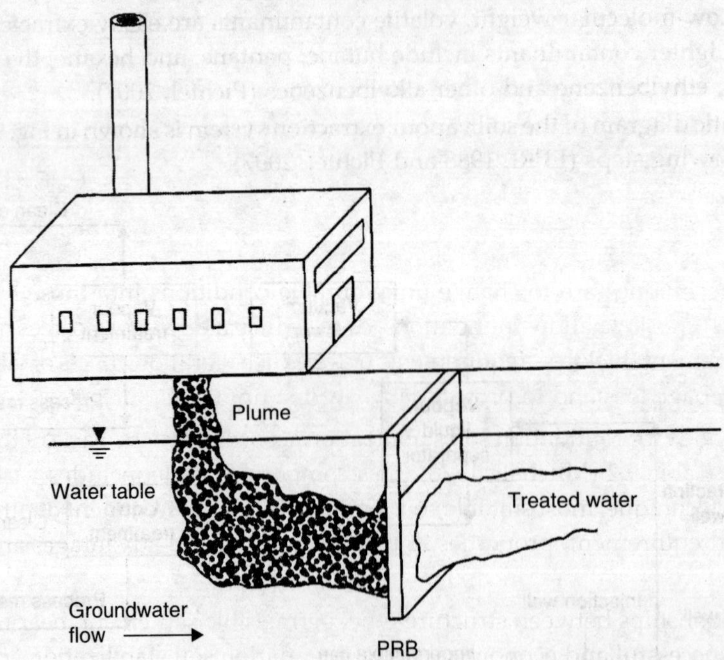

FIGURE **9.15. Principle of PRB Technology.** (*Source: Pichtel, 2007*)

Two arrangements of PRB barriers are shown in Fig. 9.16. Here the PRB is installed across the flow path of the contaminant plume. The reactive media is installed in contact with the aquifer material such that the reactive treatment zone either decomposes the contaminant or restricts its movement. Several configurations of PRB are available and the two conventional ones are shown in Fig. 9.16. In all the cases the reactive cell, called gate, portion is supplied with the necessary treatment media. Continuous reactive barriers are useful when the plume is not wide and/or containment concentrations are low.

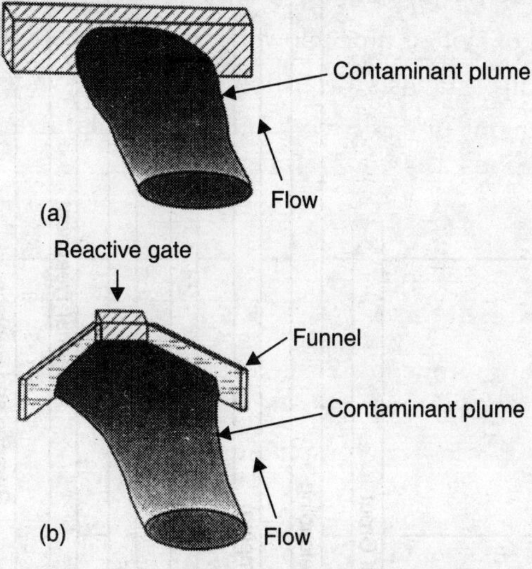

FIGURE **9.16. Two PRB Configuration.**
(*Source: Pichtel, 2007*)

9.7 GROUND IMPROVEMENT—A SUMMARY

The discussion so far made from Chapter 1 to Chapter 9 regarding ground improvement may broadly fall under three categories, *viz.*, removal of undesirable materials, control of groundwater, strengthening the geological materials and reclaimed materials. The objectives of ground improvement techniques are to change unfavourable conditions into those more suitable for the support of structures in shallow foundations, or to reduce a downdrag forces on deep foundations or to reduce pavement thickness requirements or to reduce vibration effects of vibrating machinaries, or to reduce seepage loss and to protect dams against uplift pressure, *etc*.

The methods for foundation soil stabilization according: to the basis of the soil improvement is illustrated in Table 9.2 (Mitchell, 1976). This comparison table includes various factors such as principle of the technique, most suitable soil condition, effective treatment depth and area, materials and equipment requirement, properties of treated material and advantages and limitations of the techniques.

The relationships between structure type, permissible settlement, bearing pressure, and the probability of successful and economical use of foundation soil stabilization are given in Table 9.3 (West, 1975).

Based on geologic conditions and ground strengthening techniques have been reviewed by Hunt (1986) and presented in Table 9.4.

The comparison tables of Mitchell (1976) and Hunt (1986) clearly guide to decide a suitable technique for a particular soil under a specific ground condition to attain the required property.

Most of the methods are better suited to improvement of some soil types than others. This is illustrated in Fig. 9.17 (Mitchell, 1976) which relates each stabilization method to the range of soil grain sizes for which it is most applicable.

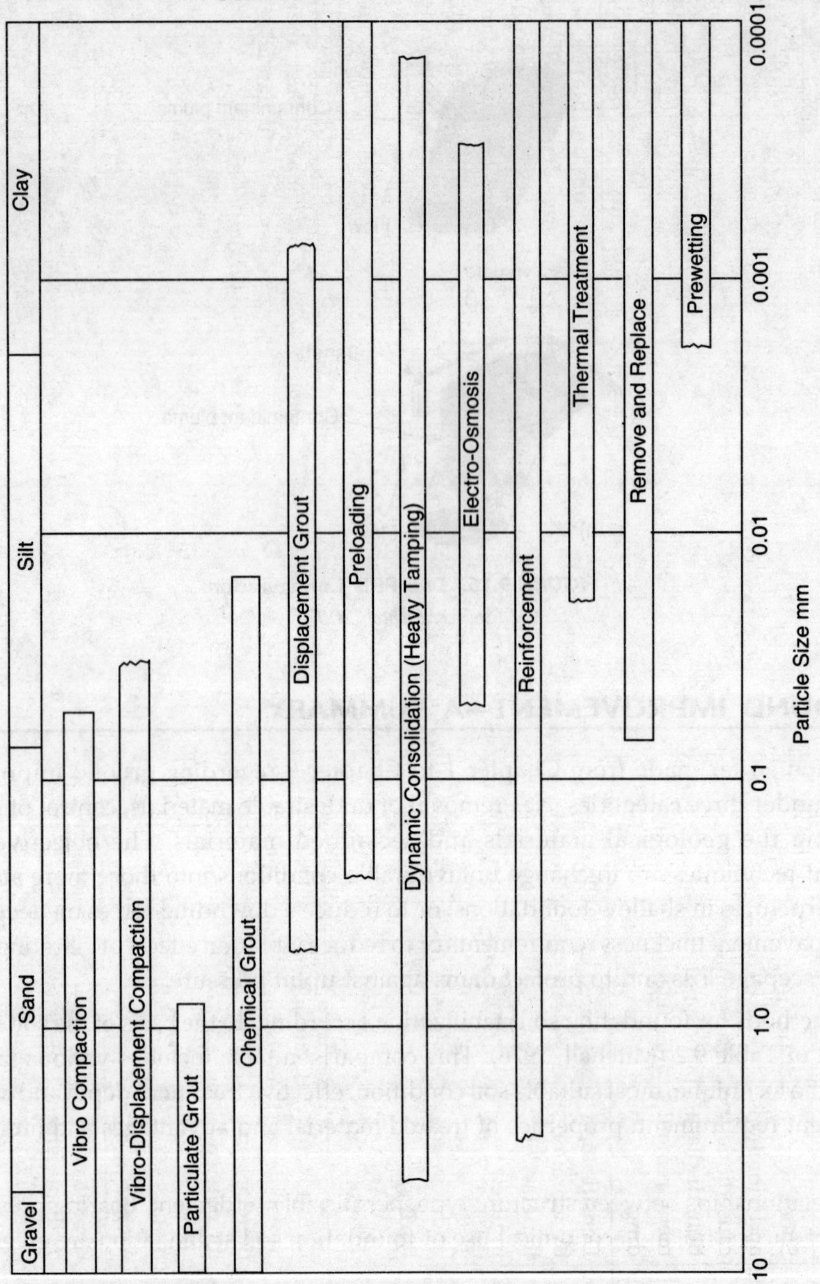

***FIGURE* 9.17. Applicable Grain Size Ranges for Different Stabilization Methods.**
(*Source: Mitchell, 1976*)

Although the discussion of various improvement techniques is directed primarily at their use for improvement of foundation soils, it should be realised that some of them are well suited to expedite or facilitate construction as well.

Table 9.2. Stabilization of Soils for Foundations of Structures
(*Source: Mitchell, 1976*)

	Method	*Principle*	*Most Suitable Soil Conditions/Types*	*Maximum Effective Treatment Depth*	*Economical Size of Treated Area*	*Special Materials Required*	*Special Equipment Required*	*Properties of Treated Materials*	*Special Advantages and Limitations*
Vibro Compaction	Blasting	Shock waves and vibrations cause liquefaction, displacement, remoulding	Saturated, clean sandy; partly saturated sands and silts after flooding	20 m	Small areas can be treated economically	Explosives, backfill to plug drill holes	Jetting or drilling machine	Can obtain relative densities to 70-80%, may get variable density	Rapid, inexpensive, can treat small areas; variable properties, no improvement near surface, dangerous
	Terraprobe	Densification by vibration; liquefaction induced settlement under over burden	Saturated or dry clean sand	20 m (Ineffective above 4 m depth)	> 1500 m²	None	Vibratory pile driver and 750 mm dia open steel pipe	Can obtain relative densities of 80%; or more	Rapid, simple, good underwater soft underlayers may damp vibrations, difficult to penetrate, stiff overlayers, not good in partly saturated soils
Vibro Displacement Compaction	Vibratory Rollers	Densification by vibration; liquefaction introduced settlement under roller weight	Cohesionless soils	2 to 3 m	Any size	None	Vibratory roller	Can obtain very high relative densities, upper few decimeters not densified	Best method for thin layer or lifts
	Compaction piles	Densification by displacement of pile volume and by vibration during driving	Loose sandy soils; partly saturated clayey soils; loess	20 m	Small to moderate	Pile material (usually sand or soil plus cement mixture)	Pile driver	Can obtain high densities, good uniformity	Useful in soils with fine, uniform compaction easy to check results; slow limited impovement in upper 1 to 2 m

(*Table Contd...*)

Table 9.2. (*Contd...*)

	Method	*Principle*	*Most Suitable Soil Conditions/Types*	*Maximum Effective Treatment Depth*	*Economical Size of Treated Area*	*Special Materials Required*	*Special Equipment Required*	*Properties of Treated Materials*	*Special Advantages and Limitations*
Vibro-displacement Compaction	Heavy temping (Dynamic consolidation)	Repeated application of high intensity impacts at surface	Cohesionless soils best, other types can also be improved	15 to 20 m	> 5000 m^2	None	Tamper of 10-40 ton high capacity crane	Can obtain high relative densities, reasonable uniformity	Simple, rapid, suitable for soils with fines, usable above and below water; requires control, must be away from existing structures
	Vibroflotation	Densification by vibration and compaction of backfill material	Cohesionless soils with less than 20% fines	30 m	> 1500 m^2	Granular backfill	Vibroflot, crane	Can obtain high relative densities, good uniformity	Useful in saturated and partly saturated soils, uniformity
Grouting and Injection	Particulate grouting	Penetration grouting-fill soil pores with soil, cement and/or clay	Medium to coarse sand and gravel	Unlimited	Small	Grout, water	Mixers, tanks, pumps, hoses	Impervious, high stength eliminate liquefaction danger	Low cost grouts high strength; limited to coarse-grained soils, hard to evaluate
	Chemical Grouting	Solutions of two or more chemicals react in soil pores to form a gel or a solid precipitation	Medium silts and coarser	Unlimited	Small	Grout, water	Mixers, tanks, pumps, hoses	Impervious, low to high strength, eliminate liquefaction danger	Low viscosity controllable gel time, good water shut-off; high cost, hard to evaluate
	Pressure injected lime	Lime slurry injected to shallow depths under high pressure	Expansive clays	Unlimited, but 2 to 3 m usual	Small	Lime, water, surfactant	Slurry tanks, agitators, injection	Lime encapsulated zones formed by channels resulting from cracks, root holes, hydraulic fracture	Rapid and economical treatment for foundation soils under light structures

(*Table Contd...*)

Table 9.2. (*Contd...*)

	Method	*Principle*	*Most Suitable Soil Conditions/Types*	*Maximum Effective Treatment Depth*	*Economical Size of Treated Area*	*Special Materials Required*	*Special Equipment Required*	*Properties of Treated Materials*	*Special Advantages and Limitations*
Grouting and Injection	Displacement grout	Highly viscous grout acts as radial hydraulic jack when pumped in under high pressure	Soft, fine-grained soils, foundation, soils with large voids or cavities	Unlimited, but a few metre usual	Small	Soil, cement water	Batching, equipment, high pressure pumps, hoses	Grout bulbs within compressed soil matrix	Good for correction of differential settlements, filling large voids; careful control required
	Electro kinetic Injection	Stabilising Chemicals moved into soil by electro osmosis	Saturated silts, silty clays	Unknown	Small	Chemical Stabilizer	DC power supply, nodes, cathodes	Increased strength, reduced compressibility	Existing soil and structures not subjected to high pressures; no good in soil with high conductivity
Precompression	Preloading	Load is applied sufficiently in advance of construction so that compression of soft soils is completed prior to development of the site	Normally consolidated soft clays, silts, organic deposits, completed sanitary landfills	—	> 1000 m^2	Earth fill or other material for loading the site; sand or gravel for drainage blanket	Earth moving equipment; large water tanks or vacuum drainage systems sometimes used; settlement markers, piezometers	Reduced water content and void ratio increased strength	Easy, theory well developed uniformity; requires long time (sand drains or wicks can be used to reduce consolidation)
	Surcharge Fills	Fill in excess of that required permanently is applied to achieve a given amount of settlmement in a shorter time; excess fill then removed	Normally consolidated soft clays, silts, organic deposits, completed sanitary landfills	—	> 1000 m^2	Earth fill or other material for loading the site; sand or gravel for drainage blanket	Earth moving equipment; settlement markers, piezometers	Reduced water content, void ratio and compressibility increased strength	Faster than pre-loading without surcharge, theory well developed; extra material handling; can use sand drains or wicks

(*Table Contd...*)

Table 9.2. (*Contd...*)

	Method	*Principle*	*Most Suitable Soil Conditions/Types*	*Maximum Effective Treatment Depth*	*Economical Size of Treated Area*	*Special Materials Required*	*Special Equipment Required*	*Properties of Treated Materials*	*Special Advantages and Limitations*
Precompression	Dynamic consolidation	High energy impacts compress and dissolve gas in pores to give immediate settlement; increased pore pressure gives subsequent drainage	Partly saturated fine grained soils, quaternary clays with 1 to 4% gas in micro bubbles	20 m	> 15000 – 30000 m^2	None	Tamper of 10 to 40 tonne, high capacity crane	Reduced water content void ratio and compressibility; increased strength	Faster than pre-loading, economical on large areas; uncertain mechanism in clays, less uniformity than preloading.
	Electro-osmosis	DC current causes water flow from anode towards cathode where it is removed	Normally consolidated silts and silty clays	10 to 20 m	Small	Anode (usually rebars or aluminium) cathodes (well points or rebars)	DC power supply, wiring, metering systems	Reduced water content and compressibility, increased strength, electro-chemical hardening	No fill loading required, can use in confined area, relatively fast; non-uniform properties between electrodes, not good in highly conductive soils
Reinforcement	Mix-in-place piles and walls	Lime, cement, or asphalt introduced through rotating auger or special in place mixer	All soft or loose inorganic soils	> 20 m	Small	Cement, lime asphalt, or chemical stabilizer	Drill rig, rotary cutting and mixing head, additive proportioning equipment	Solidified soil piles or walls of relatively high strength	Uses native soil, reduced lateral support requirements during excavation; difficult quality control
	Strips and membranes	Horizontal tensile strips or membranes buried in soil under footings	All soils	a few metre	Small	Metal or plastic strips, polyethylene, polyprophylene or polyester	Excavating earth handling, and compaction equipment	Increased bearing capacity, reduced deformations	Increased allowable bearing pressure; requires over-excavation for footing

(*Table Contd...*)

Table 9.2. (*Contd...*)

	Method	*Principle*	*Most Suitable Soil Conditions/Types*	*Maximum Effective Treatment Depth*	*Economical Size of Treated Area*	*Special Materials Required*	*Special Equipment Required*	*Properties of Treated Materials*	*Special Advantages and Limitations*
	Vibro-replacement -stone columns	Hole jetted into soft, fine-grained soil and back-filled with densely compacted gravel	Soft clays and alluvial deposits	20 m	> 1500 m^2	Gravel or crushed rock backfill	Vibroflot, crane or Vibrocat, water	Increased bearing capacity, reduced settlements	Faster than pre-compression, avoids dewatering required for remove and replace; limited bearing capacity
Thermal	Heating	Drying at low temperatures; alteration of clays at intermediate temperatures (400 to 600°C); fusion at high temperatures (1000°C)	Finegrained soils, especially partly saturated clays and silts, loess	15 m	Small	Fuel	Fuel tanks, burners, blowers	Reduced water control, plasticity, water sensitivity; increased strength	Can obtain irreversible improvements in properties; can introduce stabilizers with hot gases
	Freezing	Freeze soft, wet ground to increase its strength and stiffness	All soils	Several metres	Small	Refrigerant	Refrigeration system	Increased strength and stiffness, reduced permeability	No good in flowing ground water, temporary
Miscellaneous	Remove and replace (with or without admixtures)	Foundation soil excavated, improved by drying or admixture, and recompacted	Inorgainc soils	10 m	Small	None, unless admixture stabilizers used	Excavating and compaction equipment, dewatering system	Increased strength and stiffness, reduced compressibility	Uniform, controlled foundation soil when replaced; may require large area dewatering
	Moisture Barriers	Water access to foundation soils is prevented under footings	Expansive soils	5 m	Small	Membranes, gravel, lime, or asphalt	Excavating, and trenching equipment, compaction equipment	Initial natural or as-compacted properties retained	Best for small structures; may not be 100% effective

(*Table Contd...*)

Table 9.2. (Contd...)

	Method	Principle	Most Suitable Soil Conditions/Types	Maximum Effective Treatment Depth	Economical Size of Treated Area	Special Materials Required	Special Equipment Required	Properties of Treated Materials	Special Advantages and Limitations
Miscellaneous	Prewetting	Soil is brought to estimated final water content prior to construction	Expansive soils	2 to 3 m	Small	Water	None	Decreased swelling potential	Low cost, best for small, light structures, may still get shrinking and swelling
	Structural Fills (with or without admixtures)	Structural fill distributors loads to underlying soft soils	Soft clays or organic soils, marsh deposits	—	Small	Sand, gravel, fly ash, bottom ash, slag, expanded aggregate, clam shell or oyster shell incinerator ash	Compaction equipment	Soft subgrade protected by structural load-bearing fill	High strength, good load distribution to underlying soft soils

Table 9.3. Applicability of Foundation Soil Improvement for Different Structure and Soil Types (*Source: West, 1975*) (For Efficient use of Sallow Foundations)

Category of Structure	*Structure*	*Permissible Settlement*	*Load Intensity/ Usual Bearing pressure Required, kN/m² (tsf)*	*Probability of Advantageous Use of Soil Improvement Techniques*		
				Loose Cohesionless Soils	*Soft Alluvial Deposits*	*Old, Inorganic Fills*
Office/Apartment	High rise: More than 6 stories	Small < 25 to 50 mm	High/300 + (3 +)	High	Unlikely	Low
Frame or Load-bearing Construction	Medium rise: 3 to 6 stories	Small < 25 to 50 mm	Moderate/200 (2)	High	Low	Good
	Low rise: 1 to 3 stories	Small < 25 to 50 mm	Low/100 to 200 (1 to 2)	High	Good	High
Industrial	Large span with heavy machines, cranes; process plants; power plants	Small (< 25 to 50mm) Differential settlement Critical	Variable/high local concentrations to > 400 (> 4)	High	Unlikely	Low
	Framed ware-houses and factories	Moderate	Low/100 to 200 (1 to 2)	High	Good	High
	Covered storage, store, rack systems, production areas	Low to Moderate	Low/< 200 (< 2)	High	Good	High
Others	Water and waste water treatment plants	Moderate differentia settlement important	Low/< 150 (< 1.5)	High, if required at all	High	High
	Storage tanks	Moderate to high, but differential may be critical	High/ up to 300 (3)	High, if required at all	High	High
	Open storage areas	High	High/up to 300 (3)	High, if required at all	High	High
	Embankments and abutments	Moderate to High	High/up to 200 (2)	High, if required at all	High	High

Table 9.4. Ground-Strengthening Techniques Summarized
(*Source: Hunt, 1986*)

Conditions	*Technique*	*Application*
Low grades	Compacted Sand fill	Minimize structure settlements*
Miscellaneous fill		
Shallow	Excavate-backfill	Minimize structure settlement
Deep	Dynamic compaction	Reduce structure settlement**
	Sand columns	Reduce structure settlement
Organics		
Shallow	Excavate-backfill	Minimize structure settlement
	Geotextiles	Support low embankments
Deep	Surcharge	Reduce structure settlement
	Geotextiles	Support low embankments
	Sand columns	Reduce structure settlement
Buried	Surcharge	Reduce structure settlement
	Dynamic compaction	Reduce structure settlement
	Compaction grouting	Arrest existing structure settlement
	Sand columns	Reduce structure settlement
Soft clays		
Shallow	Excavate-backfill	Minimize structure settlement
	Geotextiles	Support low embankments
Deep	Surcharge	Reduce structure settlement
	Geotextiles	Support low embankments
	Sand columns	Reduce structure settlement
	Lime columns	Reduce structure settlement
Buried	Surcharge	Reduce structure settlement
	Dynamic compaction	Reduce structure settlement
	Compaction grouting	Arrest existing structure settlement
	Sand columns	Reduce structure settlement
	Lime columns	Reduce structure settlement
Clays, surface	Gravel admixture	Base, Subbase, low-quality pavement
	Lime admixture	Stabilize roadway base and subbase
	Freezing	Temporary arrest of settlement
Loose silts		
Shallow	Excavate-backfill	Minimize structure settlement
	Salts admixture	Dust palliative
	Surface compaction	Increase support capacity***
Deep	Surcharge	Reduce structure settlement
	Stone columns	Increase support capacity
	Electro-osmosis	Increase slope strength temporarily
Buried	Vacuum wellpoints	Improve excavation bottom stability

(Table Contd...)

Table 9.4. *(Contd...)*

Loose sands		
Shallow	Surface compaction	Increase support capacity
	Cement admixture	Base, subbase, low-quality pavement
	Bitumen admixture	Base, subbase, low-quality pavement
Deep	Vibroflotation/Terra probe	Increase support capacity
	Dynamic compaction	Increase support capacity
	Stone columns	Increase support capacity
	Wellpoints	Increase stable cut-slope inclination
	Freezing	Temporary stability for excavation
Buried	Penetration grouting	Arrest existing structure settlement
	Freezing	Temporary stability for excavation
Collapsible soils		
Shallow	Excavate-backfill	Minimize structure settlement
Deep	Hydrocompaction	Reduce structure settlement
	Dynamic compaction	Increase support capacity
	Lime stabilization	Arrest building settlement
Liquefiable soils	Dynamic compaction	Increase density
	Stone-columns	Pore-pressure relief
Expansive soils	Lime admixtures	Reduce activity in compacted fill
Rock masses		
Fractured	Compaction grouting	Increased strength
	Penetration grouting	Increase strength
	Bolts and cable anchors	Stabilize slopes and concrete dam foundations
	Shotcrete or gunite	Reinforce slopes
	Subhorizontal drains	Stabilize slopes

* Minimize structure settlement signifies that settlement will be negligible under moderate foundation loads if the technique is applied properly.

** Reduce structural settlement signifies that after application for the technique significant settlement which must be anticipated in the design of the structure may still occur.

*** Increase support capacity signifies that proper application of the technique will result in an increase in bearing capacity and a decrease in compressibility on an overall basis.

BIBLIOGRAPHY

ARCSMFE	Asian Regional Conference on Soil Mech. and Found. Engg.
ASCE	American Society of Civil Engineers, USA.
ASTM	American Society for Testing and Materials, USA.
BIS	Bureau of India Standards, India.
CGJ	Canadian Geotechnical Journal, Canada.
ECSMFE	European Conference on Soil Mech. and Found. Engg.
ENR	Engineering News Record, USA.
HRB	Highway Research Board, USA.
ICE	Institution of Civil Engineers, UK.
IGJ	Indian Geotechnical Journal, India.
INCSME	International Conference on Soil Mech. and Found. Engg.
JGED	Journal of Geotechnical Engg. Division.
JSMFD	Journal of Soil Mech. and Found. Division.

1. Aboshi, H and N. Snematzu, Sand Compaction Pile Method, State-of-the-art-paper, **3rd Int. Geotechnical Seminar on Soil Improvement Methods,** Nanyang Technological Institute, Singapore, Nov. 1985.
2. Aldrich, H.P. and E.G. Johnson, Embankment Test Section to Evaluate Field Performance of Vertical Sand Drains for Interstate 295 in Portland, Maine, **HRB Record** No. 405, pp. 50-74, 1972.
3. Azzouz, A.S., *et. al*, Regression Analysis of Soil Compressibility, **Soils and Foundations,** Vol. 16, No. 2, pp. 19-29, 1976. Bachy, SA **Traitment des Sols par Vibration Profonde,** Bachy SA, Paris, 30 pp., 1976.

4. Baker, W.H, Planning and Performing Structural Chemical Grouting, **Proc. Conf. on Grouting in Geotech. Engg.** New Orleans, pp. 515-539,1982.

5. Barron, R.A., Consolidation of Fine Grained Soils by Drain Wells, **Trans. ASCE,** Vol. 113, pp. 718-742,1948.

6. Beles, A.A., and I.I. Stanculesen, Thermal Treatment as a Means of Improving the Stability of Earth Masses, **Geotechnique.** Vol. 8, No. 4, 1958.

7. Berry, P.L. and W.B. Wilkinson, The Radial Consolidation of Clay Soils, **Geotechnique.** Vol. 19, pp. 253-284, 1969.

8. Bassett, R.H. and N.C. Last, Reinforcing Earth Below Footings and Embankments, **Proc. Symp. on Earth Reinforcement, ASCE,** Pitsburgh, Pennsylvania, pp. 202-231, 1978.

9. Bertram, G.E., Experimental Investigation of Protective Filters, **Publication of the Graduate School of Engg. Hardward University,** No. 267, 1940.

10. Binquet, J., and K.L. Lee, Bearing Capacity Analysis of Reinforced Earth Slabs, **JGED, ASCE,** Vol. 101, GT12,pp. 1257-1275,1975.

11. Bjerrum, L., Embankments on Soft Ground, **Proc. Spec. Conf. Performance of Earth and Earth-Supported Structures,** Lafeyette Vol. 2, pp. 1-54, 1972.

12. Bjerrum, L., J. Moum, and O. Eide, Application of Electro-osmosis to a Foundation Problem in a Norwegian Quick Clay, **Geotechnique,** Vol. 17, pp. 214-235, 1967.

13. Boman, P., Ny Method att Draenera tera (New Method of Clay Drainage), **Vag-och Vatten-byggaren,** Vol. 19, pp. 192-200, 1973

14. Bonazzi, D., Alluvium Grouting Proved Effective on Alpine Dam, **JSMFD, ASCE,** Vol. 91, No. SM 6, pp. 81-93, 1965.

15. Boulanger, R.W., and R.F. Hayden, Aspects of Compaction Grouting of Liquefiable Soil, **JGED, ASCE,** Vol. 121, No 12, pp. 844-855, 1995.

16. Bourges, F., M.Carissan and C. Mieussens, Le Remblai de Palavas-les-Flots, **Bull. Lias Lab. Ports Chauss,** Special No. T, pp. 119-138, 1973.

17. Bowen, R., **Grouting in Engineering Practice,** 2nd Edn., Applied Science Publishers, London, 285 pp., 1981.

18. Bowles, J.E., **Foundation Analysis and Design,** 4th ed., McGraw-Hill Book Company, New York, 1004 pp., 1988.

19. Broms, B.B. and Boman, P., Lime Columns—A New Foundation Method, **JGED, ASCE,** Vol. 105, No. GT4, pp. 539-556, 1979.

20. Bromwell, L.G., and T. Lambe, A Comparison of Laboratory and Field Values of c_v for Boston Blue Clay, **47th Annual Meeting of HRB,** 1968.

21. Brown, R.E., Vibration Compaction of Granular Hydraulic Fills, Preprint 2657, **ASCE, 'National Water Resources and Ocean Engineering Convention,** pp. 1-30, 1976.

22. Brunner, D.R., and D.J. Keller, Sanitary Landfill Design and Operation, **U.S. EPAOSWMP, Report No. SW-65,** Washington, D.C., 1972.

23. Caron, C., Applications et Essais Dans Le Domaine de L' electroosmose des Terrains Sableux et Argileux, **Bull. Tech. Suisse Romande,** No. 1. pp. 1-4, 1968.

24. Caron, C., Consolidation des Terrains Argileux par E' lectroosmose, **Ann. Inst. Tech. Batiment Travaux Publics,** No. 285, pp. 73-92, 1971.

25. Carrillo, N., Simple Two and Three Dimensional Cases in the Theory of Consolidation of Soils, **J. Math. Phys.,** Vol. 21, pp. 1-5, 1942.

26. Casagrande, A., The Determination of the Pre-consolidation Load and its Practical Significance, **1st INCSMFE,** Vol. 3, pp. 60-64, 1936.

27. Casagrande, L., Electro-osmosis in Soils, **Geotechnique,** Vol. 1, No. 3, pp. 159-177, 1949.

28. Casagrande, L., Electro-osmotic Stabilisation of Soils, **Jour. of Boston Soc. of Civil Engrs.,** Vol. 39, pp. 51-83, 1952.

29. Cassan, M., Etude des Foundations du Viaderc de la Gare de Cagnes-sur-Mer sur, L'autoroute, AB, **Chautierrs de France,** pp. 33-37, 1974.

30. Chappell, B.A., and P.L. Burton, Electro-osmosis Applied to Unstable Embankment, **JGED, ASCE,** Vol. 101, pp. 733-740, 1975.

31. Chaput, D. and Thomann, G., **Consolidation d'un Sol Avec Drains Verticaux Sons Charge Variable,** Lab. Ponts Chauss, Raport de Recherche, No. 47, 66 p., 1975.

32. Charles, J.A., D. Burford, and K.S. Watts, Field Studies of the Effectiveness of the Ground Treatment Techniques—Dynamic Consolidation, in **Proc. 10th INCSMFE,** Stockholm, 1981, Vol. 3., pp. 617-622, 1981.

33. Christopher, B.R., Update on Geosynthetics Standard Test Methods and Regulations, **Geotechnical Fabrics Report,** Jan.-Feb., pp. 50-54, 1989.

34. Coumoulos, D.G., Determination of Settlement Design Parameters from Conventional Oedometer Tests, **Proc. 7th INCSMFE,** Brighton, No. 87,1979.

35. Croce, A., C. Viggiani and G. Calabresi, In-situ Investigation of Pore Pressures in Soft Clays, **Proc. 8th INCSMFE,** Moscow, Vol. 2, pp. 53-60, 1973.

36. D'Appolonia, D.J., R.V. Whitman, and E.D' Appolonia, Sand Compaction with Vibratory Rollers, **Proc. ASCE, Spec. Conf. on Placement and Improvement of Soil to Support Structures,** New York, 1968, pp. 125-136, 1968.

37. Das, B.M., Principles of Foundation Engineering, **Cole Engineering's Division**, California, 620 pp, 2000.

38. Dastidar, AG., S. Gupta, and T.K. Ghosh, Application of Sand-Wicks in a Housing Project, **Proc. 7th INCSMFE,** Mexico City, Vol. 2, pp. 59-64, 1969.

39. Datye, K.R., Simpler Techniques for Ground Improvements, **Fourth IGS Annual Lecture, IGJ,** Vol. 12, pp. 1-82, 1982.

40. Datye, K.R. and S.S. Nagaraju, Observation Based Design Approach for Preloading, **Proc. 5th ARCSMFE,** Bengaluru, pp. 19-27, 1975.

41. Datye, K.R. and S.S. Nagaraju, Installation and Testing of Rammed Stone Columns, **Proc. IGS Speciality Session, 5th ARCSMFE,** Bangalore, 1975.

42. Datye, K.R. and S.S. Nagaraju, Practical and Economic Aspects of the Design of Ground Improvement System, **IGS Conf. Found. and Excavations in Weak Soils,** Calcutta, pp. C-9, 1976.

43. Datye, K.R. and S.S. Nagaraju, Ground Improvement, **Commemmorative Volume, IGS,** New Delhi, pp. 121-125, 1985.

44. Davis, P.H. Nelissen and A. Pladet, Mytilus, A Soil Compaction Vessel, **Proc. 10th INCSMFE.,** Stockholm, Vol. 3, pp. 641-644, 1981.

45. Dearstyne, S.C., and G.J. Newman, Subgrade Stabilization under an Existing Runway, **JATD, ASCE,** Vol. 89, ATI, pp. 1-8, 1963.

46. De Beer, E., and Welleys, Application des Drains Kjellman-Fanki et L'e' lectroosmose, **La Technique des Travaux,** Vol. 41, No. 3-4, pp. 89-102.

47. Dobson, T., and B. Slocombe, Deep Densification of Granuar Fills, **Proc. Second Geotechnical Conf. on Design and Construction,** Las Vegas, April 26-28, 21 pp., 1982.

48. Electric Power Research Institute, Remedial Technologies for Leaking Underground Storage Tanks, Chelsa, **MI:** Lewis, 1988.

49. Engelhardt, K., W.A. Flynn, A.A. Bayak, Vibro-replacement Method to Strengthen Cohesive Soils in situ, **ASCE National Struc. Eng. Meeting,** Cincinnati, 30 pp., 1974.

50. Engelhardt, K, and K. Kirsch, Soil Improvement by Deep Vibratory Technique, **Proc. 5th South-East Asian Conf. on Soil Engg.,** Bangkok, pp. 377-387,1977.

51. ENR, Weak Seabed Strata Yields Strong Support, March 10, pp. 30-31, 1983.

52. Esrig, M.I., Pore Pressures, Consolidation and Electrokinetics, **JSMFD, ASCE,** Vol. 94, pp. 899-921,1968.

53. Felt, E.J., Factors Influencing Some of the Physical Properties of Soil-Cement Mixtures, **HRB Bull.,** No. 108,1955.

54. Fetzer, C.A., Electro-osmosis Stabilization of West Branch Dam, **JSMFD, ASCE,** Vol. 93, SM4, pp. 85-106,1967.

55. Foster, C.R., Field Problems: Compaction, In **Foundation Engineering,** G.A. Leonards (ed.), Mc-Graw Hill Book Company, New York, pp. 1001-1024, 1962.

56. Fuller, W.B., and S.E. Thompson, The Laws of Proportioning Concrete, **Trans. ASCE.,** Vol. 59, pp. 67-143, Discussion, pp. 144-72, 1935.

57. Gicot, O., and J. Perfetti, **Geotextiles Conceiving and Designing Engineering Structures,** Rhone-Poulenc, France, 1982.

58. Gigan, J.P., Compactage par Pilonnage Intensif de Remblais de Comblement d' un Bras de Selne, **Bull. Liais. Lab. Ponts Chauss,** No. 90, pp. 81-102, 1977.

59. Giroud, J.P., and R.K. Frobel, **GeomebraneProducts,** Geotechnical Fabrics Reports, Fall, pp. 38-42, 1983.

60. Graf, E.D., Compaction Grouting Technique and Observations, **JSMFD, ASCE,** Vol. 95, No. SM. 5, pp. 1151-58,1969.

61. Gray, D.H., and J.K; Mitchell, Estimation of Electro-osmotic Efficiency in Soils, **ASCE Structural Engg. Conf.,** May 8-12, Preprint 497, pp. 1-46, 1967.

62. Greenwood, D.A., Mechanical Improvement of Soils below Ground Surface, **Proc. Ground Engg. Conf., ICE.,** London, 1970.

63. Griffiths, R.A., Soil-Washing Technology and Practice, **Journal of Hazardous Materials,** V.40, pp 175-189, 1995.

64. Gupta, S.N, Discussion of 'Sand Densification by Piles and Vibroflotation' by C.E. Basore and J.D. Boitano, **JSMFED, ASCE,** Vol. 95, SM6, pp. 224-226, 1969.

65. Hall, C.E., Compacting a Dam Foundation by Blasting, **JSMFD, ASCE,** Vol. 88, SM3, pp. 33-51,1962.

66. Halton, G.R., R.W. Loughney, and E. Winter, Vacuum Stabilization, of Subsoil beneath Runway Extension at Philadelphia Airport, **Proc. 6th INCSMFE,** Montreal, Vol. 2, pp. 61-65, 1965.

67. Hansbo, S., Consolidation of Clay with Special Reference to Influence of Vertical Sand Drains, **Proc. Swedish Geotech. Inst.,** No. 18, pp. 1-160, 1960.

68. Hansbo, S., How to Evaluate the Properties of Prefabricated Drains, **Proc. 8th ECSMFE,** Helsinki, No. 6.13,1983.

69. Hansbo, S. and B.A., Tornstensson, Geodrain and Other Vertical Drain Behaviour, **Proc. 9th INCSMFE,** Tokyo, Vol. 1, pp. 533-540, 1977.

70. Harris, F., **Ground Engineering Equipment and Methods,** Granda, New York, 281 pp., 1983.

71. Helmholtz, **H., Ann.Physick.** Wied., Vol. 7, p. 337, 1879.

72. Hess, K., Environmental Site Assessments, Phase-I, A Basic Guide, Boca Raton, FL, **CRC Press,** 1993.

73. Hilf, J.W., A Rapid Method of Construction Control for Embankments of Cohesive Soil, **Water Resources Technical Publications, Engg.** Monograph No. 26,1966.

74. Hodge, J., Durability Testing, Geotextile Testing and the Design Engineer, **ASTM, STP 952,** pp. 119-121, 1987.

75. Holtz, R.D. and Broms, B., Long Term Loading Tests at Ska-Ederby Sweden, **Proc. Spec. Conf. Performance of Earth and Earth-Supported Structures,** Lafayette, Vol. 1, pp. 435-464, 1972.

76. Holtz, W.G., and H.J. Gibbs, Engineering Properties of Expansive Clays, **Trans. ASCE,** Vol. 121, pp. 641-677, 1956.

77. Houlsby, A.C., Cement Grouting for Dams, **Proc. Conf. on Grouting in Geotech. Engg.,** New Orleans, pp. 1-34, 1983.

78. Houlsby, A.C., **Cement Grouting—A Compilation of Recent Papers,** Water Resources Commission, New South Wales, Australia, 1982.

79. HRB, Use of Soil-Cement Mixtures for Base Courses, Wartime Road Publications No. 7, **HRB,** Washington D.C., 1943.

80. HRB, Factors Influencing Compaction Test Results. **HRB** Bulletin No. 319, 1962.

81. Huck, P.J. and M.J. Waller, Quality Control of Grouting, **Proc. Conf. on Grouting in Geotech. Engg.,** New Orleans, pp. 781-792, 1982.

82. Hughes, J.M. N.J. Withers and D.A. Greenwood, A Field Trial of the Reinforcing Effect of a Stone Column in Soil, **Geotechnique,** Vol. 35, No. 1, pp. 34-44, 1975.

83. Hunt, R.E., **Geotechnical Engineering Analysis and Evaluation,** Mc-Graw Hill Book Company, New York, 729 pp., 1986.

84. IS: 2720—**Methods of Tests for Soils,** BIS.

85. IS: 2720—Part 7, **Determination of Water Content—Dry Density Relation using Light Compaction,** BIS, 1974.

86. IS: 2720—Part 8, **Determination of Water Content—Dry Density Relation using Heavy Compaction,** BIS, 1983.

87. IS: 2720—Part 28, **Determination of Dry Density of Soils, in-place, by the Sand Replacement Method,** BIS, 1974.

88. IS: 2720—Part 29, **Determination of Dry Density of Soils, in-place, by the Core Cutter Method,** BIS, 1975.

89. IS: 2720—Part 34, **Determination of Dry Density of the Soils, in-place, by Rubber-Ballon Method,** BIS, 1972.

90. IS: 2720—Part 38, **Compaction Control Test (Hilf Method),** BIS, 1976.

91. IS: 4332, **Methods of Test for Stabilized Soils,** BIS.

92. IS: 4332—Part 2, **Determination of Moisture Content of Stabilized Soils,** BIS, 1967.

93. IS: 4332—Part 3, **Test for Determination of Moisture Content—Dry Density Relation for Stabilized Soil Mixtures,** BIS, 1967.

94. IS: 4332—Part 7, **Determination of Cement Content of Cement Stabilised Soils,** BIS, 1973.

95. IS. 4332—Part B, **Determination of Lime Content of Lime Stabilized Soils,** BIS, 1969.

96. IS: 4332—Part 9, **Determination of the Bituminous Stabilizer Content of Bitumen and Tar Stabilized Soils,** BIS, 1970.

97. IS: 4999, **Recommendations for Grouting of Pervious Soils,** BIS, 1978.

98. IS: 6066, **Recommendation for Pressure Grouting of Rock Foundations in River Valley Projects,** BIS, 1971.

99. IS: 9759, **Guidelines for Dewatering During Construction,** BIS, 1981.

100. Iyer, T.S.R., Marine Deposits, **Proc. 5th ARCSMFE,** Bangalore, Vol. 2, pp. 35-69, 1975.

101. Iyer, T.S.R and S. Padmanabha Pillai, A Study of the In situ Stresses in Lateritic Profile, **Proc. Symp. on Strength and Deformation Behaviour of Soils,** Bengaluru, Vol. 1, pp. 159-163, 1972.

102. Jardin, J., Traitment D' alluvions Compressibles par Pieux-Colonnes Ballaste' es, **Bull. Liais. Lab. Ponts Chauss.,** No. 69, pp. 30-33, 1974.

103. John, N.W.M., **Geotextiles,** Blackie, New York, 347 pp. 1987.

104. Johnson, S. J., Foundation Precompression with Vertical Sand Drains, **JSMFD, ASCE,** Vol. 96, pp. 145-175, 1970.

105. Johnson, S.J., Foundation Precompression with Vertical Sand Drains, **Proc. ASCE, Spec. Conf. on Placement and Improvement of Soil to Support Structures,** pp. 9-39, 1968.

106. Jones, C.J.F.P., **Earth Reinforcements and Soil Structures,** Butterworths and Co., London, 183 pp. 1985.

107. Karol, R.H., Chemical Grouts and Their Properties, **Proc. Conf. on Grouting in Geotech. Engg.,** New Orleans, Wallace Hayward Baker, (ed.) ASCE, New York, pp. 359-378, 1982.

108. Katti, R.K., Kulkarni, V.S. Chandrasekaran, G. Venkatachalam and D.W. Dewaikar, Regional Soil Deposits of India, State-of-the-Art Report, **Proc. 5th ARCSMFE,** Bengaluru, Vol. 2, pp. 35-52, 1975.

109. Kjellman, W., Accelerating Consolidation of Fine Grained Soils by Means of Cardboard Wicks, **Proc. 2nd INCSMFE,** Rotterdam, Vol. 2, pp. 302-305, 1948.

110. Kelly, J.F., Performance Evaluation of Pump and Treat remediation, **U.S. Environmental Protection Agency, Superfund Ground Water Issue**; EPA/540/4-89/005, 1989.

111. Ketkar, D.J., Subsurface Conditions—Mumbay Island, **Symp. on Subsurface Exploration and Foundations of Structures in the Mumbay Region, Ind. Nat. Soc. SMFE (Mumbay Centre)** Vol. 1, 1970.

112. Koerner, R.M., **Construction and Geotechnical Methods in Foundation Engineering,** Mc-Graw Hill Book Company, New York, 496 pp. 1985.

113. Koerner, R.M. and F.K. Ko, Laboratory Studies of Long Term Drainage Capabilities of Geotextiles, **Proc. Second Int. Conf. on Geotextiles,** Vol. 1, pp. 91-95,1982.

114. Koerner, R.M., Designing with Geosynthetics, 3rd Edition, **Prentice Hall,** Eaglewood, Cliffs, USA, 1994.

115. Koppula, S.D. Statistical Estimation of Compression Index, **Geotech. Testing Journal, ASTM,** Vol. 4, No. 2, pp. 68-73, 1981.

116. Koppula, S.D., Discussion: Consolidation Parameters Derived from Index Tests, **Geotechnique,** Vol. 36, No. 2, pp. 291-292, 1986.

117. Kravetz, G.A., Cement and Clay Grouting of Fundations—The Use of Clay in Pressure Grouting, **JSMFD, ASCE,** Vol. 84, No. SM1, pp. 1546-1 to 1546-30,1958.

118. Krynine, D., and W.R. Judd, **Principles of Engineering Geology and Geotechnics,** McGraw-Hill Book Company, New York, 1957.

119. Kuesel, T.R., B. Schmidt and D. Rafaeli, Settlements and Strengthening of Soft Clay Accelerated by Sand Drains, **HRB Record,** No 457, pp. 18-26, 1973.

120. Ladd, C.C., **Strength and Compressibility of Saturated Clays,** Pan American Soils Course, Universidad Calolica Andres Bello, Caracas, 1967.

121. Ladd, C.C., J.J. Rixner, and D.C. Gifford, Performance of Embankments with Sand Drains on Sensitive Clay, **Proc. Spec. Conf. Performance of Earth and Earth-Supported Structures,** Lafayette, Vol. 1, pp. 211-242, 1972.

122. Lambe, T.W., The Engineering Behaviour of Compacted Clay, **JSMFD, ASCE,** Vol. 84, No. SM2, pp. 1665-1-35, 1958.

123. Lambe, T.W., Soil Stabilization, In **Foundation Engineering,** G.A. Leonards (ed.) Mc-Graw Hill Book Company, New York, pp. 351-437, 1962.

124. Lambe, T.W., The Integrated Civil Engineering Project, **JSMFD, ASCE,** Vol. 98, pp. 531-556, 1972.

125. Lambe, T.W., and R.V. Whitman, **Soil Mechanics,** SI Version, Wiley Eastern Limited, New Delhi, 553 pp. 1979.

126. Lancaster-Jones, P.F.F., Methods of Improving the Properties of Rock Masses, In **Rock Mechanics in Engineering Practice,** Stagg and Zienkiewicz, (eds.), John Wiley and Sons, New York, Chap. 12, 1968.

127. Lane, K.S., Discussion : Consolidation of Fine-grained Soils by Means of Drain Wells, **Trans. ASCE,** Vol. 113, p. 743, 1948.

128. Lang, T.A., Rock Reinforcement, **Bull. Assoc. Engg. Geol.,** Vol. 9, No. 3, Summer, pp. 215-239,1972.

129. Leonards, G.A., Engineering Properties of Soils, in **Foundation Engineering,** by G.A. Leonards (ed.), McGraw Hill Book Company, New York, pp. 66-240, 1962.

130. Leonards, G.A., W.A. Cutter, and R.D. Holtz, Dynamic Compaction of Granular Soils, **JGED, ASCE,** Vol. 106, No. GT1, pp. 35-43,1980.

131. Lewis, W.A., R.T. Murray, and I.F. Symoss, Settlement and Stability of Embankments Constructed on Soft Alluvial Soils, **Proc. ICE,** December, 1975.

132. Little John, G.S., Design of Cement Based Grouts, **Proc. Conf. on Grouting in Geotech. Engg.,** New Orleans, pp. 35-48, 1982.

133. Mackenna, J.M., W.A. Eyre, and D.R. Wolstenholme, Performance of an Embankment Supported by Stone Column in Soft Ground, **Geotechnique,** Vol. 25, pp. 51-59,1975.

134. Magnan, J.P. and H.T. Dang, Theoretical and Experimental Analysis of the Compressibility of Pore Fluid in a Nearly Saturated Clayey Soil, **Proc. Int. Symp. Soft Clay,** Bangkok, pp. 675-689, 1977.

135. Mahesh Varma, Construction Equipment and Its Planning and Application, **Metro-Poertan Book Co. (P) Ltd.**, 623 pp, 1979.

136. Margason, E., and I. Arango., Sand Drain Performance on a San Francisco Bay mud site, **Proc. Sepec. Conf. Performance of Earth and Earth-Supported Structure,** Lafayette, Vol. l pp. 181-210, 1972.

137. McCarthy, D.F., **Essentials of Soil Mechanics and Foundations,** Reston Publishing Company, Virgiria, 632 pp. 1982.

138. Meyerhof, G.G., Compaction of Sands and Bearing Capacity of Piles, **JSMFD, ASCE,** Vol. 85, No. SM6, 1959.

139. Me' nard, L. and Y. Broise, Theoretical and Practical Aspects of Dynamic Consolidation, **Geotechnique,** Vol. 25, pp. 3-18, 1975.

140. Michaels, A.S., and V. Puzinauskas, Chemical Additives as Aids to Asphalt Stabilization of Fine-Grained Soils, **HRB Bull.,** No. 129, 1956.

141. Mieussens, C., and P. Ducasse, Measure en Place des Coefficients de Perrneabilite et de Consolidation Horizontaux et Verticaux, **CGJ,** Vol. 14, pp. 76-90, 1977.

142. Mitchell, J.K.,In-place Treatment of Foundation Soils, **JSMFD, ASCE,** Vol. 96, pp. 73-110, 1970.

143. Mitchell, J.K., **Stabilization of Soils for Foundations of Structures,** Report prepared for U.S. Army Waterways Experiment Station, pp. 1-34,1976.

144. Mitchell, J.K. and R.K. Katti, Soil Improvement—State of the Art Report, **Proc. 10th INCSMFE.,** Stockholm, Vol. 4, pp. 567-575,1981.

145. McGown, A., K.W. Andrawes and M.H. Kabir, Load Extension Testing of Geotextiles Confined in Soil, **Proc. Second Int. Conf. on Geotextiles,** Las Vegas, Vol. 3, pp. 793-798, 1982.

146. Mohan, D., G.S. Jain, D.P. Sen Gupta and Devendra Sharma, Consolidation of Ground by Vertical Rope Drains, **IGJ,** Vol. 7, pp. 106-115, 1977.

147. Munoz, A. Jr. and R.M. Mattox, Vibroreplacement and Reinforced Earth Unite to Strengthen a Weak Foundation, **Civil Engineering, ASCE,** pp. 58-61, 1977.

148. Murry, R.T., **Embankments Constructed on Soft Foundations : Settlement Studies near Oxford** U.K., Transp. Road Res. Lab., Report LR 538, 48 p. 1973.

149. Nagaraj, T.S. and B.R.S. Murthy, **Geotech. Testing Journal, ASTM,** Vol. 8, No. 4, pp. 199-202, 1985.

150. Nagaraj, T.S. and B.R.S. Murthy, A Critical Re-appraisal of Compression Index, **Geotechnique,** Vol. 36, No. 1, pp. 27-32, 1986.

151. Nagaraj, T.S., A. Sridharan and M.V. Paul Alexander, In situ Reinforced Earth—An Approach for Deep Excavation, **IGJ,** Vol. 12, pp. 101-111, 1982.

152. Nicholls, R.A. and A.S. Barry, Vertical Drains—A Case History, **Proc. 8th ECSMFE,** Helsinki, No. 6.21, 1983.

153. OECD, **Consolidation of Roads on Compressible Soils—Road Research,** Pairs, 1979.

154. Parry, R.H.G., A Direct Method of Estimating Settlements in Sands from S.P.T. Values, **Proc. Interaction of Structure and Foundation, Symp. of Midlands SMFE,** Birmingham, pp. 29-37, 1971.

155. Paute, J.L., Consolidation d'un Sol Fin Argileux par Application de Vide, **Bull. Liais. Lab. Ponts Chauss,** Special No. T, pp. 154-161,1970.

156. Paute, J.L., Remblai de Cran. **Bull. Liais. Lab. Ponts Chauss,** Special No. T, pp. 105-118, 1973*a*.

157. Paute, J.L., Essai Oedonietrique a Drain Central, **Bull. Liais. Lab. Ponts. Chauss,** Special No. T, pp. 322-334,1973 *b*.

158. Peignaud, M., Electro-injection as Stabilization Work on Bridge Abutment. "Basse Chaine" in Anges, **Personnel Communication with G. Pilot,** 1977.

159. Pichamuttu, C.S., **Physical Geography of India,** National Book Trust, New Delhi, 223 pp. 1970.

160. Pichtel, J., Fundamentals of Site Remediation–For Metal and Hydrocarbon Contaminated Soils, Second Edition, **Government Institutes, The Scarecrow Press Inc.,** Toronto, 436 pp, 2007.

161. Pilot, G., Methods of Improving The Enginnering Properties of Soft Clay In **Soft Clay Engineering,** Brand, E. W. and R.D. Brenner, (eds.) by Elsevier Scientific Publishing Company, Amsterdam, pp. 633-696, 1981.

162. Proctor, R.R., Foundation Principles of Soil Compaction, **ENR,** August 31, September 9, 21 and 28, 1933.

163. Quast, P., Polyester Reinforcing Mats for the Improvement of Embankment Stability, **8th INCSMFE,** Rotterdam, 1983.

164. Queyroi, D., G. Pilot and J.P. Magnan, Edification d'un Remblai sur Foundation d' Argile Molle Traitee par Drains de Sable, **Proc. Int. Symp. Soft Clay,** Bangkok, pp. 245-259,1977.

165. Ramaswamy, S.D., M.A. Aziz., R.V. Subrahmanyam, M.H. Abdual Khoder, and S.L. Lee, Treatment of Peaty Clay by High Energy Impact, **JGED, ASCE,** Vol. 105, No. GT8, pp. 957-967. 1979.

166. Ramaswany, S.D., S.L. Lee, and I.U. Daulah, Dynamic Consolidation—Dramatic way to Strengthen Soil, **Civil Engineering, ASCE,** pp. 70-73, 1981.

167. Rawes, B.C., The allowable design stress available from geotextiles, **International Workshops on Geotextiles,** Bangalore, 1989.

168. Ranjan, G., Ground Treated with Granular Piles and its Response Under Load, **Eleventh IGS Annual Lecture, IGJ,** Vol. 19, pp. 1-86,1989.

169. Ranjan, G., and B.G. Rao, Skirted Granular Piles for Ground Improvement, **Proc. 8th ECSMFE,** Halsinki, 1983.

170. Rajan, G., and Rao, B.G., Granular Piles for Ground Improvement, **Proc. Int. Conf. on Deep Foundations,** China, Vol. 1, 1986.

171. Ranjan G. and A.S.R. Rao, **Basic and Applied Soil Mechanics,** Wiley Eastern Ltd., New Delhi, 640 pp. 1991.

172. Rao, B.G., Behaviour of Skirted Granular Piles, **Doctoral Thesis.** CED, Univ. of Roorkee, 1982.

173. Rao, B.G., and G. Ranjan, Closure of the paper on Settlement Analysis of Skirted Granular Piles, **JGED, ASCE,** Vol. 114, No. 1, pp. 729-736, 1988.

174. Rao, G.V. and G.V.S.S. Raju, (eds.) **Engineering with Geosynthetics,** Tata McGraw—Hill Publishing Co., Ltd., 316 pp. 1990.

175. Rao, G.V., M.P.S. Pradhan and K.K. Gupta, Hydraulic Characteristics of Geotextiles, In **Engineering with Geosynthetics** by G.V. Rao and G.V.S.S. Raju (eds.), Tata McGraw Hill Publishing Company Limited, New Delhi, pp. 101-114, 1990.

176. Rao, K.S., and B.C. Raymahashay, Influence of Clay Minerals and Iron Oxides on Selected Properties of Two Lateritic Soils, **IGJ.,** Vol. 11, pp. 255-266, 1981.

177. Rendon-Herrero, O., Universal Compression Index Equation, **JGED, ASCE,** Vol. 106, No. GT11, pp. 1178-1200,1980.

178. Rendulic, L., **Derhydrodynamische Spannunqsausgleich in Zentral entwasserten Tonzylinder, Wasserwirtschaft und Technik,** Vol. 2, pp. 250-253, 269-273, 1935.

179. Richart, F.E., A Review of the Theories of Sand Drains, **Trans. ASCE,** Vol. 124,pp. 709-739, 1959.

180. Risseeuw, P., and W. Voskamp, **Reinforcing Fabrics under Embankments on Soft Subsoils, A Calculation Method,** London : CII., 1984.

181. Rothfuchs, G., Particle size Distribution of Concrete Aggregates to obtain Greatest Density, **Zement,** Vol. 4, No. 1, pp. 8-12, 1935. **Also**: How are Asphat Bitumen Mixes of Maximum Density (Minimum Voids) to be attained? **Bitumen,** Vol. 5, No. 3, pp. 57-61, 1936.

182. Rowe, P.W., The Influence of Geological Features of Clay Deposits on the Design and Performance of Sand Drains, **ICE.,** Suppl. pp. 1-72, 1968.

183. Rowe, P.W., and L.Barden, A New Consolidation Cell, **Geotechnique,** Vol. 16, pp. 166-170, 1966.

184. Rutledge, P.C., and S.J. Johnson, Review of use of Vertical Sand Drains, **HRB, Bull. 173,** Publications 533, p. 65, 1958.

185. Samson, L., and R.Garneau, Settlement Performance of Two Embankments on Deep Compressible Soils, **CGJ,** Vol 10, pp. 211-36, 1973.

186. Schmertmann, J.M., The Undisturbed Consolidation of Clay, **Trans. ASCE,** Vol. 120, pp. 1201-1233, 1955.

187. Schmertmann, J.H., Static Cone to Compute Static Settlement Over Sand, **JSMFD, ASCE,** Vol. 96, No. SM3., 1970.

188. Schmid, G., Series of Papers in **Z. Electrochem**: Vol. 54: p. 424 (1950), Vol. 55 : p. 229 (1951), Vol. 55:p. 684 (1951), Vol. 56: p. 36 (1952), 56: 181 (1952).

189. Schultze, E., and H. Muhs, **Bodenuntersuchingen Fur Ingenieur-Bauten** Springer, Berlin, 1967.

190. Seed, H.B., Foundations and Abutment for High Dams on Rock, Committee on Embankments, Dams and Slopes, **JSMFD, ASCE,** Vol. 98, No. SM 10, pp. 1115-28, 1972.

191. Seed, H.B., and C.K. Chan, Structure and Strength Characteristics of Compacted Clays, **JSMFD, ASCE,** Vol. 85, No. SM5, 1959.

192. Seed, H.B. and I.M. Idriss, Simplified Procedure for Evaluating Soil Liquefaction Potential, **JSMFD, ASCE,** Vol. 97, No. SM. 9, 1971.

193. Seed, H.B., R.J. Woodward, Jr., and R. Lundgren, Prediction of Swelling Potential for Compacted Clays, **JSMFD, ASCE,** Vol. 88, No. SM3, pp. 53-87, 1962.

194. Sen Gupta, D.P., B.Dey and D. Rey, Substrata Treatment for a Building Using Rope Drains, **IGJ,** Vol. 10, pp. 323-331, 1980.

195. Sharma, S.C., Construction Equipment and Management, **Khanna Publishers,** New Delhi, 580 pp, 1988.

196. Shield, D.H. and P.W. Rowe, Radial Drainage Oedometer for Laminated Clays, **JSMFD, ASCE,** Vol. 91, No. SM1, pp. 15-23, 1965.

197. Shobha, K.B., Design with Geomembranes, In **Engineering with Geosynthetics** by G.V. Rao, and G.V.S.S. Raju (eds.), Tata McGraw Hill Publishing Company Limited, New Delhi, pp. 283-308, 1990.

198. Shroff, A.V. and D.L. Shah, **Grouting Technology in Tunnelling and Dam Construction,** Oxford and IBH Publishing Co., Rt. Ltd., New Delhi, 604 pp. 1992.

199. Shuster, J.A., Controlled Freezing for Temporary Ground Support, **Proc. First N. Amer. Rapid Exch. and Tunneling Conf.,** Chicago, pp. 853-894, 1972.

200. Simon, A., and M. Brigando, Le Remblai de Prechargement d'Arles, **Centre d'Etudes Technique de L Equipment, Aixen-Provence, Internal Rep.,** 60 pp. 1976.

201. Simons, N.E., Normally Consolidated and Lightly Over-Consolidated Cohesive Materials, Review Paper. In **Settlement of Structures Conference, Cambridge,** Pentech, London, pp. 500-530, 1974.

202. Sinclair, T.E. J., E.E. Rinue, and Q. Laumans, Dike Stabilization for Long Term Stage Loading and Severe Settlement, **Proc. 8th ECSMFE.** Helsinki, No. 6.24, 1983.

203. Smith, L.A., J.L. Means, A. Chen, B. Alleman, C.C. Chapman, J.S. Tixier, S.E. Brauning, A.R. Gavaskar and M.D. Royer, Remedial Options for Metals–Contaminated Soils, Boca Raton, FL: **CRC Press**, 1995.

204. Sole' tanche, S.A., **Cokerie d'Alfortville, France ; Foundation du Gazometre No.** 4, Sole' tanche SA, Paris, 4 pp. 1960.

205. Som, N., Regional Deposits, **Proc. 5th ARCSMFE,** Bangalore, Vol. 2, pp. 70-81, 1975.

206. Somanathan, C., Neyveli Fly ash as an Admixture in Cement Grout, **Ind. JSMFE,** Vol. 7, No. 3, 1968.

207. Sowers, G.F., Engineering Properties of Residual Soils Derived from Ingneous and Metamorphic Rocks, **Proc. Second Pan. Am. Conf. SMFE.,** Brazil, Vol. 1, p. 39, 1963.

208. Sowers, G.F., Soil and Foundation Problems in the Southern Piedmont Region, **Proc. ASCE,** Vol. 80, Separate No. 416, 1953.

209. Sowers, S.F., **Introductory Soil Mechanics and Foundations: Geotechnical Engineering,** 4th ed., Macmillan Publishing Co. Inc., New York, 621 pp. 1979.

210. Smith, J.C., The-Chrome-Lignin Process and Ion Exchange Studies, **Proc. MIT Soil Stabilization Conference,** 1952.

211. Sridharan, A., Bearing Capacity Improvement, In **Engineering with Geosynthetics,** by G.V. Rao and G.V.S. Raju (eds.), Tata McGraw- Hill Publishing Company Limited, New Delhi, pp. 175-196, 1990.

212. Sridharan, S., Application of Electrokinetics for Piles in Black Cotton Soil, **Indian Concrete Journal**, Vol. 46, No. 6, pp. 245-248, 1972.

213. Stamatopoulos, A.C., and P.C. Kotzias, Settlement—Time Predictions in Preloading, **JGED, ASCE.,** Vol. 109, No. 6,1983.

214. Stamatopoulos, A.S. and P.C. Kotzias, **Soil Improvement by Preloading,** John Wiley and Sons, Inc., New York, 261 pp. 1985.

215. Su, H., J.C. Chang and R.A Forsyth, Shear Strength Increase in a Soft Clay Foundation, **HRB, Record,** No. 463, pp. 28-34, 1973.

216. Subbaraju, B.H., T.K Natarajan and R.K. Bhandari, Field Performance of Drain Wells Designed Especially for Strength Gain in Soft Marine Clays, **Proc. 8th INCSMFE,** Moscow, Vol. 2.2, pp. 217-220,1973.

217. Tadashi, Electro-osmotic Dewatering and Distribution of the Pore Water Pressure, **Proc. 5th INCSMFE,** Vol. 1, p. 255, 1961.

218. Tallard, G.R. and C. Caron, **Chemical Grouts for Soils : Available Materials,** Vol. 1 and **Engineering Evaluation of Available Materials,** Vol. 2, Federal Highway Administration, U.S. Dept. of Transportation, Washington, DC, 1977.

219. Taylor, D.W., **Fundamentals of Soil Mechanics,** Asia Publishing House, New Delhi, 700 pp. 1948.

220. Terrafigo, A.B., **Geotechnical Reports about Geodrains,** Terrafigo AB, Gothenberg, Collected Papers, 52 pp. 1976.

221. Terzaghi, K., **Erdaumechanik,** Frauz Deuticke, Vienna, 1925.

222. Terzaghi, K., **Theoretical Soil Mechanics,** John Wiley and Sons, New York, 510 pp. 1943.

223. Terzaghi, K., and R.B. Peck, **Soil Mechanics in Engineering Practice,** 2nd ed., John Wiley and Sons, New York, 727 pp. 1967.

224. Thorburn, S., and R.S.L. McVicar, Soil Stabilization Employing Surface and Depth Vibrations, **Structural Engineer,** Vol. 96, No. 10, pp. 117-120, 1968.

225. Tomblinson, M.J., **Foundation Design and Construction,** Fifth Edition, English Language Book Society/Longman, Essex, 842 pp. 1986.

226. Torstensson, B.A., Pore Pressure Sounding Instrument, **Proc. Spec. Conf. In situ Measurement of Soils Properties,** Raleigh, Vol. 2, pp. 48-54, 1975.

227. Ueda, Y., Case Histories of Highway Structures on Soft Ground: Box Type Culverts and Embankments, **Proc. 4th ARCSMFE,** Bangakok, Vol. 1, pp. 323-28, 1971.

228. U.S. Environmental Protection Agency, Design and Construction for Solid Waste Land Fills, **Publication No. EPQ-600/2-79-165,** Ohio, USA, 1979.

229. U.S. Environmental Protection Agency, Handbook Remedial Action at Waste Disposal Sites (Revised), EPA/625/6-85/006, Washington D.C., **Office of Solid Waste and Emergency Response**, 1985.

230. U.S. Environmental Protection Agency, Cover for Uncontrolled Hazardous Waste Sites, **Publication No.EPA-540/2-85-002,** Ohio, 1986.

231. U.S. Environmental Protection Agency, Requirements for Hazardous Waste Landfill Design, Construction and Closure, **Publication No.EPA-625/4-89-022,** Ohio, 1989.

232. Ibid., Corrective Action: Technologies and Applications, **Seminar Publication**, EPA/625/4-89/020, 1989.

233. USBR, **Pressure Grouting,** U.S. Bureau of Reclamation, Tech. Memo. No. 646, Denver, June, 1957.

234. USBR, **Earth Manual,** U.S. Department of the Interior Bureau of Reclamation, Washington, D.C., 209 p. 1974.

235. U.S. Waterways Experiment Station, The Unified Classification, Appendix A—Characeristics of Soil Groups Pertaining to Airfields and Appendix B—Characteristics of Soil Groups Pertaining to Embankment and Foundations, **Technical Memorandum No. 357,**1953.

236. Van Wanbeke, A. and E. De Beer, Etude sur Aire D'essai du Pilonnage Intensif, **Proc. 8th INCSMFE,** Moscow, Vol. 4.3, pp. 362-65, 1973.

237. Varadarajulu, G.H. and A.K.D. Gupta, Ground Strengthening by Vibroflotation, **Civil Engg. and Construction Review,** March, pp. 33-37, 1993.

238. Vautrain, J., Mur en Terre Arme'e sur Colonnes Ballaste'es, **Proc. Int. Symp. Soft Clay,** Bangkok, pp. 613-626, 1977.

239. Veeraraghavan, N., Geotextiles in India—An overview, **Jour. Int. of Engineers, (India), Civil Engg.,** Vol. 68, CII, pp. 10-14, 1987.

240. Verfel, J. Soil Strengthening by Grouting for the Metro of Prague (in Czech) In **Methods of Foundation Engg.** by **Z. Bazant,** Elsevier, Amsterdam, 1979.

241. Vidal, H., **The Principle of Reinforced Earth,** Highway Research Record, No. 282, pp. 1-16, 1969.

242. Wahlstrom, E.E., **Dams, Dam Foundations and Reservoir Sites,** Elsevier, New York, 1974.

243. Welsh, J.P., Dynamic Deep Compaction: A Ground Modification Technique, **Ground/Expo. 82,** Houston, TX, March 22, 1982.

244. West, J.M., The Role of Ground Improvement in Foundation Engineering, **Geotechnique,** Vol. 25, No. 1, pp. 72-78, 1975.

245. Wheeless, L.D., and G.T. Sowers, Mat Foundation and Preload Fill, VA Hospital, Tampa, **Proc. Spec. Conf. Performance of Earth and Earth-supported Structures,** Lafayette, Ind., Vol. 1, pp. 939-952, 1972.

246. Westergaard, H.M., A Problem of Elasticity Suggested by a Problem in Soil Mechanics : Soft Material Reinforced by Numerous Strong Horizontal Sheets in Contributions to the Mechanics of Solids, **Stephen Timosharko 60th Anniversary Volume,** New York, 1938.

247. Wilson, S.D., Small Compaction Apparatus Duplicates Field Results Closely, **Engineering News Record,** Nov. 2, pp. 34-36, 1950.

248. Zanten, V.V., **Geotextiles and Geomembranes in Civil Engineering,** A.A. Balkema, Rotterdam, 658 pp. 1986.

INDEX

E

F

G

H

I

J

K

L

M

N

O

P